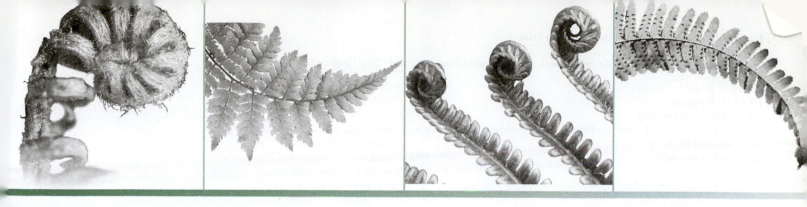

Quantitative Reasoning and the Environment
Mathematical Modeling in Context

Greg Langkamp
Seattle Central Community College

Joseph Hull
Seattle Central Community College

The material in this text was developed, in part, with the support of the National Science Foundation. Any opinions, findings, and conclusions or recommendations expressed in this material are those of the authors and do not necessarily reflect the views of the Foundation.

PEARSON
Prentice Hall

Upper Saddle River, NJ 07458

Library of Congress Cataloging-in-Publication Data

Langkamp, Greg.
    Quantitative reasoning and the environment: mathematical modeling in context / Greg Langkamp,
Joseph Hull.—1st ed.
       p. cm.
    Includes index.
    ISBN 0-13-148527-X
      1. Environmental sciences—Mathematics—Textbooks. I. Hull, Joseph. II. Title.
    GE45.M38L36 2007
    628.01'51—dc22   2006043203

Editor in Chief: *Sally Yagan*
Acquisitions Editor: *Chuck Synovec*
Project Manager: *Michael Bell*
Production Editor: *Raegan Keida Heerema*
Assistant Managing Editor: *Bayani Mendoza DeLeon*
Senior Managing Editor: *Linda Mihatov Behrens*
Executive Managing Editor: *Kathleen Schiaparelli*
Manufacturing Buyer: *Maura Zaldivar*
Manufacturing Manager: *Alexis Heydt-Long*
Director of Marketing: *Patrice Jones*
Marketing Manager: *Wayne Parkins*
Marketing Assistant: *Jennifer de Leeuwerk*
Editorial Assistant: *Jennifer Urban*
Art Director: *Heather Scott*
Interior Design: *Judith Matz-Coniglio, Dina Curro*
Cover Designer: *John Christiana*
Art Editor: *Thomas Benfatti*
Creative Director: *Juan R. López*
Director of Creative Services: *Paul Belfanti*
Cover Image: *FotoSearch / Design*
Manager, Visual Research and Permissions: *Karen Sanatar*
Director, Image Resource Center: *Melinda Patelli*
Manager, Rights and Permissions: *Zina Arabia*
Manager, Visual Research: *Beth Brenzel*
Image Permission Coordinator: *Nancy Seise*
Permission Researcher: *Melinda Alexander*
Art Studio: *Laserwords*
Part Openers, Chapter Openers: *Corbis*©

© 2007 by Pearson Education, Inc.
Pearson Prentice Hall
Pearson Education, Inc.
Upper Saddle River, NJ 07458

Pearson Prentice Hall™ is a trademark of Pearson Education, Inc.

Printed in the United States of America

ISBN 0-13-148527-X

Pearson Education, Ltd., *London*
Pearson Education Australia PTY. Limited, *Sydney*
Pearson Education Singapore, Pte., Ltd
Pearson Education North Asia Ltd, *Hong Kong*
Pearson Education Canada, Ltd., *Toronto*
Pearson Education de Mexico, S.A. de C.V.
Pearson Education – Japan, *Tokyo*
Pearson Education Malaysia, Pte. Ltd

To Laura

— G.L.

To Mom

—J.H.

Contents

Part 2   Function Modeling   77

Chapter 4   Linear Functions and Regression   78

Chapter 5   Exponential Functions and Regression   103

Chapter 6   Power Functions   135

Part 3   Difference Equation Modeling   163

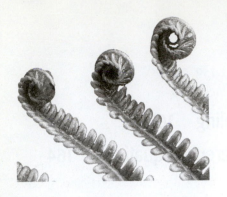

Preface

As the Earth's human population sprints toward the 7 billion mark, humanity is faced with a number of serious environmental problems: global warming, extinction of species, acid rain, vanishing forests, atmospheric ozone holes, polluted drinking water, over-fishing, mountaintop strip mining, soil erosion, and depleted freshwater aquifers. The list goes on and on. "Staying the course" can only lead to ruin, as many readers will agree. But if we take actions to prevent further environmental degradation, which actions are best? Most environmental issues are quite complex scientifically (not to mention politically, economically, and socially). How can we better understand these topics?

Let's take, for example, the issue of transportation energy. Gasoline-burning automobiles release greenhouse gases, which contribute to global warming and emit toxins and particulates that compromise human health. Gasoline from petroleum is also a finite resource, soon to be depleted forever. We can switch to hydrogen-powered vehicles, which have been billed as pollution free. But the hydrogen necessary to meet our transportation needs will have to be manufactured using other sources of energy, such as coal energy or nuclear energy. Both of these energy sources have their own set of environmental problems. Are we really better off switching to hydrogen-based transportation? How can we analyze this problem in more depth?

If you scratch beneath the surface to investigate environmental issues, you will quickly encounter a lot of quantitative information in the form of facts, figures, and formulas. Math is the key to sorting through this quantitative information. One of our goals is to convince the reader that critical issues, such as the choice among transportation energy sources, can be better understood using number sense, basic algebra, simple models, and introductory statistics.

This text is intended primarily for nonscience majors who need to complete a first-year college mathematics course—often referred to as a *liberal arts mathematics* course or a *quantitative reasoning* course. A prerequisite course for this text is intermediate algebra, although many of the skills presented in intermediate algebra are not necessary to comprehend and master the material in this text. A good sense of working with numbers, the ability to read and write clearly, a curiosity to learn—these skills are just as important. A hallmark of this text is that it introduces elementary mathematical concepts and techniques and asks students to apply them in sophisticated ways.

The text fully integrates the mathematics with environmental science topics and illustrates how mathematics is used as a tool to describe, model, and analyze data. This approach is similar to what science students might experience in advanced undergraduate-level courses (e.g., a calculus-based ecology course), only at an introductory level. The integrated nature of this text may also appeal to beginning undergraduate science students seeking a first introduction to applied mathematics. A third possible audience for this text is advanced high school students who have already completed intermediate algebra but seek a capstone experience with real-world mathematics.

TOPICAL COVERAGE

Part 1 of the text, *Essential Numeracy*, helps build many of the quantitative skills that are indispensable in mathematics and science, let alone daily life. Topics such as units of measure, scientific notation, ratios, and percentages are commonly presented in developmental mathematics courses and will already be somewhat familiar to students; our examples focus on environmental applications and emphasize the utility and pervasiveness of these topics in the real world. We expect that other topics in this section will be new for students (logarithms, normalization, etc.), so we explore them from an intuitive viewpoint. This section concludes with an overview of the many different types of graphical displays: Pie and bar charts, histograms, and scatterplots.

Part 2 of the text, *Function Modeling*, explores what we feel are the three most important functions used in analyzing environmental data and modeling environmental problems: Linear, exponential, and power-law functions. For students with an intermediate algebra background, this section bridges familiar territory (slope and linear functions) with more exotic lands (fractals and power-law distributions). Basic features of each of these functions are examined via graphs, tables, and equations. Special topics such as balancing units in linear equations and finding the doubling time of exponential functions are also included. The statistical topic of curve fitting is also explored; a paper-and-pencil technique (the straightedge method) helps students comprehend a more advanced method (least squares regression).

Part 3 of the text is titled *Difference Equation Modeling*. Why do we include an entire section on a subject rarely taught in any mathematics class, let alone in a book for liberal arts majors? Because of the simple yet powerful equations that can model complex behavior, and because of the rich applications to environmental science. We begin by introducing sequence notation so that students learn to read and write difference equations and develop an intuitive sense of recursive formulas. Linear and exponential difference equations are explored and compared to functional equations. A discussion on affine difference equations and equilibrium values rounds out the first half of this section. The second half covers logistic difference equations and system models. Applications touch on environmental issues related to harvesting, periodic and chaotic behavior, population dynamics and stable age distributions, and epidemics.

Part 4 of the text, *Elementary Statistics*, introduces topics such as measures of center, deviation, skew, and displays of data, all in an environmental context. The many practical examples and exercises will help students become adept at analyzing and summarizing data. Discussions on sampling and inference will help students develop an intuitive understanding of the practice of statistics. This section states and explores Chebychev's Rule and then introduces normal distributions and the Empirical Rule. The reader will find a classic approach to determining areas below the standard normal curve, with the problems couched in terms of environmental issues and problems. This section concludes with an intuitive explanation of confidence intervals and illustrates how to compute 95% confidence intervals for means and proportions.

USING THE TEXT

The chapter organization and contents of *Quantitative Reasoning and the Environment* allows instructors to build courses with lengths varying from a 10-week quarter to a full academic year. There are several paths that can be followed through the chapters, with prerequisite material determining the paths diagrammed below. Some chapter topics can be skipped before proceeding to the next chapter indicated. For example, scatterplots can be skipped in Chapter 3 if the statistics path (Chapters 11–13) is chosen. Also, the topic of regression in Chapters 4–5 can be avoided while proceeding into difference equations (Chapters 7–10). Further suggestions for using this text are provided in the Instructor's Resource Manual.

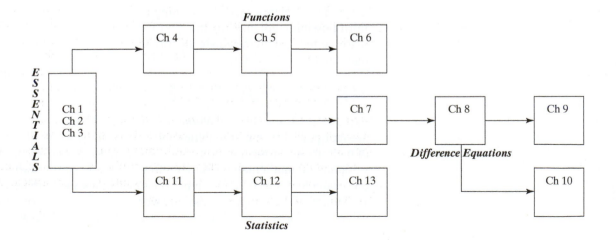

TEXT WEB SITE

The Web site for this book is found at **enviromath.com**. Featured at the Web site are chapter **projects** and all supporting materials, chapter **data sets** in several electronic formats, and **online applications** to support students who are not using the TI-83/84 graphing calculator. Directions for using the online applications are provided at the Web site.

Visitors to the Quantitative Environmental Learning Project (QELP) Web site (**http://www.seattlecentral.edu/qelp/**) will find many data sets not provided in the text, as well as additional background material and links.

SPECIAL FEATURES OF THE TEXT

- **Clear, concise presentation**, with an emphasize on understanding the mathematics as well as using the mathematics in context to examine environmental information.
- A **modular approach** involving four parts: Numeracy, Functions, Difference Equations, and Statistics. This approach allows for multiple ways in which instructors can design a course (see the preceding flow chart).
- **Real-world examples and data sets** drawn from all facets of environmental science: Ecology and population studies, agriculture and food, air pollution and global warming, water resources and pollution, energy and transportation, environmental hazards, and solid waste.
- A just-in-time approach to **using technology**. The text illustrates the essential keystrokes for the **TI-83/84 graphing calculator** to display scatterplots and histograms, create tables and graphs for functions and difference equations, and generate descriptive statistics. As an alternative, our Web site provides directions on using **online applications** to accomplish the same tasks.
- **Extensive student exercises** that help students master essential skills but emphasize both the application of the mathematics and the interpretation of answers.
- A *Science in Depth* article at the end of each chapter that expounds on critical environmental issues that face the world today. These articles also introduce the reader to the chapter projects.
- **Chapter projects** that encourage students to use their mathematical skills and problem-solving capabilities in more sophisticated applications. The projects are the text's zenith of integrating mathematics with environmental science topics. Be sure to check these out at the text Web site: **enviromath.com**.
- **Answers** to odd-numbered exercises are provided at the end of the text.

HISTORY

This textbook resulted from a common vision—the desire to make our course materials more meaningful and interesting to students. We felt that elementary college mathematics would be more compelling if driven by real-world, socially relevant examples and exercises, such as those found in the field of environmental science. We were not satisfied by contrived and tiresome problems about mixing Brazil nuts and cashews, or about one train leaving Boston and another Los Angeles.

We began teaching an integrated mathematics and environmental science course at Seattle Central Community College in 1998. Our course notes developed into a preliminary version of this text, as no text was available to meet our needs. Hundreds of students at our college have now used early versions of this text to complete their college mathematics graduation requirement. Many students say that they really *had fun* in our course and that they thought mathematics and science was *useful and interesting*. We wrote this text to share this integrated approach, and the fun and excitement, with you. We hope you enjoy it!

ACKNOWLEDGMENTS

We would like to thank the National Science Foundation for funding the Quantitative Environmental Learning Project (QELP) from 2000 to 2002. Through QELP we collected many of the data sets for this text, developed the QELP Web site, wrote many of the chapter projects found in this text, and hosted an exciting workshop in the summer of 2002. Through that workshop and the travel promoting QELP, we met numerous educators who shared our common interest in bridging mathematics and environmental science instruction. A special thanks goes to Robert Cole of The Evergreen State College for his substantial role as project evaluator, mentor, and friend. We also would like to thank Lawrence Morales for his exceptional work in designing the QELP Web site.

We would like to thank the students, faculty, administration, and staff at Seattle Central Community College for their continued support of our efforts over the past six years. In particular, we would like to thank our colleagues Sanford Helt, Mike Pepe, Bobby Righi, and Janet Ray for their wonderful suggestions and steadfast encouragement while we labored on both QELP materials and this textbook. You are more than colleagues, you are friends!

A special thanks also goes to our friend Erik Neumann of Seattle, Washington, who graciously accepted the challenge of designing and programming our functions and difference equations Java applets in the summer of 2005. Erik's awesome physics simulation Web site is found at **www.myphysicslab.com**. We would also like to acknowledge Webster West from the University of South Carolina for his work in creating StatCrunch, the online statistics applet. StatCrunch is found at **www. StatCrunch.com**. Webster's applet integrates beautifully with the QELP Web site and our text's data sets. Thanks Erik and Webster!

To all the researchers who have contributed data to either the text or the QELP Web site, either directly or indirectly, we express our utmost gratitude. Without your devotion and accomplished work, many of our examples and exercises would not exist.

We would like to thank the editorial, design, and production staff at Prentice Hall, especially Sally Yagan and Petra Rector, for making this text possible as you now read it. Your encouragement to write this text and your skill, professionalism, and dedication in publishing it are deeply appreciated.

Finally, a special thanks is given to the numerous reviewers and site testers of this text—your insightful comments have been most helpful. These people include

Peter Benedict—Colorado Rocky Mountain School
David Buchthal—Cascadia Community College
Dave Cole—Spokane Community College
MaryAnn Firpo—Seattle Central Community College
Mike Fletcher—Davidson High School
Ben Fusaro—Florida State University
Barbara Grover—Salt Lake Community College
Gregg Harbaugh—University of Washington
Ed Harri—Whatcom Community College
William Long—Northland College
Andrew Long—Northern Kentucky University
Kim Luna—Eastern New Mexico University
Derek Ogle—Northland College
Nicole Pfeiffer—Spokane Community College
Bobby Righi—Seattle Central Community College
Catherine Roberts—College of the Holy Cross
Wm. D. Stone—New Mexico Tech
Jim Wright—Green Mountain College
Kevin Yokoyama—College of the Redwoods
Sarah N. Ziesler—Dominican University

Greg and Joe
Spring 2006

PART ONE

Essential Numeracy

1

Measurement and Units

Real environmental information and data are used to introduce, develop, and illustrate the mathematics throughout this textbook. Following a case study of heavy metals in the diets of Greenland Inuit, this chapter introduces a few topics related to measuring: measurement strategies, accuracy and precision, and estimation. We then discuss units of measurement and conversion of one unit of measurement to another. The chapter ends with powers of 10, scientific notation, and logarithmic scales of representing measurements.

MERCURY AND THE INUIT OF GREENLAND

The traditional diet of Greenland Inuit is dominated by marine mammals such as small whales, seals, walrus, the occasional polar bear (yes, it's a marine mammal, *Ursus maritimus*!), fish such as cod and halibut, and seabirds. This meaty diet is rich in fats, proteins, and vitamins and is well suited to life in the cold Arctic. Inuit that adhere to the traditional diet are nearly free of heart disease and diabetes. Besides providing sustenance, marine species are an essential component of the culture of Greenland Inuit (Figure 1-1).

One might surmise that the Greenland Inuit, living so far north, would be isolated from the ill effects of an industrialized world. Scientists studying toxic contaminants used blood and breast milk from Arctic peoples as an uncontaminated "control" group.[1] The scientists reasoned that Arctic natives, living so far from industrialized society, would have few contaminants in their bodies. They were shocked to discover that the Inuit of Greenland have one of the highest concentrations of mercury (a heavy metal) in their blood and breast milk of any group of people on Earth. One out of six Greenland Inuit had mercury levels in their blood exceeding 200 parts per billion (200 ppb), a level classified as "acute mercury poisoning."[2] High levels of mercury in humans have been linked to neurological damage.

Figure 1-1 In Northwest Greenland, Kigutikak Duneq cuts a piece of *muktuk* (whale skin) to eat. *Photo by Bryan & Cherry Alexander Photography.*

China is the second largest energy consumer on the planet, and 75% of its energy comes from coal. Chinese coal-fired power plants and smaller-scale heaters emit about 300 metric tonnes (over 600,000 pounds) of mercury each year into the atmosphere. Winds carry this metal from Asia north and east to the Arctic, where the mercury drops out of the atmosphere with rain and snow. Mercury enters the food chain in oceanic microorganisms, small land plants, and animals, and then **bioaccumulates**, becoming more and more concentrated in the bodies of animals with each step up the food chain. Inuit, at the top of the food chain, eat predators such as seals that carry a potentially toxic load of mercury.

The story of mercury and the Greenland Inuit raises many questions. How much will China's demand for energy increase as its population grows and its economy becomes more industrialized? Will China develop cleaner energy sources that emit less mercury? How will global climate change affect atmospheric deposition of mercury in the high Arctic? Should the Inuit, especially pregnant women and the young, limit their consumption of seals and whales? If so, how will the Inuit pay for store-bought food and what impact will a changing diet have on their health? How will the problem of mercury contamination affect the culture of the Inuit?

To answer these questions, scientists must begin with careful, comprehensive measurements, such as China's population growth, the amount of mercury in Asian coal, the amount of metal transferred to the atmosphere, the percentage of mercury deposited from the atmosphere, the concentration of metals in humans and their food, and the effects of mercury on human health. In this chapter, we explore some of the basic issues scientists face with measurements.

MEASURING

We quantify environmental information using a wide variety of tools, such as a ruler to determine the length of a clam shell, an inclinometer to measure the angle to the top of a tree, a mass spectrometer to quantify the amount of mercury in tissue, or a satellite-mounted fluorometer to measure the amount of chlorophyll in the ocean. The choice of tool depends on what is available, how much money we are willing to spend, how accurate and precise the measurement must be, what tool is appropriate for the task, and so on.

EXAMPLE 1-1   Measuring Stream Velocity

The velocity or speed of a stream is an important part of the stream's ecology. Stream velocity influences the kinds of organisms that live on the stream bed. And stream velocity determines the sizes of particles that can be carried downstream. How can we measure stream velocity?

Solution One low-tech method is to toss a piece of wood into the stream, let the wood float downstream for a fixed distance (60 feet, for example), and measure the time it takes to traverse this distance (20 seconds, for example). Then the velocity is calculated by dividing distance by time:

$$\frac{60\,\text{feet}}{20\,\text{seconds}} = \frac{3\,\text{feet}}{1\,\text{second}}$$

Scientists often take multiple measurements and average the results. Always use the same or similar piece of wood; various shapes and sizes might influence the result. Place the wood in the same part of the stream for consistent and comparable results.

Figure 1-2 Student measuring stream velocity. The velocity probe is attached to a data collection device. *Photo: Langkamp/Hull*

There are more sophisticated and expensive devices, such as electronic current meters, that allow rapid measurement of velocity at any place and at any depth in the stream (Figure 1-2). With automated and computerized equipment, a more complete picture of stream velocities can be produced.

ACCURACY AND PRECISION OF MEASUREMENT

Imagine a student project quantifying the amount of lake habitat in central Florida. One of the first tasks involves measuring diameters of lakes on a map of central Florida. For example, Dinner Lake is a reasonably circular lake. The *actual* diameter of Dinner Lake is 322 meters, measured to the nearest meter. Five students in Group A measure the diameter of Dinner Lake. The results are shown in Table 1-1. Group

	Group A	Group B
TABLE 1-1	310 m	400.1 m
	320 m	399.2 m
	320 m	401.7 m
	330 m	398.3 m
	310 m	399.8 m

A's measurements, taken together, are accurate though not very precise. Five other students in Group B measure the diameter of Dinner Lake (Table 1-1). These measurements are precise, though not very accurate.

Accuracy refers to how close the measurement is to the actual value. Group A's measurements were fairly accurate; their results, taken together, were close to the actual diameter of 322 meters. But Group A's measurements had low precision; they only measured to the nearest 10 meters. **Precision** refers to the detail or refinement of the measurement. Group B was very precise, measuring to the nearest tenth of a meter. However, the second group was not accurate; none of their measurements were close to the actual value. Their effort to be precise was somewhat wasted, in this case, because of the low accuracy. A practical rule might be as follows: **accuracy first, precision second**.

EXAMPLE 1-2 MTBE in Gasoline

Suppose that groundwater beneath a gasoline station was contaminated with methyl tert-butyl ether (MTBE), a gasoline additive used to increase gas mileage by increasing combustion. Unfortunately, MTBE is a suspected cancer-causing agent. A sample of the groundwater right at the gasoline station was analyzed for MTBE, and its concentration in the groundwater was measured at 455 parts of MTBE in a billion parts of water (often phrased as 455 parts per billion). The threshold of measurement below which MTBE cannot be detected is about 1 part per billion. What can we say about the precision and accuracy of these measurements?

Solution The precision seems quite high. The procedure can detect values of MTBE as small as 1 molecule out of a billion molecules, which seems very precise. Nothing can be said about the accuracy, as we do not know the real value.

How many digits to the left or right of the decimal point should we report? A general rule is to report the same precision as measured. For example, the length of a clam shell would probably be measured to the nearest millimeter using a metric ruler; 66 millimeters or 6.6 centimeters (6.6 cm). We would not write 66.0 mm, because that implies a measurement to the nearest tenth of a millimeter. And we would not write 7 cm either, because we actually measured more precisely.

How many decimal places should we use when making calculations? One general suggestion is to carry one extra decimal place during calculations (one more decimal place than was present in the original values) and then round off the result when the calculations are completed. For example, consider a rectangular agricultural plot that is 12.3 meters long and 7.7 meters wide, with each side measured to the nearest tenth of a meter. The plot's area is therefore 94.71 m² (length times width), written to the nearest one hundredth of a square meter. We might carry this extra decimal place into the next calculation and then round off when calculations are finished.

How many decimal places should we round off to when reporting our final answer? This is a fairly complicated subject, whose mathematics is beyond the scope of this text. In applied environmental science problems, the number of decimal places reported also may depend on how the answer is to be used. Common sense derived from experience should play an important role. We will provide many examples throughout the text that illustrate rounding of answers.

EXAMPLE 1-3 Sagebrush Habitat

Students were asked to determine the amount of sagebrush cover on a **transect** (a linear traverse) by setting out a tape measure and measuring the amount of sagebrush intersected by the tape measure. Three sagebrush were crossed by a 13.00-meter tape, with lengths of 2.13 meters, 0.74 meters, and 1.33 meters (Figure 1-3). Calculate the fraction of sagebrush cover.

To calculate the fraction of sagebrush habitat on this transect, we add the three sagebrush lengths together and then divide by the total length of the traverse. The

Figure 1-3 Sagebrush along a 13-m tape measure (heavy line), looking down from above. Three sagebrush are intersected by the tape measure.

sum of the three sagebrush lengths is 4.2 meters. However, because we measured to the nearest centimeter, we would report this sum as 4.20 meters and retain the extra decimal place for the moment. The fraction of cover is

$$\frac{\text{sagebrush length}}{\text{total length surveyed}} = \frac{4.20 \text{ meters}}{13.00 \text{ meters}} = 0.3230769231$$

The sagebrush covers a bit less than a third of the ground. What number do we report? The students did not measure precisely enough to justify 10 decimal places. Reporting 0.323 as the fraction of sagebrush cover, with one more decimal place than originally measured, is reasonable.

ESTIMATION AND APPROXIMATION

If measurements and calculations appear to be complicated or lengthy, it is often a good strategy to make an **estimate** or an **approximation**. An estimate can be very useful as an independent check on intricate measurements or calculations. Estimates guard against some of the minor mistakes that can have large consequences. One common problem when working with very large or very small numbers is dropping or adding a zero or two; a small error with nasty consequences.

Suppose the task is to measure the length of the Little Knifepoint Glacier in Wyoming, as part of a study of global warming. Glaciologists have measured the length of the Little Knifepoint Glacier using sophisticated technology and have found the length to be 561 meters (to the nearest meter). We can make our own estimate of the length of the glacier by using the graphical scale at the bottom of the glacier's map (Figure 1-4). Hold your fingers apart over the length of the glacier and estimate the distance (between the arrows) using the scale bar. The Little Knifepoint Glacier appears to be about 0.6 km or 600 meters long, an estimate that is close to the actual value of 561 meters.

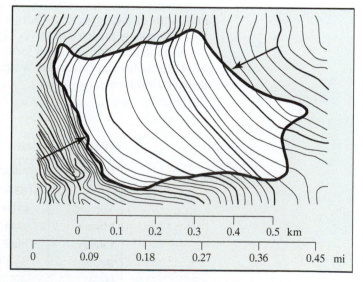

Figure 1-4 Topographic map of Little Knifepoint Glacier, Wyoming. The glacier is shown in white.

We can also estimate the result of a calculation. Glaciologists have determined the average width of the Little Knifepoint Glacier as 382 meters. Assuming a rectangular shape, the glaciologists then calculated the area of the glacial ice:

$$\text{area} = \text{length} \times \text{average width} = 561\,\text{m} \times 382\,\text{m} = 214{,}302\,\text{m}^2$$

The area of the Little Knifepoint Glacier is 214,302 square meters. Is this number reasonable? To make an estimate of the calculation, first round off the two distances:

$$\text{area} = 561\,\text{m} \times 382\,\text{m} \approx 500\,\text{m} \times 400\,\text{m}$$

We rounded the length down and rounded the width up. Rounding some numbers up and some numbers down should give us a better estimate.

$$\text{area} \approx 500\,\text{m} \times 400\,\text{m} = 200{,}000\,\text{m}^2$$

The Little Knifepoint Glacier covers about 200 thousand square meters of area. The estimate is very close to the actual value of 214,302 square meters and provides a quick check on the detailed calculation.

It's very common in environmental science to make "back of the envelope" estimates for many types of questions. For example, how do we estimate the volume of water an average household uses each year to wash dishes?

Let's assume an average household washes dishes once a day and that it uses one sinkfull of water for washing and rinsing. We can mentally compare the volume of the sink with a gallon of milk (for example) to estimate that a sink holds about 6 to 10 gallons of water:

$$\text{gallons per year} = 6 \text{ to } 10 \text{ gallons per day} \times 365 \text{ days per year}$$

The calculation can be rounded off for this estimate:

$$\begin{aligned} \text{gallons per year} &\approx 8 \text{ gallons per day} \times 400 \text{ days per year} \\ &= 3{,}200 \text{ gallons per year} \end{aligned}$$

The "back of the envelope" estimate of water use for dishwashing is about 3,000 gallons of water per year per household.

UNITS OF MEASUREMENT

Units are the conventions or "yardsticks" of measurement for various quantities such as dimension and size. For example, distance can be measured in familiar units such as millimeters, inches, feet, meters, and miles. Tables 1-2 through 1-4 give a few common units of measurement. There is a larger and more explanatory table of units in the Appendix at the back of this book; take a quick glance at it now. Tables 1-2 through 1-4 also include abbreviations for the different units listed. It is useful to memorize a few key abbreviations and to watch out for similar-appearing units; for example, m is meter, but mi is mile.

TABLE 1-2 **Metric Units**

Distance	millimeter (mm)	meter (m)
Area	square meter (m^2)	hectare (ha)
Volume	cubic cm (cc or cm^3)	liter (l or L)
Mass	gram (g or gm)	tonne
Energy	joule (J)	kilojoule (kJ)
Power	watt (W)	gigawatt (GW)
Heat	calorie (cal)	kilocalorie (Cal)

TABLE 1-3 English Units

Distance	inch (in)	mile (mi)
Area	acre	square mile (mi^2)
Volume	quart (qt)	gallon (gal)
Mass	slug (no kidding)	
Energy	foot-pound (ft-lb)	erg
Power	ft-lb/sec	horsepower (hp)
Heat	British thermal unit (BTU)	

TABLE 1-4

Time	second (sec or s)	minute (min)	year (yr or a)
Temperature	Celsius ($°C$)	Fahrenheit ($°F$)	Kelvin (K)

Most scientists use the metric system. In this text, you will have plenty of opportunity to use metric units; however, we will provide both metric and English units as often as is reasonable.

There are a few simple rules for the arithmetic of units. Units that are the same can be added or subtracted without changing the unit. For example,

$$12 \text{ centimeters } + 9 \text{ centimeters } = 21 \text{ centimeters}$$

$$4 \text{ quarts } - 3 \text{ quarts } = 1 \text{ quart}$$

When multiplying and dividing the same units, treat the units like numbers. For example,

$$3 \text{ feet} \times 3 \text{ feet} = \underbrace{3 \times 3}_{\text{numbers}} \underbrace{\text{feet} \times \text{feet}}_{\text{units}} = 3^2 \text{ feet}^2 = 9 \text{ feet}^2$$

$$\frac{10 \text{ cm}^3}{2 \text{ cm}^2} = \frac{10}{2} \frac{\text{cm}^3}{\text{cm}^2} = 5 \frac{\text{cm} \times \cancel{\text{cm}^2}}{\cancel{\text{cm}^2}} = 5 \text{ cm}$$

Units can cancel completely during division. For example,

$$\frac{6 \text{ miles}^2}{12 \text{ miles}^2} = \frac{6}{12} \frac{\text{miles}^2}{\text{miles}^2} = 0.5 \frac{\cancel{\text{miles}^2}}{\cancel{\text{miles}^2}} = 0.5$$

This last result has no units.

UNIT CONVERSION

The Amazon River in South America is approximately 3,900 miles long. How many kilometers is that? In order to solve this problem, we need the **conversion identity** that relates miles to kilometers. A few common identities are given in Tables 1-5 through 1-7. More extensive tables are found in the Appendix.

TABLE 1-5 Metric to Metric

1 cm	=	10 mm	1 cm^3	=	1,000 mm^3
1 meter	=	100 cm	1 liter	=	1,000 cm^3
1 kilometer	=	1,000 m	1 m^3	=	1,000 liters
1 m^2	=	10,000 cm^2	1 kilogram	=	1,000 grams
1 hectare	=	10,000 m^2	1 tonne	=	1,000 kg
1 km^2	=	1,000,000 m^2	1 km^2	=	100 hectare

TABLE 1-6 English to English

1 foot	= 12 inches	1 quart	= 2 pints
1 yard	= 3 feet	1 quart	= 57.75 in^3
1 mile	= 5,280 feet	1 gallon	= 4 quarts
1 acre	= 43,560 ft^2	1 ft^3	= 7.4805 gallons
1 mile^2	= 640 acres	1 pound	= 16 ounces
1 quart	= 4 cups	1 ton (short)	= 2,000 pounds

TABLE 1-7 English to Metric

1 inch	= 2.54 cm	1 mile^2	= 2.59 km^2
1 foot	= 0.3048 m	1 quart	= 0.9464 liters
1 yard	= 0.9144 m	1 gallon	= 3.7854 liters
1 mile	= 1.609 km	1 pound	= 0.4536 kg
1 acre	= 4,046.9 m^2	1 ton (short)	= 0.9072 tonnes
1 acre	= 0.4047 ha	1 BTU	= 251.996 calories

From Table 1-7, we can see that 1 mile equals 1.609 kilometers (approximately). Because "1 mile" and "1.609 km" are equivalent, we can simply substitute for the length of the Amazon River in miles and evaluate:

$$3,900 \text{ mi} = 3,900 \ (1 \text{ mi}) \approx 3,900 \ (1.609 \text{ km}) \approx 6,300 \text{ km}$$

The previous unit change is the easiest kind of unit conversion problem; just substitute the equivalent unit and simplify to get the desired result. We refer to this approach as the **substitution method** of unit conversion.

EXAMPLE 1-4   *Prestige* Oil Spill Volume

In November 2002, the 25-year-old, single-hulled tanker *Prestige* broke in half off the Spanish coast (Figure 1-5).[3] This tanker was carrying about 19.6 million gallons of heavy fuel oil. How many liters of oil is this?

Solution From the conversion tables, we see that 1 gallon is 4 quarts, and 1 quart is approximately 0.9464 liters. We first convert gallons to quarts:

$$19.6 \text{ million gallons} = 19.6 \text{ million } (4 \text{ quarts}) = 78.4 \text{ million quarts}$$

We then convert quarts to liters:

$$78.4 \text{ million quarts} = 78.4 \text{ million } (0.9464 \text{ liters}) \approx 74.198 \text{ million liters}$$

Therefore, 19.6 million gallons of oil is about 74.2 million liters.

Figure 1-5 Oil from the 2002 *Prestige* spill. This heavy oil did not spread out as a thin film. Liencres, Cantabria, Spain. *Photo by Eduardo Cando (www.costaquebrada.com).*

Many conversion problems are *slightly* more complicated because there is no ready equivalency in a table. For example, suppose the average spacing between old-growth mahogany trees in a tropical rainforest is 36 meters. How many feet is this? From the conversion tables, we find that 1 foot = 0.3048 meters. This equivalency is useful but it is not in the right form for this problem. Therefore, we divide both sides by 0.3048, and simplify:

$$1 \text{ foot} = 0.3048 \text{ meters}$$

$$\frac{1 \text{ foot}}{0.3048} = \frac{0.3048 \text{ meter}}{0.3048}$$

$$3.281 \text{ feet} = 1 \text{ meter}$$

This new equivalency looks correct; 1 meter is a bit longer than 1 yard, which is 3 feet. We can now substitute this equivalency to get the spacing between old mahogany trees in feet:

$$36 \text{ meters} = 36 \ (3.281 \text{ feet}) \approx 118 \text{ feet}$$

EXAMPLE 1-5   *Prestige* Oil Spill Thickness

If all the fuel oil from the *Prestige* spilled out onto the surface in a layer 4 inches thick, the resulting oil slick would cover about 741,946 square meters in area. How many acres is this?

Solution From Table 1-7, 1 acre = 4,046.9 m². But we're converting a known number of square meters to an unknown number of acres. We need to rearrange the identity such that 1 m² = *xxx* acres. We divide both sides of the equality by 4,046.9, and simplify:

$$1 \text{ acre} = 4{,}046.9 \text{ m}^2$$

$$\frac{1 \text{ acre}}{4{,}046.9} = \frac{4{,}046.9 \text{ m}^2}{4{,}046.9}$$

$$0.0002471 \text{ acres} = 1 \text{ m}^2$$

This last result is a new equality in a form that is appropriate for the substitution method of unit conversion. Now substitute this identity to convert square meters to acres:

$$741{,}946 \text{ (m}^2) = 741{,}946 \text{ (0.0002471 acres)} \approx 183.3 \text{ acres}$$

The fuel oil carried by the *Prestige* would cover 183 acres (about a third of a square mile) to a depth of 4 inches.

Another common approach to unit conversion uses **conversion ratios**. The equivalency is expressed as a ratio of the two related units. This ratio is equal to 1 and therefore can be substituted readily into equations. For example, there are two conversion ratios between feet and meters:

$$\frac{1 \text{ foot}}{0.3048 \text{ meters}} = 1 \quad \text{or} \quad \frac{0.3048 \text{ meters}}{1 \text{ foot}} = 1$$

Let us try the mahogany spacing problem again, using a conversion ratio. We start with a 36-meter spacing and use the first of the two conversion ratios:

$$36 \text{ meters} = 36 \text{ m} \times \frac{1 \text{ ft}}{0.3048 \text{ m}} = \frac{36 \text{ m} \times 1 \text{ ft}}{0.3048 \text{ m}} = \frac{36}{0.3048} \frac{\cancel{m} \times \text{ ft}}{\cancel{m}} \approx 118 \text{ ft}$$

Once again we find that the mahogany trees are spaced about 118 feet apart.

The conversion ratio used in the mahogany problem can be expressed as feet per meter, or meters per foot. How do we know which ratio to use? There are two hints to assist you: (1) The units should cancel, and (2) The numbers should make sense.

1. *Watch the units.* We start with meters and want to convert to feet; the meters must cancel out. Therefore, multiply by the conversion factor that has meters in the denominator:

$$\text{meters} = \text{meters} \times \frac{\text{feet}}{\text{meters}} = \frac{\cancel{\text{meters}} \times \text{feet}}{\cancel{\text{meters}}} = \text{feet}$$

Notice how the meters in the numerator and denominator cancel out.

2. *Should the number go up or down?* Meters are larger units of distance than feet. When converting from meters to feet, the number of feet should be greater than the number of meters. Therefore, multiply by the conversion ratio that makes the number increase:

$$36 \text{ meters} \times \frac{1 \text{ foot}}{0.3048 \text{ meters}} \approx 118 \text{ feet}$$

EXAMPLE 1-6 Hectares in a Section of Land

A **section** of land in the section-township-range system used in the United States is 1 square mile in area. How many metric hectares are there in a section?

Solution Using the tables, we see that there are 640 acres in one square mile and that there are approximately 2.47 acres in one hectare. Therefore,

$$1 \text{ mile}^2 = 640 \text{ acres} = 640 \text{ acres} \times \frac{1 \text{ hectare}}{2.47 \text{ acres}} \approx 259 \text{ hectares}$$

The first part of this solution is a simple identity ($1 \text{ mile}^2 = 640$ acres). The second part involves multiplying by the correct conversion ratio. We may not know, in advance, whether the number of hectares should be smaller or larger than the number of acres. But we do know that the acres must cancel, leaving hectares:

$$\text{acres} = \text{acres} \times \frac{\text{hectare}}{\text{acres}} = \cancel{\text{acres}} \times \frac{\text{hectare}}{\cancel{\text{acres}}} \approx \text{hectares}$$

COMPOUND UNITS

Basic units of time, distance, mass, and so on can be combined to produce **compound units** for quantities such as velocity, density, and energy. For example, typical units of velocity are miles per hour (mi/hr), a unit of distance divided by a unit of time. Another unit of velocity is kilometers per second (km/sec).

Converting one compound unit to another is straightforward but can involve a large number of steps. For example, to convert miles per hour to kilometers per second, we have to convert both the distance in the numerator (miles to kilometers) and the time in the denominator (hours to seconds):

$$70 \frac{\text{mi}}{\text{hr}} = 70 \frac{\text{mi}}{\text{hr}} \times \left(\frac{1.6 \text{ km}}{1 \text{ mi}} \right) \times \left(\frac{1 \text{ hr}}{60 \text{ min}} \times \frac{1 \text{ min}}{60 \text{ sec}} \right) \approx 0.031 \frac{\text{km}}{\text{sec}}$$

Study this example to confirm that the units cancel correctly to yield $\frac{\text{km}}{\text{sec}}$. Also check whether the number makes sense. Seventy miles in one hour should be a small distance in one second.

We also recommend writing these conversions with the units and values "stacked" on top of each other, as shown in the preceding examples. For example, we recommend $\frac{50 \text{ gallons}}{6 \text{ minutes}}$, not 50 gallons/6 minutes. When the units and values are written in a "stack," it is easy to see how the units cancel, and it is easy to see how to arrange the conversion ratios. Make sure the units cancel properly and check the result.

Another recommendation for avoiding mistakes in a complicated unit conversion is to arrange the numbers on the left and units on the right. This recommendation is illustrated in Example 1-7.

EXAMPLE 1-7 Acre-Feet of Water

A cotton farmer applied 2 acre-feet of irrigation water to his cotton crop. How many cubic feet is this volume?

Solution The first challenge is to recognize that the unit "acre-foot" means "acre times foot." The hyphen is often used instead of a multiplication symbol in this particular compound unit. A unit of area times a unit of length is a volume. To convert acre-feet to cubic feet, we must know the number of square feet in an acre. From Table 1-6 we see that

$$1 \text{ acre} = 43{,}560 \text{ ft}^2$$

This equivalency is shown in Figure 1-6.

We can substitute this equality to change acres to square feet:

$$2 \text{ acre-feet} = 2 \text{ acre} \times 1 \text{ foot} = 2(43{,}560 \text{ ft}^2) \times 1 \text{ foot}$$

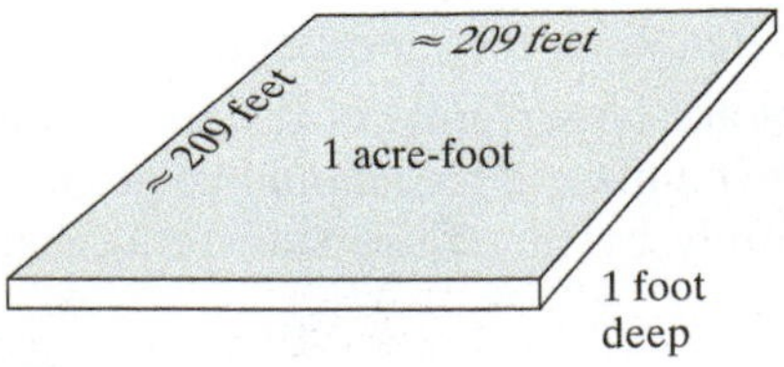

≈ 209 feet
≈ 209 feet
1 acre-foot
1 foot deep

Figure 1-6 1 acre-foot of volume. The 1-foot depth has been exaggerated.

We can rewrite the rightmost portion of the previous calculation so that the numbers are on the left and the units are on the right:

$$2\,(43{,}560\,\text{ft}^2) \times 1\,\text{foot} = \underbrace{2 \times 43{,}560 \times 1}_{\text{numbers together}} \quad \underbrace{\text{ft}^2 \times \text{ft}}_{\text{units together}} = 87{,}120\,\text{ft}^3$$

There are $87{,}120\,\text{ft}^3$ of water in 2 acre-feet of water. Rearranging the computation with the numbers on the left and the units on the right can help with complicated conversions.

Converting compound units can be tedious, but these conversions arise all the time. In 2002, the U.S. Coast Guard seized a fishing vessel carrying 32 tons of shark fins, destined for the Asian soup market.[4] Sharks are a very important part of marine ecosystems, yet they are routinely killed just for the fins. How many sharks were killed to produce 32 tons of fins? An average shark might have 1.5 pounds of fin. Using this estimate, we can convert tons of shark fins to number of sharks in two steps:

$$32\,\text{tons of shark fins} \times \frac{2{,}000\,\text{pounds}}{1\,\text{ton}} = 64{,}000\,\text{pounds of shark fins}$$

$$64{,}000\,\text{pounds of shark fins} \times \frac{1\,\text{shark}}{1.5\,\text{pounds of shark fins}} \approx 43{,}000\,\text{sharks}$$

Over 40,000 sharks were killed just for their fins. The rest of the body parts were tossed overboard.

UNITS IN EQUATIONS AND FORMULAS

Units are sometimes more important than numbers in understanding reality-based mathematics problems and getting the computations right. Here is an example where units aid hydrologists studying stream flow.

EXAMPLE 1-8   Owen's River Discharge

The size of a river is often characterized by its **discharge**, the volume of water passing by a measuring point on the river in a certain amount of time. Units of discharge might be gallons per minute, cubic feet per second, or cubic meters per second. Imagine standing in a small stream, with a 5-gallon bucket in hand, trying to scoop out all of the water passing by every second. This cannot be done, even with a tiny stream.

The Owens River originates on the east side of the Sierra Nevada near the California-Nevada border. A 200-mile-long aqueduct brings the river water around the mountains to Los Angeles (Figure 1-7). In 1884, J. M. Keeler declared that

Figure 1-7 The Owens River is diverted through a large pipe to Los Angeles. *Courtesy of Eric and Avery Wilmanns.*

the "Owens River carries a volume of water, carefully measured, fifty feet wide, average depth six feet, flow [7 feet per second]; and . . . will by irrigation give to the county 50,000 acres of fine agricultural land."[5] How is the discharge of the Owens River determined using this information?

Solution Discharge cannot be measured directly but can be *calculated* from other measurements. The measurements of width, depth, and velocity can be combined to calculate discharge:

$$\text{width} \times \text{average depth} \times \text{average velocity}$$

We can check the mathematics with English units:

$$\text{feet} \times \text{feet} \times \frac{\text{feet}}{\text{sec}} = \frac{\text{ft}^3}{\text{sec}}$$

The result is volume per time. Unit analysis shows that discharge can be calculated by multiplying width, average depth, and average velocity. The discharge of the Owens River in 1884 was

$$50 \text{ feet} \times 6 \text{ feet} \times \frac{7 \text{ feet}}{\text{sec}} = \frac{2{,}100 \text{ ft}^3}{\text{sec}}$$

You will see more examples of units in equations in Chapter 4 on linear functions.

UNIT PREFIXES

The Darlington nuclear power facility in Ontario Province, Canada, can produce up to 3,524,000,000 watts of power when operating at peak capacity.[6] Cryptosporidium is a microbe that can contaminate municipal water supplies. Cryptosporidium is usually about $\frac{6}{1{,}000{,}000}$ of a meter in length; a very fine mesh must be used to filter out this microbe.

These very large and very small numbers can be difficult to convey in writing and speech and can easily lead to mathematical errors during computation. To simplify these numbers, scientists and others use **unit prefixes** (Table 1-8). A larger table of unit prefixes is given in the Appendix.

TABLE 1-8

Prefix	Symbol	Value	Exponential Equivalent	Example Using Prefix
giga	G	1 billion	10^9	gigawatt (GW)
mega	M	1 million	10^6	megayear (Ma)
kilo	k	1 thousand	10^3	kilopascal (kPa)
centi	c	1 hundredth	10^{-2}	centimeter (cm)
milli	m	1 thousandth	10^{-3}	milliliter (mL)
micro	μ	1 millionth	10^{-6}	micrometer (μm)

The Darlington nuclear facility generates 3.524 billion watts or 3.524 gigawatts of electricity, abbreviated 3.524 GW. Cryptosporidium is usually about $\frac{6}{1{,}000{,}000}$ of a meter in length, or 6 one-millionths of a meter, or 6 micrometers, abbreviated 6 μm.

The prefix **milli** is abbreviated "m," which is also the abbreviation for **meter**. "Milli" is a prefix and so will always precede another unit. A milliliter (ml or mL) is a thousandth of a liter, and a millimeter (mm) is a thousandth of a meter.

As can be seen in Table 1-8 and the Appendix, there's a prefix for some of the powers of 10 but certainly not all of them. With one exception (centi), the prefixes in Table 1-8 and the Appendix correspond to powers of 10 whose power or superscript is a multiple of 3.

EXAMPLE 1-9   Heat from Coal

Coal provides between 9,500 and 14,000 British thermal units (BTUs) of heat energy per pound. Express this range of values in calories. Then rewrite both the BTUs and calories using unit prefixes.

Solution There are 251.996 calories per BTU; therefore, the range of heat production for coal is 2,393,962 to 3,527,944 calories per pound. Coal generates 9.5 to 14 kiloBTUs or approximately 2.393 to 3.528 megacalories of energy per pound.

SCIENTIFIC NOTATION AND ORDER OF MAGNITUDE

The Darlington nuclear power facility produces about 3,524,000,000 watts of electricity at peak capacity, or 3.524 billion watts. The word *billion* represents the number 1,000,000,000, which can be written as 10^9. In other words, the Darlington nuclear power facility can produce up to 3.524×10^9 watts of electricity. Numbers expressed in **scientific notation** have a decimal value between 1 and 10 multiplied by 10 raised to some integer power. Some examples are 2×10^3, 5.7×10^{-3}, and 9.4×10^{11}.

EXAMPLE 1-10   Size of Cryptosporidium Cysts

Cryptosporidium cysts have a diameter of approximately 0.000006 meters or $\frac{6}{1,000,000}$ meters. What is that value in scientific notation?

Solution We know that 1 one-millionth is equal to 10^{-6}. Therefore, 6 one-millionths of a meter can be expressed as 6×10^{-6} meters.

USING TECHNOLOGY: SCIENTIFIC NOTATION

The size of a cryptosporidium cyst in scientific notation (6×10^{-6}) is typically entered into the calculator as $6 \times 10 \wedge -6$. Many calculators will have a shortcut key that can be used for scientific notation such as **E** or **EE** (e.g., 6 E −6), **10^x** (e.g., 6×10^{-6}), or **y^x**, where 10 can be substituted for y. More details for the TI-83/84 graphing calculator are given next.

Number Display on the TI-83/84

Press **MODE** for three choices: **Normal** (typical display), **Sci** (scientific notation) and **Eng** (engineering notation). Using each choice, multiply 12,345 times 12,345. "E8" stands for "$\times 10^8$". The number of decimal places is controlled by the second line under **MODE**.

```
12345*12345
           152399025
12345*12345
      1.52399025E8
12345*12345
      152.399025E6
```

Using scientific notation, it is much easier to compare measurements or quantities when those quantities show a wide range of values. For example, imagine three different electricity generating plants of different sizes; 3,524,000,000 watts, 17,700,000 watts, and 5,500 watts (Table 1-9). Using scientific notation, we can express these three values as 3.524×10^9 watts, 1.77×10^7 watts and 5.5×10^3 watts. Scientific notation draws our attention to the power of 10 or **order of magnitude** of a number. The first electrical plant is approximately 2 orders of magnitude (10^2 times) greater in capacity than the second, and the second plant is a bit under 4 orders of magnitude (10^4 times) greater in capacity than the third.

TABLE 1-9

Plant Size	Scientific Notation
3,524,000,000 watts	3.524×10^9 watts
17,700,000 watts	1.77×10^7 watts
5,500 watts	5.5×10^3 watts

Scientific notation is also very useful in simplifying the arithmetic of very large or very small numbers.

EXAMPLE 1-11   U.S. Per Capita Energy Consumption, 2001

The United States consumed approximately 97,000,000,000,000,000 British thermal units of energy in 2001.[7] In the same year, the U.S. population was approximately 285,000,000 people.[8] What was the per capita energy consumption?

Solution   Energy consumption and U.S. population in scientific notation are 9.7×10^{16} BTU and 2.85×10^8 people, respectively. Therefore, per capita energy use was

$$\frac{9.7 \times 10^{16} \text{ BTU}}{2.85 \times 10^8 \text{ people}} = \frac{9.7}{2.85} \times 10^8 \frac{\text{BTU}}{\text{person}} \approx 3.40 \times 10^8 \frac{\text{BTU}}{\text{person}}$$

POWERS OF 10 AND LOGARITHMS

Not satisfied with scientific notation, mathematicians and scientists have developed an even more compact method of expressing large and small numbers. Let's return to the Darlington electrical facility, which has a 3,524,000,000-watt or 3.524×10^9–watt capacity. This generating capacity is somewhere between 10^9 and 10^{10} watts, some *noninteger power* of 10. What is that fractional power? In other words, what value of x makes $10^x = 3,524,000,000$?

To estimate the answer, we start by using the calculator to "guess and check," a process more formally known as **successive approximation**. Most scientific calculators will have a **10^x** key. We plug in some numbers between 9 and 10 and track the results (Table 1-10).

TABLE 1-10

1st guess	10^9	1,000,000,000
2nd guess	$10^{9.1}$	1,258,925,412
3rd guess	$10^{9.5}$	3,162,277,660
4th guess	$10^{9.55}$	3,548,133,892
5th guess	$10^{9.54}$	3,467,368,505
6th guess	$10^{9.545}$	3,507,518,740
7th guess	$10^{9.548}$	3,531,831,698
8th guess	$10^{9.547}$	3,523,708,710
9th guess	$10^{9.5471}$	3,524,520,168

With a few more guesses we find that

$$3,524,000,000 \text{ watts} \approx 10^{9.54704} \text{ watts}$$

The Darlington nuclear complex can produce about $10^{9.54704}$ watts of electricity.

There has to be a more direct way to get the power of 10, and there is: the **LOG** or **LOG$_{10}$** key on your calculator:

$$\log(3,524,000,000) \approx 9.54704$$

"Log" is an abbreviation for **logarithm**. In simple terms, the logarithm of a number gives the power of 10 that represents that number. In other words, the logarithm gives the precise order of magnitude based on 10. One hundred can be represented as 10^2, and so the logarithm of 100 is 2 (the power of 10 that makes 100).

$$\log(100) = \log(10^2) = 2$$

One one-hundredth $\left(\frac{1}{100}\right)$ can be represented as 10^{-2}; therefore, the logarithm of 10^{-2} is -2:

$$\log\left(\frac{1}{100}\right) = \log(10^{-2}) = -2$$

Table 1-11 gives some numbers and their logarithms so you can see the pattern. Try computing these logarithms with your calculator.

TABLE 1-11

Number	Scientific Notation	Logarithm
1/1,000,000,000	10^{-9}	-9
1/1,000	10^{-3}	-3
1/10	10^{-1}	-1
1	10^0	0
10	10^1	1
10,000	10^4	4
1,000,000	10^6	6

It is assumed that the word *log* refers to logarithms based on 10. For this reason, log is often written as $\log_{10}$ and pronounced "log base 10." Anytime you see "log" it means $\log_{10}$, and vice versa.

Any positive real number can be a base of a log. Other logarithms include $\log_2$, $\log_7$, and $\log_e$, where e is Euler's constant (approximately 2.718281828). When working with logs that have a base not equal to 10, it is necessary to indicate the base using a subscript. These logs usually do not have their own keys on the calculator, with the exception of $\log_e$, which is typically the **LN, Ln,** or **ln** key on the calculator. The logarithms $\log_{10}$ and $\log_e$ are used so often that they have been given special names. $\log_{10}$ is called the **common logarithm** and $\log_e$ is called the **natural logarithm**. As mentioned above, "log" means $\log_{10}$ and LN means $\log_e$. In this book we will be using $\log_{10}$ exclusively.

LOGARITHMIC SCALES

In the previous section, we learned how to represent a single value (3,524,000,000) by its logarithm (≈ 9.54704), the power of 10 that represents that number. When values of size range over many orders of magnitude, scientists may substitute the logarithms for the original numbers and focus on the differences in orders of *magnitude* rather than the differences among the numbers themselves. This approach creates a **logarithmic scale** of the values.

EXAMPLE 1-12   pH

One measure of the chemical composition of liquids is **pH**, which stands for the "power of hydrogen." pH is a logarithmic scale based on powers of 10 and is a measure of the concentration of hydrogen ions (H^+) in a liquid. In distilled water at a temperature of 25°C, there are 1.0×10^{-7} **moles** of H^+ in a liter. A mole is a very large number of ions or molecules (see the Appendix). By definition, a liquid with this concentration of hydrogen ions, 1.0×10^{-7} moles/liter, is said to be **neutral**.

To calculate the pH of a liquid, we take the negative of the logarithm of the H^+ concentration in moles/liter:

$$pH = -\log (\text{concentration of } H^+)$$

For example, the pH of a neutral liquid is 7 because

$$-\log (1.0 \times 10^{-7} \text{ moles/liter}) = 7$$

Lemon juice has a much higher concentration of hydrogen ions than distilled water; 1.0×10^{-2} moles of H^+ per liter of liquid is a typical concentration in lemon juice. What is the pH of lemon juice?

$$\text{lemon juice pH} = -\log(1.0 \times 10^{-2}) = 2$$

Lemon juice has a pH of 2; liquids with low pH (that is, high concentrations of H^+) are **acidic** (Table 1-12).

TABLE 1-12

Liquid	Hydrogen Ions	pH	
Hydrochloric acid	1×10^{0} moles/liter	0.0	acidic
Lemon juice	1×10^{-2} moles/liter	2.0	
Distilled water	1×10^{-7} moles/liter	7.0	neutral
Blood	4×10^{-8} moles/liter	7.4	
Ammonia	1×10^{-11} moles/liter	11.0	basic
Sodium hydroxide	1×10^{-14} moles/liter	14.0	

Normal human blood has a hydrogen concentration of about 4×10^{-8} moles/liter. Thus the pH of blood is

$$\text{blood pH} = -\log(4 \times 10^{-8}) = 7.4$$

Liquids with pH greater than 7 are **basic** in composition; ammonia has a pH of 11 (Table 1-12). The pH for common liquids ranges from a low of zero (1 mole/liter of hydrogen atoms, as found in pure hydrochloric acid) to a high of 14 (10^{-14} moles/liter of hydrogen atoms, as found in pure sodium hydroxide).

Each integer *increase* in pH corresponds to a 10-fold *decrease* in hydrogen concentration; a liquid with a pH of 8 is ten times more acidic than a liquid with a pH of 9. You can understand why scientists sometimes use a logarithmic scale as opposed to the plain old numbers. Common liquids from apple juice to household cleaners show an enormous range in hydrogen concentration; therefore, a logarithmic scale based on order of magnitude is much more convenient.

EXAMPLE 1-13 Sizes of Earthquakes

Earthquakes are the shaking of the ground caused by two blocks of rock moving past each other, relatively quickly, along a break in the rocks, called a **fault**. When blocks of rock move past each other, the irregular movement of the two blocks causes waves to radiate out in all directions. The more movement along the fault during an earthquake event, the bigger the waves produced and the more likely the ground shaking will cause significant damage. Earthquakes come in a wide variety of sizes, from the gigantic earthquake in Alaska in November 2002, which rattled the Trans-Alaska Pipeline, to tiny vibrations that can only be measured by ultrasensitive machines. Earthquakes measured by a typical set of earthquake recorders range over 9 orders of magnitude in size.

Scientists measure the *size* of an earthquake event today with something called the **moment**, which is the product of the rock strength, the area of the fault between the two blocks, and the average amount of movement or slip of one block past the other. But rather than reporting the moment or actual size of the earthquake, scientists report the **moment magnitude**, which uses the logarithm of the moment:

$$\text{moment magnitude} = \log\left(\frac{\text{moment}^{2/3}}{10^{10.7}}\right)$$

In the formula above, the moment has units of dyne-centimeters, which is a very small unit of force multiplied by a small unit of distance.

The January 17, 1995, Kobe earthquake in south-central Japan that killed over 5,000 people and caused $200 billion of damage had a moment of 2.5×10^{26} dyne-centimeters.[9] The moment magnitude is therefore

$$\text{Kobe magnitude} = \log\left(\frac{(2.5 \times 10^{26})^{2/3}}{10^{10.7}}\right)$$

In a complicated expression like the one above, it's a good strategy to break the computation into steps. Start by entering and evaluating the numerator inside the large parentheses on the calculator. Remember to enclose the 2/3 power in parentheses. Then divide the result by $10^{10.7}$ and take the logarithm. The moment magnitude of the 1995 Kobe earthquake is 6.9. See the following calculator screen.

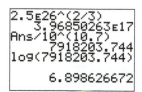

The November 3, 2002, Denali earthquake in central Alaska that rattled the oil pipeline had a moment of 8.0×10^{27} dyne-centimeters and a moment magnitude of 7.9.[10] The Denali earthquake was 32 times larger than the Kobe earthquake, but the magnitude between these two earthquakes differs by a value of only 1. Each unit of magnitude on the moment scale for earthquakes represents a 32-fold difference in size. A moment magnitude 6 earthquake is *not* twice as big as a moment magnitude 3 earthquake; it is 32^3 or 33,000 times larger.

CHAPTER SUMMARY

Measurements such as the weight of radioactive waste and the rate of immigration involve quantifying physical, chemical, and biological variables. Environmental scientists are always concerned about the accuracy and precision of their measurements. **Accuracy** refers to how close the measurement is to the actual value, whereas **precision** refers to the degree of detail or refinement of the measurement.

Units are the conventions of measurement; for example, distance can be measured in meters or inches. Units can be converted from metric to English or vice versa using **conversion identities** that relate one system to another. A mile of stream-side habitat is approximately 1.6 kilometers, for example. **Compound units** are created by multiplying or dividing one unit by another; for example, a whooping crane flying 30 miles per hour or a farmer applying an acre-foot of irrigation water.

Unit prefixes are employed in order to express large or small quantities in a more succinct form. A billion watts of electric power is a gigawatt of power, whereas a water filter with holes a millionth of a meter in diameter has micrometer-sized holes.

Another shorthand system uses **scientific notation** or **powers of 10**; in 2005, there were approximately 6.45×10^9 people on Earth. The size difference between two objects, based on powers of 10, is often called the **order of magnitude**. The rhinovirus (2×10^{-8} meters long) is two orders of magnitude smaller than the *E. coli* bacterium (2×10^{-6} meters). **Logarithms** can be used for precise calculation of powers of 10. The logarithm of 6.5×10^3 is 3.813; therefore, 6.5×10^3 hectares of nature preserve is equivalent to $10^{3.813}$ hectares of preserve.

When objects such as earthquakes or wildfires vary in size over many orders of magnitude, scientists may use a **logarithmic scale**, often based on powers of 10. For example, the pH scale is based on the logarithm of the hydrogen concentration; a liquid with a pH of 3 has 10 times the hydrogen concentration of a liquid with a pH of 4.

END *of* CHAPTER EXERCISES

Measuring

1. Measure the length (long axis) and width (perpendicular to length) of the butter clam shell shown in Figure 1-8. Alternatively, collect your own bivalve shell (clam, mussel, oyster, etc.) and measure its length and width. What problems did you encounter making these measurements?

Figure 1-8 Butter clam shell. *Photo: Langkamp/Hull.*

2. Measure the length (long axis) and width (perpendicular to the length) of the oak leaf shown in Figure 1-9. Alternatively, collect your own leaf and measure its length and width. What problems did you encounter making these measurements?

Figure 1-9 Red oak leaf. *Photo: Langkamp/Hull.*

Accuracy and Precision

3. The rate of evaporation of water from a reservoir was measured using microwave technology at 0.56 inches per day. The actual rate of evaporation was 0.49 inches per day. Discuss the accuracy of the microwave technology.

4. The thickness of soil in an agricultural plot was measured with ground-penetrating radar. The radar gave a thickness of 67 centimeters; the actual value, measured by

coring, was 126 centimeters. Discuss the accuracy of the ground-penetrating radar method.

5. The concentration of the gasoline additive methyl tert-butyl ether (MTBE) in groundwater beneath a leaking storage tank was measured at 3,144 parts of MTBE in a billion parts of groundwater (3,144 parts per billion parts). Discuss the precision of this measurement.

6. Researchers measured the concentration of water vapor in warm air using a laser system. The concentration of water vapor was found to be 28.3 grams of water vapor in 1,000 grams of air. Discuss the precision of this measurement.

Estimation

7. Estimate the gallons of gasoline consumed per year by a typical U.S. driver. How many miles are driven per year? What is a typical gas mileage?

8. Estimate the volume of water used per person per day in a typical household. Make separate estimates for bathing, drinking, cooking, washing clothes, and using the toilet.

9. Make an estimate of the amount of food (in pounds) one U.S. resident consumes each year. Include a brief explanation of the steps in this estimate and the result.

10. Make an estimate of the number of automobile tires discarded each year in the United States. Include a brief explanation of the steps in this estimate and the result.

Units

11. Give a common metric unit for mass and energy.

12. Give a common metric unit for volume and power.

13. "Acre-foot" is a unit of which quantity?

14. "Watt-hour" is a unit of which quantity?

Unit Conversion

15. A transect (a linear survey) through a tropical forest measured 13,267 feet in length. How many miles long was this survey?

16. Three and a half barrels of used oil were dumped next to a stream. How many gallons of oil was this?

17. 1.45 km^2 of wetlands were preserved for migrating waterfowl. How large is this area in hectares (ha)?

18. A committee on rail freight transport recommended increasing the mass carried by a trailer from 42 tonnes to 46 tonnes. How big are these masses, in kilograms?

19. A pickup truck was tested using both conventional diesel fuel and **biodiesel** made from vegetable oil. The pickup generated 137 horsepower running on regular diesel and 133 horsepower using biodiesel.[11] Convert these horsepower values to kilowatts.

20. A river otter traveled 60 miles in one season. How many kilometers is 60 miles?

21. The circumference of the Earth is approximately 40,000 km. Convert this distance to miles.

22. According to the Environmental Protection Agency (EPA), the total amount of toxic pollutants released to the U.S. environment in the year 2000 equaled 3.2205 million metric tonnes.[12] How many U.S. tons is this?

Compound Units

23. A small well in a rural village produces 3.5 gallons per minute. What is this discharge in $\frac{m^3}{hour}$?

24. How much mass is 1 m^3 of water, in kilograms? Recall that the density of water is $\frac{1\,gram}{1\,cm^3}$.

25. The energy utility PacifiCorp wants to remove its hydroelectric dam on the White Salmon River in Washington State.[13] By 2002, an estimated 2.4 million cubic yards of river sediment (mud, sand, and gravel) had accumulated in the Condit Dam reservoir since the dam was built in 1913. Calculate the average rate of sediment accumulation behind the dam over the 89-year period, both per year and per day.

26. Rain that falls on the Cascade Pole and Lumber Company in Tacoma, Washington, runs into a storm drain and then into the Puyallup River. About 24 million liters of rainwater per year are discharged by this storm drain. This rainfall runoff is contaminated with various chemicals, including arsenic, a toxic heavy metal. The concentration of arsenic in the storm drain water is about 170 micrograms for every liter of stormwater.[14] How many grams and kilograms of arsenic are being dumped into the Puyallup River each year by this storm drain?

Units in Equations

27. The distance water flows down a stream in a certain amount of time is given by

$$\text{distance (meters)} = 1.31\frac{?}{?} \times \text{time (seconds)}$$

What are the units at the question marks?

28. The temperature of the lower atmosphere decreases steadily as the height above the ground increases. The equation is

$$\text{temperature (°C)} = 18°C - (6.8\frac{?}{?}) \times \text{height (km)}$$

What are the units at the question marks?

Unit Prefixes

29. What is the abbreviation for a million watts of electricity?
30. What is the abbreviation for a thousandth (1/1,000) of a gram of mass?
31. A μm is how many meters in distance?
32. A GPa is how many pascals of pressure?

Scientific Notation and Order of Magnitude

33. In 2001, there were 4,650 species of named mammals compared with 1,025,000 named species of insects.[15] Express each of these abundances in scientific notation. About how many orders of magnitude do insect species exceed mammal species?

34. Brazoria County in Texas produced 12,961,149 tons of toxic waste in 1997, whereas Midland County in Texas produced 6,981 tons of toxics in the same period.[16] Express each of these amounts in scientific notation. By how many orders of magnitude do these counties differ in toxic waste output?

35. The average amount of **PM10** (particulate matter smaller than 10 microns) in Chicago's air was found to be 0.000025 grams of PM10 in a cubic meter of air.[17] The U.S. air quality standard for PM10 is 0.00015 grams per cubic meter of air. Express each of these values in scientific notation using units of grams per cubic meter. By how many orders of magnitude do these two measurements differ?

36. Soil erosion during rainfall events following wildfires in western Montana was measured. Erosion rates during intense, short-duration thunderstorms were as high as 40 tonnes of soil per hectare of land. Erosion rates during low-intensity, long-duration rainfalls averaged about 0.01 tonnes per hectare.[18] Express each of these erosion rates in scientific notation. By how many orders of magnitude do these two rates differ?

Powers of 10 and Logarithms

37. Without using a calculator, what is the log of 10^{-6}? the log of 10^8? the log of 100,000? the log of 0.001?

38. Without using a calculator, what is the log of 10^{-3}? the log of 10^{11}? the log of 10,000,000? the log of 0.000001?

39. Report the answers to the following, rounded to three decimal places. What is the log of 1,731,124? the log of 0.000352? Now express both of the original (unlogged) values in the form 10^x.

40. Report the answers to the following, rounded to three decimal places. What is the log of 71,333? the log of 0.489? Now express both of the original (unlogged) values in the form 10^x.

41. In 2000, lightning-caused wildfires burned 101,013 acres of National Park land and 1,676,414 acres of National Forest land in the United States.[19] Calculate the logarithms of these two acreages. By how many orders of magnitude do these two acreages differ?

42. Fort Lauderdale had a 2003 population of 162,917 residents, whereas Tarpon Springs had a 2003 population of 22,240.[20] Calculate the logarithms of these two populations. By how many orders of magnitude do these two populations differ?

Figure 1-10 A tsunami warning sign along the Oregon coast. Residents should run to high ground following a coastal earthquake or siren warning. *Photo: Langkamp/Hull.*

Logarithmic Scales

43. The December 2004 Indonesian earthquake triggered a giant *tsunami* (translated as "harbor wave"); the earthquake and tsunami killed around 250,000 people (Figure 1-10). The earthquake had a moment of 3.5×10^{29} dyne-centimeters.[21] What is the corresponding magnitude of this earthquake?

44. An earthquake took place off the coast of Mexico in June 2004. The moment for this earthquake is 1.0×10^{24} dyne-centimeters.[22] Is this a large, medium, or small earthquake?

45. A lake at the Wangaloa coal mine in New Zealand has a hydrogen ion concentration of 1.6×10^{-5} moles/liter.[23] What is the pH of this lake water? Is the lake acidic or basic?

46. The hydrogen concentration of Big Moose Lake in the Adirondack Mountains of New York has been monitored since 1992.[24] The average hydrogen concentration during this time period has been about 5.0×10^{-6} moles/liter. What is the pH of the lake water? How does this pH compare with neutral water?

SCIENCE IN DEPTH
Global Warming

With over 6 billion people on Earth, the impacts of humans on the natural environment are beginning to reach planetary proportions. Humans have become a major agent in rearranging the world's landscape, on the same order as landslides, rivers, and wind. About 100 billion metric tonnes of earth materials are moved every year by people.[25] Humans are a potent geologic force!

Humanity has also grown into an important agent of global climate change. Since the beginning of the Industrial Revolution, gases in the atmosphere created by human activities have increased exponentially along with the exponential increase in the human population. Two of these gases, carbon dioxide and methane, are accumulating rapidly in the lower atmosphere. Carbon dioxide in the Northern Hemisphere has increased from about 280 parts per million before the Industrial Revolution to about 380 parts per million today. Methane has increased in the atmosphere from 750 parts per billion to 1,750 parts per billion in this same time period.[26]

Carbon dioxide and methane play an important role in the **energy budget** of the lower atmosphere. Radiation from the Sun passes through the atmosphere, where it is absorbed by dark surfaces such as vegetation and parking lots, and is then reemitted in the form of infrared radiation or heat. Carbon dioxide and methane are both very efficient at absorbing this outgoing infrared radiation. Therefore, any increase in these "greenhouse gases" should produce an increase in the temperature of the lower atmosphere. And indeed, direct measurements around Earth show an average increase in the temperature of the lower atmosphere of $1°$ Celsius or so over the last 100–150 years.

What are the consequences of human-induced global warming? Global warming has, in some cases, benefited species. For example, pied flycatchers in Germany have had more reproductive success due to an increase in spring temperatures, when these birds lay their eggs and rear their young.[27] Wildflowers in New York State have bloomed earlier, in concert with warming temperatures.[28]

Human-induced global warming has also increased the temperature of the sea surface, with both physical and biological consequences. As sea surface temperatures increase, diseases among coral reefs have become more prevalent, producing bleaching of corals. Coral bleaching has been documented in many of the world's oceans.

In the Arctic, **tree line** has moved northward, with evergreen trees colonizing previously unforested areas, perhaps as a consequence of slightly warmer temperatures. The increased amount of carbon dioxide in the atmosphere may also benefit trees, as plants absorb carbon dioxide from the atmosphere and convert the carbon dioxide to carbohydrates for nourishment. Both extra nourishment of plants and an increase in temperatures may be the result of excess carbon dioxide in the atmosphere.

CHAPTER PROJECT:
MELTING OF THE ICE CAPS

Another consequence of global warming is the shrinking of the Earth's glaciers. Warming of the atmosphere generally results in both less precipitation as snowfall and more melting of glacial ice. Glaciers are the most important storehouse of fresh water, containing three times as much fresh water, in frozen form, than all the lakes, rivers, and groundwater combined. As glaciers shrink in size due to global warming, sea level will rise and threaten coastal communities.

In the companion project for this chapter found at the text's Web site (**enviromath.com**) you will determine the current volume of ice in the Greenland ice cap, the second largest glacier system on Earth. You'll determine how much sea level would rise if all of the Greenland and Antarctic glaciers were to disappear. And you'll investigate the impact on coastal communities. Check it out!

NOTES

[1] Marla Cone, *Silent Snow: The Slow Poisoning of the Arctic* (New York: Grove Press, 2005).

[2] Arctic Monitoring and Assessment Programme, http://www.amap.no/

[3] John Pickrell, "Oil Spills Pollute Indefinitely and Invisibly, Study Says," *National Geographic News*, November 22, 2002, http://news.nationalgeographic.com/news/2002/11/1122_021122_OilSpill.html

[4] Janet Raloff, "No Way to Make Soup—Thirty-Two Tons of Contraband Shark Fins Seized on the High Seas," *Science News Online* 162, no.10, 2002, http://www.sciencenews.org/articles/20020907/food.asp

[5] The Borax Museum, Death Valley National Park.

[6] Ontario Power Generation, http://www.opg.com/ops/N_darlington.asp

[7] Lawrence Livermore National Laboratory, U.S. Energy Flow Trends–2001, http://eed.llnl.gov/flow/01flow.php

[8] U.S. Census Bureau, National Population Estimates, http://www.census.gov/popest/archives/2000s/vintage_2002/NA-EST2002-01.html

[9] Dept of Civil and Environmental Engineering, University of California-Berkeley, "The Kobe Earthquake," http://www.ce.berkeley.edu/geo/research/Kobe/kobe.html

[10] U.S. Geological Survey Fact Sheet, FS014-03, http://pubs.usgs.gov/fs/2003/fs014-03/

[11] University of Idaho, Dept of Biological and Agricultural Engineering, http://www.uidaho.edu/bae/biodiesel/

[12] U.S. Environmental Protection Agency, Toxics Release Inventory (TRI), http://www.epa.gov/tri/tridata/index.htm

[13] Andrew Engelson, "Condit Dam Removal Hits Snags," *High Country News* 34, no. 23, December 9, 2002.

[14] Washington State Dept of Ecology, Water Quality Program, http://www.ecy.wa.gov/programs/wq/permits/permit_pdfs/cascade_pole/

[15] W. W. Gibbs, "On the Termination of Species," *Scientific American* 285, no. 5, November 2001, 40–50.

[16] U.S. Environmental Protection Agency, Toxics Release Inventory (TRI), http://www.epa.gov/tri/tridata/index.htm

[17] U.S. Environmental Protection Agency, AirData Program, http://www.epa.gov/air/data/index.html

[18] P. R. Robichaud, "Wildfire And Erosion: When to Expect the Unexpected," *Geological Society of America, Abstracts with Programs*, Paper 143-10, 2002.

[19] National Fire Information Center, National Fire News, http://www.nifc.gov/fireinfo/nfn.html

[20] U.S. Census Bureau, http://www.census.gov

[21] U.S. Geological Survey, Earthquake Hazards Program, http://neic.usgs.gov/neis/eq_depot/2004/eq_041226/

[22] U.S. Geological Survey, Earthquake Hazards Program, http://earthquake.usgs.gov/eqinthenews/2004/ci14065544/

[23]University of Otago (New Zealand), Dept of Geology, http://www.otago.ac.nz/geology/features/restoration/wangaloa/amdandph.html

[24]Adirondack Lakes Survey Corporation, http://www.adirondacklakessurvey.org/monthly2.html

[25]Roger LeB. Hooke, "On the Efficacy of Humans as Geomorphic Agents," *GSA Today* 4 (1994), 217, 224–225.

[26]U.S. Dept of Commerce, National Climate Data Center, http://lwf.ncdc.noaa.gov/oa/ncdc.html

[27.]J. P. McCarty, "Ecological Consequences of Recent Climate Change," *Conservation Biology* 15, no. 2 (2001), 320–331.

[28]Ibid.

2
Ratios and Percentages

The ivory-billed woodpecker (*Campephilus principalis*) once occupied much of the southeastern United States and Cuba, but logging and farming destroyed almost all of its habitat (Figure 2-1). The woodpecker was thought to be extinct, until 2004, when it was spotted in a swamp in Arkansas.[1] As of 2005, there have been seven sightings of both males and females, but how many ivory-bills are there?

It's neither possible nor desirable to trap and count all the animals in a population. Without a complete census, how do biologists estimate the size of a population? One approach is the **capture-recapture** method, which is based on ratios and proportions and involves tagging a small sample of the population. Ratios and proportions are discussed in this chapter, along with normalization, percentages, concentrations, probability, and recurrence intervals.

RATIOS

Figure 2-1 A pair of ivory-billed woodpeckers. The male has a red crest. *Original painting by Mark Bowers.*

In 1998, Tillamook County, Oregon, disposed of 30,700 tons of municipal garbage whereas Clatsop County disposed of 15,060 tons of garbage.[2] What is the **ratio** of Tillamook County garbage disposal to that of Clatsop County? The word *ratio* implies division. A simple estimate suggests that Tillamook County disposed of twice as much garbage as Clatsop County:

$$\frac{30,700 \text{ tons}}{15,060 \text{ tons}} = \frac{30,700 \text{ tons}}{15,060 \text{ tons}} = \frac{30,700}{15,060} \approx 2.04$$

The ratio of Tillamook to Clatsop garbage is indeed about 2. Notice that the units cancel in this problem; ratios of like quantities have no units.

The lack of units in a ratio can be very handy in some circumstances. The **map scale** (a ratio) is often given at the bottom of a standard United States Geological Survey topographic map (Figure 2-2).

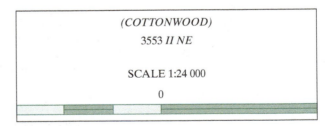

(COTTONWOOD)

3553 *II NE*

SCALE 1:24 000

0

Figure 2-2 Map scale

A common U.S. map scale is 1:24,000 or $\frac{1}{24,000}$. The map scale is the ratio of the *distance on the map* to the *distance in reality*. For the map scale of 1:24,000, one unit of distance measured on the map is equal to 24,000 of the same units in reality. For example, let us choose inches for the unit of distance. Then 1 inch measured on the map corresponds to 24,000 inches in reality, which is 2,000 feet. But we can use any kind of ruler or unit of measurement we want to! Suppose we choose centimeters. Then 1 cm measured on the map corresponds to 24,000 cm in reality. The distance scale, because it is a ratio of like quantities, is independent of units:

$$\frac{1 \text{ in}}{24,000 \text{ in}} = \frac{1 \text{ cm}}{24,000 \text{ cm}} = \frac{1}{24,000}$$

The ratio does not provide as much information as the original two values and therefore can be a problem under some circumstances. For example, suppose you read that Tillamook County disposed of twice as much garbage as Clatsop County in 1998. How much garbage did Tillamook County dispose? That question cannot be answered with the information given. Both counties may have been incredible polluters, disposing of millions of tons of garbage, or both may have been squeaky clean:

$$\frac{30,700 \text{ million tons}}{15,060 \text{ million tons}} = \frac{30,700 \text{ ounces}}{15,060 \text{ ounces}} \approx 2.04$$

Ratios appear frequently in environmental issues. The **energy payback ratio** of an electrical power generating plant is the ratio of the electrical energy produced to the energy costs of building, operating, and decommissioning the facility.[3] A natural gas–fired electrical plant might produce 11,350,000 gigajoules of electricity per year at a "cost" of 2,800,000 gigajoules of energy per year (a giga is one billion). The energy payback ratio is therefore

$$\frac{\text{electrical energy generated}}{\text{energy consumed by facility}} = \frac{11,350,000 \dfrac{\text{GJ}}{\text{yr}}}{2,800,000 \dfrac{\text{GJ}}{\text{yr}}} \approx 4.05$$

The energy payback ratio for this gas-fired plant is approximately 4; wind-driven electrical facilities have energy payback ratios of approximately 25.

NORMALIZATION

Ratios are one way of **normalizing** or standardizing environmental data in order to express relative values rather than actual values. One quantity is often normalized by a second quantity to eliminate the second quantity as a variable.

EXAMPLE 2-1   Prairie Dog Colonies

White-tailed prairie dogs (*Cynomys leucurus*) live in large colonies in the Rocky Mountain states and are the sole source of food for endangered black-footed ferrets (Figure 2-3). Sylvatic plague has decimated some prairie dog colonies, thinning out the populations. The Little Snake colony in Colorado has 36,875 prairie dogs in 31,624 hectares of area, whereas the Wolf Creek colony has 20,009 prairie dogs in 3,174 hectares.[4] Normalize the populations to decide which colony is more robust.

Solution The Little Snake colony appears healthier than the Wolf Creek colony. However, the Little Snake colony is spread out over 31,624 hectares of land, whereas the Wolf Creek colony covers only 3,174 ha. While the Wolf Creek colony is smaller in population size, it has a much higher density or concentration of prairie dogs (Table 2-1).

Figure 2-3 Prairie dogs. *U.S. Fish and Wildlife Service.*

TABLE 2-1

Colony Name	Prairie Dogs	Hectares of Land	Prairie Dogs per Hectare
Little Snake	36,875	31,624	1.17
Wolf Creek	20,009	3,174	6.30

Source: U.S. Dept of Interior

The Wolf Creek colony has a prairie dog density almost six times that of the Little Snake colony.

| EXAMPLE 2-2 | Chicken Production in the United States |

The United States produced 7 billion pounds of broiler chicken in 1965 and 34.2 billion pounds in 1995, a very large increase in chicken production.[5] The U.S. population grew from 194 million to 263 million in this same time period.[6] Has chicken production increased simply because of the population increase, or has demand for chicken gone up?

Solution We can normalize the chicken production to the U.S. population (Table 2-2). The rate of chicken production went from about 36 pounds per person in 1965 to about 130 pounds per person in 1995. U.S. residents are indeed consuming chicken at a much higher rate than in the past.

TABLE 2-2

	1965	1995
Chicken Production	7,000 million lb	34,200 million lb
U.S. Population	194 million people	263 million people
Chicken per Capita	36 lb per person	130 lb per person

Source: U.S. Dept of Interior and U.S. Census Bureau

Normalizing data to the human population is prevalent in environmental science, and therefore has its own special phrase. The **per capita** chicken production increased from 36 pounds to 130 pounds. *Per capita* means "per person."

PERCENTAGE AS A TYPE OF RATIO

A **percentage** is a ratio expressed as a part of 100 or per hundred. Suppose that in an environmental science class, there were 21 female students out of 35 total students. What percentage of this class was female? To calculate the percentage, take the ratio of the two values and multiply by 100%:

$$\frac{21 \text{ female students}}{35 \text{ total students}} \times 100\% = 0.6 \times 100\% = 60\%$$

Sixty percent of the students were women. The % sign is shorthand for "the number of parts per hundred parts." One hundred percent is 100 parts out of 100 parts or $\frac{100 \text{ parts}}{100 \text{ parts}}$ or 1. Multiplying by 100% is the same as multiplying by 1. Thus $0.6 = 60\%$.

If all the students were female (35 out of 35), the percentage of female students would be 100%, and if none of the students were female, the percentage of female students would be 0%. Notice the lack of units in percentages. The unit *students* is found in both the numerator and denominator and therefore cancels.

| EXAMPLE 2-3 | Recycling in Seattle |

According to Chuck Clarke, director of Seattle Public Utilities, Seattle residents recycled 72,000 tons of material in 2001, but 52,000 tons of recyclable materials were dumped in the garbage.[7] What percentage of potentially recyclable materials were actually recycled by Seattle residents in 2001?

Solution In order to calculate the percentage recycled, we must divide by the total of all recyclable materials. Therefore,

$$\frac{72,000 \text{ tons}}{72,000 \text{ tons} + 52,000 \text{ tons}} \times 100\% = \frac{72,000 \text{ tons}}{124,000 \text{ tons}} \times 100\%$$

$$= 0.5806 \times 100\%$$

$$\approx 58\%$$

Approximately 58% of the recyclable materials disposed of by Seattleites were actually recycled.

The previous examples used percentages to represent a part or fraction of a whole (21 female students out of 35 total students, for example). Percentages can also be used to express the ratio of *any* like quantities, as illustrated in the next two examples.

EXAMPLE 2-4 Texas Wetlands

The Texas Parks and Wildlife Department estimated in 1980 that there were 95,342 acres of inland swamp wetlands in Texas, compared with 611,760 acres of coastal marsh wetlands.[8] What was the ratio of inland wetland acreage to coastal wetland acreage?

Solution The ratio of inland wetlands to coastal wetlands is

$$\frac{95{,}342 \text{ acres}}{611{,}760 \text{ acres}} \times 100\% = 0.1558 \times 100\% \approx 15.6\%$$

EXAMPLE 2-5 Gasoline Mileage

The Ferrari 550 Maranello sports car gets about 14 miles per gallon of gasoline while driving on the freeway, while the Chevrolet Metro gets about 43 miles per gallon.[9] What percentage of the Metro's gas mileage is the Ferrari's? In other words, the Ferrari's mileage is what percentage of the Metro's?

Solution Take the ratio of the Ferrari's mileage to the Metro's, and multiply by 100%:

$$\frac{14\,\dfrac{\text{miles}}{\text{gallon}}}{43\,\dfrac{\text{miles}}{\text{gallon}}} \times 100\% = \frac{14\,\dfrac{\text{miles}}{\text{gallon}}}{43\,\dfrac{\text{miles}}{\text{gallon}}} \times 100\% \approx 32.6\%$$

The Ferrari gets only 33% (about a third) of the gas mileage of the Chevy Metro.

PARTS PER THOUSAND

We've seen that the percent symbol, %, means "parts per hundred parts." In a similar fashion, the "‰" symbol stands for "parts per thousand parts" or "ppt." Thus the shorthand for "17 parts out of 1,000" is 17 ppt or 17‰.

Ratios related to human populations are often expressed in parts per thousand. Suppose a birth rate averages 20 live births per 1,000 people. This birth rate is therefore

$$\frac{20 \text{ people born}}{1{,}000 \text{ people}} = \frac{20}{1{,}000} = 20 \text{ ppt} = 20\text{‰}$$

With a quick calculation you should find that 20 ppt is the same as 2.0%. Either expression is correct, but human population ratios are often expressed in parts per thousand.

EXAMPLE 2-6 Salinity of Seawater

Ions are charged elements or charged molecules. Common ions in seawater include sodium, chlorine, and bicarbonate. The **salinity** of seawater is typically expressed as the number of grams of extra ions in 1 liter or 1 kilogram of water. Because seawater typically has about 35 grams of extra ions in a liter of water, the typical salinity is

$$\frac{35 \text{ grams}}{1 \text{ liter}} = \frac{35 \text{ grams}}{1 \text{ kilogram}} = \frac{35 \text{ grams}}{1{,}000 \text{ grams}} = \frac{35}{1{,}000} = 35 \text{ ppt} = 35\text{‰}$$

The salinity of common seawater is 35 ppt.

It can be seen from this calculation that 35 grams of ions per kilogram of water is the same as 35 parts per thousand parts or 35 ppt. Indeed, there are usually several different ways of expressing the same ratio.

PARTS PER MILLION AND PARTS PER BILLION

When the ratio of two values is small, the ratio is often expressed as the number of parts out of a million parts or the number of parts out of a billion parts. Parts per million is abbreviated as "ppm" and parts per billion as "ppb." 170 ppm means that there are 170 things out of a total of 1 million things. 5 ppb means that there are 5 objects out of 1 billion objects, a very low ratio. Small ratios such as 170 ppm or 5 ppb are quite common in environmental science, especially in discussions of the **concentrations** of contaminants. For example, the concentration of lead in drinking water is often expressed as the number of grams of lead in a billion grams of contaminated water:

$$\frac{3 \text{ grams lead}}{1{,}000{,}000{,}000 \text{ grams total}} = \frac{3 \text{ parts}}{1 \text{ billion parts}} = 3 \text{ parts per billion} = 3 \text{ ppb}$$

Three grams is about a tenth of an ounce. A billion grams of water is a million kilograms, or about 2,200 U.S. tons. A concentration of 3 ppb is therefore equivalent to a tenth of an ounce out of 2,200 tons, a very low concentration.

Were a billion grams or 2,200 tons of contaminated water analyzed to discover that 3 grams were lead? Not likely! Analytical machines in a modern chemistry laboratory only need small quantities for detailed analysis. We reemphasize that 3 ppb is just a ratio and that the ratio by itself does not tell us which two values were used in calculating that ratio.

EXAMPLE 2-7 Arsenic in Fish

Arsenic is a heavy metal that readily accumulates in tissue and is passed up the food chain. Suppose the concentration of arsenic in a contaminated fish was given in a report as "3.5 mg/kg." What is the concentration of arsenic in this fish in parts per million?

Solution This problem combines unit manipulation with ratios. Our goal is to eliminate the units and end up with 1 million in the denominator. The units in the numerator of 3.5 mg/kg are milligrams or 0.001 grams. The units in the denominator are kilograms or 1,000 grams. Using powers of 10, we rewrite the original ratio:

$$3.5 \frac{\text{mg}}{\text{kg}} = 3.5 \frac{10^{-3} \text{g}}{10^3 \text{g}} = 3.5 \frac{10^{-3} \cancel{g}}{10^3 \cancel{g}} = 3.5 \frac{10^{-3}}{10^3}$$

Recall that $x^{-n} = \frac{1}{x^n}$ (more on the rules of exponents in Chapter 5). Therefore, we can move 10^{-3} into the denominator as 10^3, then combine powers of 10:

$$3.5 \frac{10^{-3}}{10^3} = 3.5 \frac{1}{10^3 \times 10^3} = \frac{3.5}{10^6} = 3.5 \text{ ppm}$$

We see that 3.5 mg/kg is the same as 3.5 ppm. When reading environmental literature, you will find that scientists use mg/kg and ppm interchangeably.

PERCENTAGE AS A MEASURE OF CHANGE

The amount of change with time can also be expressed as a percentage. The percent increase or decrease in a value from one time to the next is given by the **percentage change formula:**

$$\frac{\text{final value} - \text{initial value}}{\text{initial value}} \times 100\% = \% \text{ change}$$

This equation calculates the ratio of the *difference* between the two values compared to the initial or starting value and expresses this ratio in percentage form.

EXAMPLE 2-8 Colonies of Bacteria in a Lake

Suppose that one week, bacteria in a lake were averaging 720 bacterial colonies for every liter of lake water or 720 col./L. The next week, the concentration increased to 1,260 col./L. What was the percentage change with time?

Solution The initial concentration of bacterial colonies in the lake water sample was 720 col./L, while the final value was 1,260 col./L. The percent change from the first week to the second week was

$$\frac{1{,}260\,\frac{col.}{L} - 720\,\frac{col.}{L}}{720\,\frac{col.}{L}} \times 100\% \ = \ \frac{540\,\frac{col.}{L}}{720\,\frac{col.}{L}} \times 100\% \ = \ \frac{540}{720} \times 100\% \ = \ 75\%$$

There was a 75% increase in fecal coliform bacteria from one week to the next. In other words, the second week's concentration was 75% *greater than* the first week's concentration. The phrase *greater than* means "added onto" in this context. Adding 75% of the starting value onto the starting value yields

$$720 + (75\% \text{ of } 720) = 720 + (0.75 \times 720) = 720 + 540 = 1{,}260$$

Thinking about some simple cases involving percentage change might help to clarify this topic. If there is no change between the final and initial values, the percentage change is zero (0%). If the final value is greater than the initial, the percentage change is positive. If the final value is less than the initial, the percentage change is negative. If the final value is twice that of the initial, the percentage change is 100%. If the final value is half that of the starting value, the percentage change is −50%.

Percentage increases can be larger than 100%, but there is no percentage decrease less than −100%. Try substituting zero for the final value in the formula, and see what happens. If we are analyzing environmental data that show a change with time, and the starting value is zero, we cannot calculate the percentage increase because zero in the denominator is mathematically undefined. This is a special situation that comes up frequently.

EXAMPLE 2-9 Lead in Gasoline

Several decades ago, lead was added to gasoline to prevent "knocking" in vehicle engines and to increase fuel efficiency. A small amount of tetraethyl lead increases the combustion of gasoline. But ingestion of lead by people can cause health problems, especially among children. Regulations adopted by the U.S. government reduced lead in gasoline from 12 grams/gallon before 1985 to 0.5 grams/gallon after 1985 (more recent regulations have lowered the concentration even further).[10] What was the percentage reduction in lead concentration in gasoline?

Solution The percentage decrease can be calculated using the percentage change formula:

$$\frac{0.5\,\frac{g}{gal} - 12\,\frac{g}{gal}}{12\,\frac{g}{gal}} \times 100\% \ = \ \frac{-11.5\,\frac{g}{gal}}{12\,\frac{g}{gal}} \times 100\% \ \approx \ -95.8\%$$

There was about a 95.8% decrease of lead in gasoline following the new regulation. In other words, the second concentration is 95.8% *less than* the first concentration. The phrase *less than* in this context means "subtracted from." Subtracting 95.8% of the starting value from the starting value yields

$$12 - (95.8\% \times 12) = 12 - 11.5 = 0.5$$

In the lake bacteria example, the number of bacterial colonies increased by 75% from the first to the second week. Suppose that the bacteria increased by the same percentage from the second to the third week, and then again from the third to the fourth week. Values that grow or decay by a constant percentage with time are said to grow or decay **exponentially**. The topic of exponential growth and decay is discussed thoroughly in Chapter 5.

A final word of caution about percentages is appropriate. Suppose the population of wild condors in California (56 birds in June 2005[11]) remains unchanged for the next decade or so into the future. Then the percentage change will be

$$\frac{56 \text{ condors then} - 56 \text{ condors now}}{56 \text{ condors now}} \times 100\% = \frac{0 \text{ condors}}{56 \text{ condors}} \times 100\% = 0\%$$

However, the ratio of condors then to now (in percentage) is

$$\frac{56 \text{ condors then}}{56 \text{ condors now}} \times 100\% = 100\%$$

The condor population will show a 0% increase from the present to the future. The condor population in the future will be 100% of the present population. These two statements are equivalent but can be easily misread or misinterpreted.

PERCENTAGE DIFFERENCE AND PERCENTAGE ERROR

The percentage change formula can be used in situations where there is no change with time, that is, no initial and final values. We can use the formula to calculate the **percentage difference** between any two comparable values. For example, the volume of water in Lake Baikal is 23,600 km^3, whereas Lake Superior has 12,000 km^3 of water.[12] Let's compare the volume of Lake Baikal to the volume of Lake Superior, choosing Lake Superior as the reference. The **percentage difference formula** is

$$\frac{\text{comparison value} - \text{reference value}}{\text{reference value}} \times 100\% = \% \text{ difference}$$

Inserting the volumes of the two lakes appropriately, we get

$$\frac{23{,}600 \text{ km}^3 - 12{,}000 \text{ km}^3}{12{,}000 \text{ km}^3} \times 100\% = \frac{11{,}600 \text{ km}^3}{12{,}000 \text{ km}^3} \times 100\%$$
$$\approx 0.967 \times 100\%$$
$$= 96.7\%$$

Lake Baikal is about 96.7% more voluminous than Lake Superior, meaning that Lake Baikal is nearly twice as big as Lake Superior. Now let us switch the two lakes around and compare Lake Superior to Lake Baikal:

$$\frac{12{,}000 \text{ km}^3 - 23{,}600 \text{ km}^3}{23{,}600 \text{ km}^3} \times 100\% = \frac{-11{,}600 \text{ km}^3}{23{,}600 \text{ km}^3} \times 100\%$$
$$\approx -0.492 \times 100\%$$
$$= -49.2\%$$

Lake Superior is about 49.2% smaller than Lake Baikal. In other words, Lake Superior is approximately half of Lake Baikal's volume.

The percentage difference formula can also be used to calculate the **percentage error** associated with a measurement. Percentage error indicates how close a measurement is to the true or correct value. The **percentage error formula** is

$$\frac{\text{measured value} - \text{true value}}{\text{true value}} \times 100\% = \% \text{ error}$$

The "General Sherman" sequoia tree growing in Sequoia National Park in California is 83.82 meters tall.[13] Assume that this value for the tree's height is correct. Suppose we visit the tree and measure its height using technology. Our result is 84.71 meters. We can calculate the percentage error in our measurement as follows:

$$\frac{84.71\ m - 83.82\ m}{83.82\ m} \times 100\% = \frac{0.89\ m}{83.82\ m} \times 100\%$$
$$\approx 0.0106 \times 100\%$$
$$= 1.06\%$$

There was a 1.06% error in attempting to measure the height of this giant sequoia. Not bad!

The three percentage formulas (percentage change with time, percentage difference, and percentage error) all have a similar structure. In each, a reference value was chosen, and the difference from that reference value was expressed as a percentage.

PROPORTIONS

A **proportion** is an equality between two ratios. Suppose an 18-hole golf course in Arizona uses 100 million gallons of water per year for irrigation. A 27-hole course will therefore require 150 million gal/yr, because both have the same ratio of gallons per hole:

$$\frac{100\ \text{million}\ \frac{gal}{yr}}{18\text{-hole course}} = \frac{150\ \text{million}\ \frac{gal}{yr}}{27\text{-hole course}} \approx \frac{5.6\ \text{million}\ \frac{gal}{yr}}{1\ \text{hole}}$$

In many proportions, a value for a numerator or denominator is unknown, and we must solve for the unknown quantity. For example, consider the proportion

$$\frac{x}{10} = \frac{6}{4}$$

To solve for x, we multiply both sides of the equation by 10. The result is

$$x = \frac{6}{4} \times (10) = 15$$

In other proportions, the unknown resides in the denominator:

$$\frac{5}{3} = \frac{20.5}{x}$$

One simple way to solve for x is to first cross-multiply, then isolate the unknown:

$$5x = (20.5)(3)$$
$$x = \frac{(20.5)(3)}{5}$$
$$x = 12.3$$

Our result tells us that the ratio 5/3 is the same as the ratio 20.5/12.3, which seems reasonable.

EXAMPLE 2-10 Spacing of Turtle Nests

Green sea turtles dig their nests near the high-tide mark (Figure 2-4). Suppose a detailed map shows the location of turtle nests along a stretch of beach. The scale of the map reads "one centimeter equals 2.5 meters." The map distance between two adjacent turtle nests is 6.8 map centimeters. What is the actual separation on the beach in meters?

Solution We first set up the correct proportion:

$$\frac{2.5\ \text{real meters}}{1\ \text{map centimeter}} = \frac{x}{6.8\ \text{map centimeters}}$$

Figure 2-4 Green sea turtle. *U.S. Fish and Wildlife Service.*

Now multiply both sides of the equality by 6.8 map centimeters to isolate x, and simplify:

$$\frac{2.5 \text{ real m} \times 6.8 \text{ map cm}}{1 \text{ map cm}} = x$$

$$17.0 \text{ real meters} = x$$

The two turtle nests are 17.0 meters apart on the beach. Notice that the units "map cm" cancel when solving for x, yielding the correct units for x.

EXAMPLE 2-11 Ions in Seawater

Seawater contains charged elements or molecules (called **ions**) such as sodium, chlorine, bicarbonate, and sulfate. Table 2-3 gives the typical concentrations of 7 ions in seawater in units of grams of ion per kilogram of seawater.[14]

TABLE 2-3

Ion	Grams per kg
Chlorine	19.35
Sodium	10.76
Sulfate	2.71
Magnesium	1.29
Calcium	0.41
Potassium	0.39
Bicarbonate	0.14

Source: *Chemical Oceanography*

The amounts of these ions vary from place to place, but the **Principle of Constant Proportion** states that the *ratios* of these ions remains constant. For example, the ratio of magnesium to calcium in seawater is always

$$\frac{\text{magnesium}}{\text{calcium}} = \frac{1.29 \frac{\text{g}}{\text{kg}}}{0.41 \frac{\text{g}}{\text{kg}}}$$

Suppose a sample of seawater has a magnesium concentration of 1.55 g/kg, which is higher than usual. What is the concentration of calcium in that sample, assuming the Principle of Constant Proportion holds?

Solution We can set up the proportionality:

$$\frac{\text{typical magnesium}}{\text{typical calcium}} = \frac{\text{enriched magnesium}}{\text{enriched calcium}}$$

Substituting values yields

$$\frac{1.29\,\frac{g}{kg}}{0.41\,\frac{g}{kg}} = \frac{1.55\,\frac{g}{kg}}{x}$$

To solve, cross-multiply and then isolate the unknown x:

$$\left(1.29\,\frac{g}{kg}\right)x = \left(0.41\,\frac{g}{kg}\right)\left(1.55\,\frac{g}{kg}\right)$$

$$x = \frac{\left(0.41\,\frac{g}{kg}\right)\left(1.55\,\frac{g}{kg}\right)}{\left(1.29\,\frac{g}{kg}\right)}$$

$$\approx 0.49\,\frac{g}{kg}$$

There should be about 0.49 grams of calcium for every kilogram of ion-rich seawater.

The Principle of Constant Proportion for seawater may seem a little odd. Why should the abundance of one ion be "linked" to all the others? Evaporation of seawater involves liquid H_2O (water) converting to gaseous H_2O (water vapor). Evaporation removes some of the water from the sea but leaves the ions behind. Therefore, the ions increase in abundance as the water evaporates, but the *ratios* of ions remain the same. A similar, but opposite, effect results from dilution of seawater by fresh water.

Proportions can be used to solve percentage problems. For example, in March 2004, the U.S. Coast Guard boarded a New England fishing vessel and discovered 1,100 lobsters, of which 1,050 were females that should have been released upon capture.[15] The percentage of illegal lobsters can be calculated using a proportion:

$$\frac{x}{100\%} = \frac{1{,}050\text{ illegal lobsters}}{1{,}100\text{ total lobsters}}$$

Solving for x yields

$$x = \frac{(100\%)(1{,}050\text{ illegal lobsters})}{1{,}100\text{ total lobsters}} \approx 95\%$$

Proportions can also be used to solve unit conversion problems (see Chapter 1). For example, a common loon (*Gavia immer*) took five days to travel from California to Montana, a distance of 950 miles.[16] To convert this distance to kilometers, we can set up the proportion as

$$\frac{x}{950\text{ mi}} = \frac{1.609\text{ km}}{1\text{ mi}}$$

After multiplying each side by 950 miles, we get

$$x = \frac{(950\text{ mi})(1.609\text{ km})}{1\text{ mi}} \approx 1{,}530\text{ km}$$

Capture-Recapture and Proportions

Determining the size of a population can be very challenging. One obvious approach is to collect every individual and take a "headcount," but this strategy is usually either impossible or impractical. Biologists often take advantage of the mathematical rules of proportions and estimate the population size using a method called **capture-recapture** or **catch and release**.

Imagine a small island populated with nonnative rats (this is an imaginary, but not unrealistic, example). To better understand the impact these invasive rats have on ground-nesting birds, we must determine how many rats are on the island (the total population of rats). In the capture-recapture method, we first live-trap a large number of rats in a short period of time and tag these rats by implanting a harmless microchip beneath their skin. The tagged rats are then released and given some time to mix back into the island population. Then a sample of the population of rats is trapped, and the number of marked rats in the sample is determined (Figure 2-5).

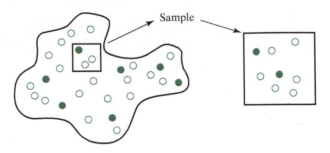

Figure 2-5 Marked rats (solid circles) and unmarked rats (open circles) on an island and in a sample (square) from that island.

Let's denote the total number of rats in the population by N and the number of rats in the sample by n. Let T equal the number of tagged rats in the population, and let t be the number of tagged rats in the sample. If the sample is representative of the population, then the ratio of tagged rats in the population to tagged rats in the sample should be the same as the ratio of the population size to the sample size:

$$\frac{T}{t} = \frac{N}{n}$$

We know the number of tagged rats in the sample (t) and the number in the population (T), and we know the total sample size (n). Therefore, we can calculate the total population N of rats on the island. Suppose $T = 250$ rats were originally tagged and released into the population, and a later sample $n = 320$ rats had $t = 11$ tagged rats in that sample. The proportionality is

$$\frac{250 \text{ tagged rats in population}}{11 \text{ tagged rats in sample}} = \frac{N}{320 \text{ rats in sample}}$$

Multiply both sides by 320 to solve for N:

$$N = \frac{250 \text{ tagged} \times 320 \text{ rats}}{11 \text{ tagged}} \approx 7{,}273 \text{ rats}$$

There are over 7,200 rats on the island, according to this experiment.

There are some assumptions about the capture-recapture estimate of population size. As noted before, the sample must be representative of the population; good mixing of the originally tagged rats is important. But the original capture/release and the later recapture must be fairly close together in time, otherwise the tagged rats will die off and be replaced by untagged rats. A **closed population** as might be found on an isolated island is ideal for this method; however, most populations are open to migration.

PROBABILITY

Probabilities are used to express the chance of occurrence of some event, such as an earthquake, a rain shower, or global warming. Determining probabilities of natural phenomena can be complex, so we start with a simple example. What is the chance that an even number turns up when we roll one standard die (singular of dice)? We know that there are three ways to get an even number (2, 4, and 6) out of the six possible numbers (1, 2, 3, 4, 5, and 6), so the probability or chance of rolling an even number is 3/6 = 0.5 or 50%. Probabilities can be expressed in either fraction, decimal, or percentage form.

More formally, we define the **probability of an event** as the number of successful outcomes divided by the total number of possible outcomes.

$$P(\text{event}) = \frac{\text{number of succesful outcomes}}{\text{total number of possible outcomes}}$$

Each of the possible outcomes must be *equally likely;* that is, they must have the same chance of occurring. For the simple example above, the event is rolling an even number. There are three successful outcomes (ways of getting an even number) out of the six possible outcomes (the numbers 1, 2, 3, 4, 5, and 6). Each possible outcome is equally likely if the die is fair. So again we compute that

$$P(\text{rolling an even number}) = \frac{3}{6} = 0.5 = 50\%$$

EXAMPLE 2-12 Flipping Coins

Suppose that you want to determine the probability that exactly two heads will occur when you flip three coins in sequence. Using H for heads and T for tails, the equally likely outcomes are the following:

HHH   HHT   HTH   HTT

THH   THT   TTH   TTT

There are eight total possible outcomes, all equally likely (each has probability of 1/8 of occurring). Out of the eight outcomes, there are three outcomes in which exactly two heads occur. Thus

$$P(\text{getting 2 heads when flipping 3 coins}) = 3/8 = 0.375 = 37.5\%$$

Notice that the probability of getting three heads is 1/8 = 0.125 or 12.5%. This means that getting two heads is three times as likely as getting three heads.

The following rules about probabilities are extremely useful:

- The probability of any event is always between 0 and 1 (i.e., between 0% and 100%). Using inequalities: $0 \leq P(\text{event}) \leq 1$.
- The probability that an event will *not* occur is equal to 1 minus the probability of the event occurring: $P(\text{not event}) = 1 - P(\text{event})$.

Making use of the second rule, we find that the probability of *not* getting two heads is the following:

$$
\begin{aligned}
P(\text{not 2 heads}) &= 1 - P(\text{2 heads}) \\
&= 1 - 0.375 \\
&= 0.675 \\
&= 67.5\%
\end{aligned}
$$

The probability of rolling one die and getting an even number is 50%. Likewise, the probability of getting two heads when flipping three coins is 37.5%. Both of these probabilities were obtained using a theoretical approach. We could have also estimated either of these probabilities by conducting experiments. For example, suppose we flip 3 coins a total of 200 times and get 2 heads in 83 of the cases. The **experimental probability** of flipping exactly two heads is 83/200 = 0.415 = 41.5%. We would use this experimental percentage to estimate that the **theoretical probability** of getting two heads is 41.5% (which is not too different from 37.5%). It's important to understand that as we flip three coins more and more times, our estimate of the true (theoretical) probability improves. The **law of large numbers** states that as a procedure is repeated over and over, the experimental probability will approach the true probability.

Determining the probability of occurrence of natural events such as earthquakes, landslides, and tornadoes using only a theoretical approach is usually next to impossible. Often the best that scientists can do is estimate these probabilities. What makes estimation difficult is that earth systems cannot be manipulated like a coin-tossing experiment. Instead, scientists often rely on historical data to estimate probabilities.

EXAMPLE 2-13 U.S. Hurricanes

Over the 101-year period from 1900 to 2000, there were only 2 years in which a Category 5 hurricane (sustained winds greater than 155 mph) struck the coast of the United States (Figure 2-6).[17] From these data, we estimate that there is about a $2/101 \approx 2\%$ probability that in any year, the United States will be struck by such a large hurricane. Stated differently, there is roughly a 98% chance that the United States will *not* be struck by a hurricane of this size in any given year.

Figure 2-6 Hurricane Katrina devastated the southeast coast of the United States in August 2005. Residents of Navarre Beach, Florida, pump flood waters away from their home. *Photo courtesy of Marvin Nauman/FEMA*.

EXAMPLE 2-14 Earthquake Probability

Earthquakes are ground shakings, mostly caused by rather rapid movement or slip along breaks in the Earth called **faults.** Earthquakes are often rated according to their **magnitude,** which is the logarithm of their size (see Chapter 1 for more details). From 1900 through 1999, there have been 10 earthquakes with magnitude greater than or equal to 7 along the San Andreas fault (Table 2-4).[18] Earthquakes of this magnitude have the potential for causing a lot of damage. The historical earthquake with the largest magnitude on the San Andreas fault is the great San Francisco earthquake of 1906 (Figure 2-7).

TABLE 2-4

Year	Mag.	Year	Mag.	Year	Mag.	Year	Mag.	Year	Mag.
1906	8.3	1923	7.2	1940	7.1	1980	7.2	1992	7.3
1922	7.3	1927	7.3	1952	7.7	1989	7.1	1999	7.1

Source: U.S. Dept of Interior

What is the probability that a magnitude 7 or greater earthquake will take place in any given year in the twenty-first century along the San Andreas fault?

Figure 2-7 Statue of geologist Louis Agassiz at Stanford University, 1906. The great San Francisco earthquake of 1906 caused much damage. *Historical photo courtesy of U.S. Geological Survey.*

Solution We start by calculating the probability for the twentieth century, for which we have data. The number of possible outcomes is 100, corresponding to every year in the twentieth century. There were 10 years with large earthquakes in the last century; therefore, the probability of having an earthquake during any year of the twentieth century was

$$P(\text{quake in any year in 20th century}) \ = \ \frac{10 \text{ actual years}}{100 \text{ possible years}} \ = \ 0.1 \ = \ 10\%$$

There was a 0.1 or 10% probability of a large earthquake in any given year of the twentieth century along the San Andreas fault. We might argue that the same probability holds true for the twenty-first century, but some caution would be appropriate. Was the twentieth century *representative* of the earthquake activity along the San Andreas fault, which has been active for millions of years? Does earthquake occurrence vary from century to century?

RECURRENCE INTERVALS

The **recurrence interval** is the time between events or how often an event occurs. The recurrence interval is also referred to as the **return period.** Using the same data as in the previous example, the recurrence intervals between large earthquakes on the San Andreas fault system in the twentieth century are shown in Table 2-5.

TABLE 2-5

Years	Int.	Years	Int.	Years	Int.
1906–1922	16	1927–1940	13	1980–1989	9
1922–1923	1	1940–1952	12	1989–1992	3
1923–1927	4	1952–1980	28	1992–1999	7

The intervals between large earthquakes vary from 1 year to 28 years. We can calculate an average recurrence interval for the twentieth century:

$$\frac{16 \text{ yr} \ + \ 1 \text{ yr} \ + \ 4 \text{ yr} \ + \ 13 \text{ yr} \ + \ 12 \text{ yr} \ + \ 28 \text{ yr} \ + \ 9 \text{ yr} \ + \ 3 \text{ yr} \ + \ 7 \text{ yr}}{9 \text{ intervals}}$$

$$= \frac{93 \text{ years}}{9 \text{ intervals}}$$

$$\approx \frac{10.3 \text{ years}}{\text{interval}}$$

Large earthquakes on the San Andreas fault system take place about every decade, on average. The average recurrence interval can also be calculated using the time interval between the first and last event:

$$\frac{1999 - 1906}{9 \text{ intervals}} = \frac{93 \text{ years}}{9 \text{ intervals}} \approx \frac{10.3 \text{ years}}{\text{interval}}$$

An average recurrence interval of 10 years does not imply that earthquakes will occur precisely every decade. As can be seen from the data in the table, the time intervals between successive large earthquakes are quite variable. Nonetheless, knowing the average recurrence interval of damaging earthquakes could be useful in disaster planning.

EXAMPLE 2-15 Large Landslides in Central America and the Caribbean

The International Landslide Center maintains a global database of large landslides, mudflows, debris flows, and other **mass movements.** In 2003, nine different mass movements in the Caribbean and Central America killed an estimated 171 people (Table 2-6).[19] What was the average recurrence interval, in days, of large mass movements?

TABLE 2-6

Date	Day of Year	Country	Fatalities
Jan. 22	22	Bermuda	1
Mar. 04	63	Mexico	1
Mar. 31	90	Bolivia	117
Apr. 24	114	Guatemala	23
May 06	126	Mexico	5
Jul. 17	198	Mexico	9
Oct. 06	279	Haiti	13
Oct. 07	280	El Salvador	1
Nov. 17	321	Dominican Republic	1

Source: International Landslide Centre

Solution We calculate that there were 299 days between the first landslide (January 22) and the last event (November 17) in 2003. There were nine events or eight intervals between events. Therefore, the average recurrence interval was

$$\frac{299 \text{ days}}{8 \text{ intervals}} \approx \frac{37.4 \text{ days}}{\text{interval}}$$

The average recurrence interval for large landslides and mudflows in Central America and the Caribbean in 2003 was approximately 37 days.

The recurrence interval of an event and the probability of that event taking place are closely related. The longer the recurrence interval, the lower the probability of that event happening.

CHAPTER SUMMARY

Ratios are comparisons of two quantities by division. Ratios of like quantities are dimensionless, as the units cancel. Ratios of unlike quantities have compound units.

Ratios can be used to **normalize** or standardize information, to eliminate a second quantity as a variable. Two nearby habitats have different numbers of an endangered specie of herbivore, but the amount of edible plant material in the two habitats is not the same. Therefore, we divide the animal abundance by the food mass available, to compare the two habitats without the influence of food supply.

A common type of ratio is a **percentage,** where the ratio is expressed as a part of 100. Percentages can refer to parts of a whole; the volume of carbon monoxide compared to the total volume of gases emitted by an automobile, for example. Percentages can be calculated for any ratio of like quantities.

A percentage can be thought of as the number of parts per 100 parts (parts per hundred). Other similar expressions of **concentration** include parts per thousand (ppt), parts per million (ppm), and parts per billion (ppb). These expressions are used when concentrations are low. The amount of pesticide contamination in runoff from irrigation is often given in parts per million or parts per billion.

The **percentage change formula** can be used to quantify the amount of change with time:

$$\frac{\text{final value} - \text{initial value}}{\text{initial value}} \times 100\% = \% \text{ change}$$

For example, checkerspot butterflies in a small area increased from 500 to 600 butterflies, an increase of 20%. Similar formulas can be used to calculate the **percentage difference,** where time is not involved, and **percentage error** between two measurements.

A **proportion** is an equality between two ratios. City A has 75 electric buses for its population of 100,000 people. If City B has the same ratio of buses to people, and City B has a population of 200,000, then City B has 150 buses:

$$\frac{75 \text{ buses}}{100,000 \text{ people}} = \frac{150 \text{ buses}}{200,000 \text{ people}}$$

Proportions can be used to solve numerical problems such as calculating percentages and converting units. Proportionality is also the basis for the **capture-recapture** method of estimating a population.

A **probability** expresses the chance of occurrence of an event:

$$P(\text{event}) = \frac{\text{number of succesful outcomes}}{\text{total number of possible outcomes}}$$

The probability of rolling a die and getting a 5 is 1/6 or $0.1\overline{6}$ (about 17%). The probabilities of events such as earthquakes, wildfires, and landslides occurring can be calculated from historical data.

The **recurrence interval** is the time between successive events. The **average recurrence interval** can also be calculated from historical data on a large number of events, such as the number of hurricanes during the last century.

END *of* CHAPTER EXERCISES

Ratios

1. The World Conservation Union compiles a "Red List" of threatened plants and animals worldwide.[20] Organisms that are critically endangered or vulnerable to extinction are included on the Red List. In 2002, 10,731 species were on the Red List; this total increased to 15,042 organisms in 2004. What is the ratio of the number of species on the 2004 list to the 2002 list? Give the ratio to 2 decimal places.

2. The groundwater in Pensacola, Florida has been contaminated by **radium,** a radioactive element, from the old Agrico Chemical Company plant (now a Superfund site). Eleven monitoring wells have been installed and the concentration of radium in groundwater has been measured in these wells. Data for 2002 are given in Table 2-7.[21] The federal limit for radium in drinking water is 5 picocuries per liter of water (5 pC/L). A picocurie is a measure of radioactivity (see the Appendix).

   a. Calculate the ratio of the measured radioactivity compared to the federal limit for drinking water for each well.

   b. How many monitoring wells have water whose contamination exceeds the federal limit?

   c. For the well with the highest contamination, how many times does its radium concentration exceed the federal limit?

TABLE 2-7

Well	pC/L	Ratio	Well	pC/L	Ratio
AC-2D	3.1		NWD-4D	10.2	
AC-3D	19.1		AC-25D	8.0	
AC-29D	18.2		AC-35D	10.5	
AC-30D	15.9		AC-12D	10.5	
AC-13D	7.0		AC-8D	3.5	
AC-36D	2.9				

Source: *Pensacola News Journal*

3. The Amazon rain forest covers about 4 million square kilometers of area, about 10 times the surface area of California. Using this ratio, estimate the area of California in square kilometers.

4. The 2004 U.S. population was approximately 294 million people. The ratio of the world population to the U.S. population for that year was 21.77.[22] Using this ratio, calculate the 2004 world population.

Normalization

5. Many power plants consume energy in the form of fossil fuels and emit carbon dioxide to the atmosphere. A small sample of California power plant values is given in Table 2-8, with energy input in trillions of BTUs and carbon dioxide output in tons.[23]

TABLE 2-8

Plant Name	Energy Input (10^{12} BTU)	CO_2 Output (tons)	Normalized CO_2 Output
Santa Clara	833,261	48,213	
SCA	4,487,554	2,340,088	
Scattergood	9,889,331	537,922	
South Bay	23,038,851	1,353,523	
SPA	2,460,438	145,701	
Walnut	40,594	2,349	

Source: U.S. Environmental Protection Agency

a. Normalize the carbon dioxide output by the energy input. Include units.

b. Are these power plants similar in their normalized CO_2 output, or are there wide variations?

c. Which plant produced the most CO_2 gas per energy input?

6. Table 2-9 shows the 1990 production of electricity by hydropower plants (in terawatt-hours) along with the amount of water used (in gigagallons per day) to generate that hydroelectricity, for several different states.[24]

TABLE 2-9

State	Electricity (TWh)	Water (Ggal/day)	Ratio
Alabama	10.340	0.218199	
Arizona	8.180	0.031801	
Montana	10.688	0.066797	
Wisconsin	1.148	0.043972	

Source: U.S. Dept of Interior

a. Which state produces the most hydroelectricity? Which state uses the most water to produce hydroelectricity?

b. Normalize the electricity produced to the amount of water used for each state. What are the units of the ratio?

c. Which state is the most *efficient* in producing hydroelectricity? Which state is the least efficient?

7. The number of automobiles in the San Francisco Bay area, along with the population, are given in Table 2-10 for each decade from 1930 to 2000.[25]

TABLE 2-10	Year	Automobiles	Population	Normalized
	1930	461,800	1,578,000	
	1940	612,500	1,734,300	
	1950	1,006,400	2,681,300	
	1960	1,620,600	3,638,900	
	1970	2,503,100	4,630,600	
	1980	3,281,800	5,179,800	
	1990	3,953,200	6,023,600	
	2000	4,799,300	6,875,400	

Source: San Francisco Transportation Commission

   a. Calculate the per capita automobile ownership. Include units.

   b. Has the per capita automobile ownership increased, decreased or stayed the same over the last 70 years? Discuss briefly.

8. The weights of disposed garbage in 1998 in four Oregon counties, along with the populations of those counties, are given in Table 2-11.[26]

TABLE 2-11	County	Tons Disposed	Population	Ratio
	Marion	237,166	271,750	
	Lane	261,958	313,000	
	Umatilla	52,484	60,600	
	Lake	6,361	7,400	

Source: Oregon Dept of Environmental Quality

   a. Calculate the per capita disposal rates. Include units.

   b. Is the per capita disposal rate higher, lower, or the same for populous counties compared to nonpopulous counties? Discuss briefly.

Percentages

9. The total area of tropical forest and the area of protected forest in countries of Central Africa (in 2000) are given in Table 2-12.[27]

TABLE 2-12	Country	Forest Area (km^2)	Protected Forest (km^2)	% Protected
	Cameroon	289,965	17,854	
	Central African Republic	199,018	41,608	
	Congo	278,797	12,935	
	Equatorial Guinea	23,540	0	
	Gabon	239,369	8,975	
	Zaire	1,439,178	93,160	

Source: U.N. Environment Programme

   a. Calculate the percent protected forest in each country to one decimal place.

   b. Which country has the most protected forest?

   c. Which country has the highest percentage of protected forest?

10. In 2004, Germany consumed about 582 terawatt-hours (TWh) of electricity. About 167 TWh of electricity was generated by nuclear power plants and consumed domestically.[28] What percentage of Germany's electricity consumption is dependent upon nuclear power facilities?

11. According to a newspaper article, Californians "guzzle more than 1 billion water bottles a year. Just 16 percent of the water bottles sold in California are being recycled."[29] Calculate the number of water bottles tossed into the garbage each year in California.

12. The Union of Concerned Scientists tested commercial plant seeds such as corn, canola, and soybean, and discovered small quantities of genetically modified (GM) seeds in bags of seed marked "GM free."[30] This contamination comes about by farmers mixing GM and non-GM seeds together after harvesting. Suppose you purchase a 50-pound bag of corn seed. There are 85 kernels per ounce of seed.

    a. How many corn seeds are there in a 50-pound bag?

    b. If 0.1% of those seeds are GM corn, how many GM corn seeds are there in a 50-pound bag?

Parts Per Thousand, Million, and Billion

13. The total midyear U.S. population in 2000 was 281,422,000 people. There were 4,137,000 live births in the United States in the same year.[31] Express the birth rate in terms of parts per thousand.

14. The fertility rate is based on the number of live births to women of child-bearing age. For Sweden in 1980, there were 97,061 live births by 1,694,447 women of child-bearing age.[32] Express the fertility rate in terms of parts per thousand. How did Sweden's fertility rate compare to the rate of 68 for the United States and 59 for Japan?

15. The mean concentration of lead in whole trout from the Spokane River is about 1 milligram of lead in every kilogram of trout.[33] Express this concentration of lead in parts per million (ppm) and parts per billion (ppb).

16. A cubic meter of air has a mass of approximately 1.25 kilograms. About 2 milligrams of that mass is methane. What is the concentration of methane in air, in ppm and ppb?

Percentage as Change with Time

17. Suppose that a population of prairie dogs in a small area tripled from 10 to 30 prairie dogs. What was the percentage change? Another population in a larger area also tripled, from 100 to 300 prairie dogs. What was the percentage change?

18. Suppose that the concentration of ozone in a city's air decreased from a high of 2.29 ppm to a low of 0.88 ppm from Friday evening to Saturday morning. What was the percentage change?

19. The Environmental Protection Agency's Toxic Release Inventory records the amount of toxic chemicals reported by industry to have been released to the environment. In 1999, 7.7 billion pounds of toxics were reported released, whereas in 2000, 7.1 billion pounds were reported released.[34] What was the percentage decrease?

20. The amount of carbon dioxide in the lower atmosphere before the Industrial Revolution was about 280 ppm and in 2004, after burning lots of fossil fuels, the concentration of carbon dioxide was about 380 ppm.[35] What is the percentage change between these two values?

21. In England and Wales, air pollution incidents are ranked from Category 1 (most severe impact on air quality) to Category 3 (small impact on air quality). A total of 225 Category 1 and 2 incidents damaged the air quality in 2002, a 54% decrease from 2001.[36] How many Category 1 and 2 incidents did England and Wales suffer through in 2001?

22. Ethanol, made from fermented corn, is usually mixed with gasoline and touted as a "renewable" fuel. In 2003, the U.S. House of Representatives proposed to increase U.S. ethanol production by 100%, to 5 billion gallons a year by the year 2012.[37] What was the ethanol production in 2003? You should be able to write down the answer just by thinking about the problem. Then calculate the answer, using the formula for percentage change.

Percentage Difference and Percentage Error

23. Idaho is divided into five wheat production districts. Wheat farmers in each of these districts were surveyed, and the average size of farms in each district was calculated (Table 2-13).[38] Using District 1 as a reference, what is the percentage difference in farm size between District 1 (north Idaho) and District 3 (southwest Idaho)? Between District 1 and District 5 (southeast Idaho)?

TABLE 2-13

District 1	District 2	District 3	District 4	District 5
1,259 acres	1,040 acres	789 acres	996 acres	1,690 acres

Source: Univ. of Idaho Pest Management Center

24. Carbon is stored in various "reservoirs" on planet Earth: in gases in the atmosphere, in vegetation on land, in ocean waters, and in fossil fuels such as coal and petroleum. The amount of carbon in these reservoirs is often measured in petagrams (Pg), where 1 petagram is equal to 10^{15} grams. The amount of carbon in the six main storehouses is given in Table 2-14.[39]

TABLE 2-14

Atmosphere	Fossil Fuels	Vegetation	Shallow Ocean	Deep Ocean	Ocean Mud
590 Pg	3,700 Pg	2,300 Pg	900 Pg	37,100 Pg	150 Pg

Source: *Physics Today*

    Using fossil fuels as a reference, what is the percentage difference between carbon stored in fossil fuels and carbon stored in vegetation? Between carbon in fossil fuels and carbon in the deep ocean?

25. Suppose that a scientist sends some tissue samples to a laboratory for analysis. In order to determine the accuracy of its machines, the laboratory first analyzes a **reference sample** that contains a known amount of the chemical of interest. The reference sample has a known concentration of 150 ppb, whereas the lab's analysis of the reference sample yields a concentration of 159 ppb. Calculate the percentage error.

26. Automated sampling of rivers and streams for water quality produces high-quality results but is very expensive. The siphon sampler is a much cheaper alternative, but does it produce meaningful results? Scientists from the U.S. Geological Survey and the Wisconsin Department of Natural Resources compared these two samplers. In one test, the water captured by the automated sampler contained 1,020 ppb of phosphorus whereas the water captured by the siphon sampler at the same time and place contained 880 ppb of phosphorus.[40] Assuming that the results from the automated sampler are correct, what is the percentage error by the siphon sampler?

Proportions

27. Suppose that an average person uses about 50 gallons of water per day. What is the daily water use of a city of 115,000 people? Solve this problem using a proportion.

28. Urban planners decide that each city block should have at least 6,000 square feet of public green space (planters, beauty strips next to sidewalks, etc.). How many total square feet of green space should there be in a neighborhood of 35 blocks? Solve this problem using a proportion.

29. 100 hectares of land is approximately equal to 0.386 square miles. How many hectares are in 1 square mile? Solve this problem using a proportion.

30. 100 gallons is approximately the same as 13.37 cubic feet. How many gallons are in 100 cubic feet? Solve this problem using a proportion.

31. **Brownfields** are former industrial areas whose redevelopment is often complicated by pollutants in the soil and groundwater. Suppose that 3.2 acres of a 28.7-acre brownfield have been cleaned of pollutants. What percentage of this brownfield has been decontaminated? Solve this problem using a proportion.

32. The Susquehanna River basin in Pennsylvania contains 36,100 miles of streams and rivers, of which 26,840 miles were examined for environmental impairment in 2002. Impairment could be caused by chemical contamination, excessive water temperatures, lack of oxygen, or other reasons. In 2002, there were 4,140 miles of impaired streams.[41] What percentage of assessed streams were impaired? Solve this problem using a proportion.

33. Polybrominated diphenol ether (PBDE), a chemical flame retardant associated with reproductive disorders and cancer, was measured in a variety of foodstuffs. One sample of salmon contained 3,078 parts of PBDE per trillion parts of fish.[42] Express this concentration in parts per billion (ppb). Solve this problem using a proportion.

34. Different layers of ice from glaciers in the Alps were analyzed for heavy metals to determine the history of European air pollution dating back to Roman times. The maximum amount of copper in twentieth-century ice was 0.0382 parts per million (ppm).[43] Express this concentration in parts per billion (ppb). Solve this problem using a proportion.

35. A biological survey of the Ironwood Forest National Monument near Tucson, Arizona, was conducted from 2001 to 2003. One of the study plots in the Roskruge Mountains in the southeast corner of the monument contained three large plants: the saguaro cactus, the ironwood tree, and the foothill palo verde tree. In the Roskruge plot, biologists counted 315 ironwoods per hectare, 270 foothill palo verdes per hectare, and 165 saguaros per hectare. A different plot had 120 ironwoods per hectare.[44] If the same proportions of large plants are found in both plots, what are the densities of foothill palo verdes and saguaros in the second plot?

36. Mercury (chemical symbol Hg) was measured in bass captured from lakes in Washington State. As these fish age, mercury accumulates in their fatty tissue (Table 2-15).[45] Calculate the concentration of mercury in older fish for each lake, assuming the same ratio of mercury and age as in younger fish.

TABLE 2-15

Lake Name	Age (yr)	Hg (ppb)	Age (yr)	HG (ppb)
Kitsap	2	185	12	
Terrell	3	115	13	
Newman	2	68	7	
Moses	2	33	11	

Source: Washington State Dept of Ecology

Capture-Recapture

37. Wildlife biologists in Big Bend National Park (Texas) captured 61 tadpoles of the canyon treefrog (*Hyla arenicolor*) at Boot Spring. They marked the tadpoles with a harmless red dye, released them into the spring, and then captured 68 tadpoles two hours later. There were 36 dyed tadpoles in the new sample.[46] Estimate the size of the tadpole population in Boot Spring. How well did this study match the assumptions behind the capture-recapture method? Explain briefly.

38. In a small area in Ohio, 22 gray squirrels (*Sciurus carolinensis*) were captured, marked, and released back into the population. The next day, 13 squirrels were captured, of which 7 were marked.[47] Estimate the size of the gray squirrel population.

39. Dryden Lake in central New York State has a large population of pickerel (*Esox lucius*), a kind of fish. In 1970, biologists captured 232 pickerel from Dryden Lake, marked their fins, and released the fish. A few weeks later, a sample of 329 pickerel contained 16 fish with marks on their fins.[48] Estimate the size of the pickerel population in Dryden Lake in 1970. Are the pickerel in Dryden Lake a closed population, and does this help or hinder the population estimate? Explain briefly.

40. Suppose a state agricultural agency surveyed farmers in their state and identified about 50,000 farms statewide. However, the agricultural agency felt that a number of farms had been missed by this survey and hypothesized that the total number of farms was greater than 50,000. In order to test this idea, the agency turned to its aerial photograph collection. Agents selected a sample area, identified every farm on the aerial photos in that sample area, and then visited each of those farms on the ground to determine the actual number of farms in the sample area. The agency discovered that out of 2,000 farms in the sample area, 1,750 were on the survey list, and 250 were not. Estimate the total number of farms in the state.

Probability

41. Suppose that you roll two dice, each with six sides. Consider the two numbers rolled.
    a. List the 36 equally likely outcomes (each outcome has 2 numbers).
    b. What is the probability that the sum of the 2 numbers equals 7?
    c. What is the probability that the sum of the 2 numbers equals 9?
    d. What is the probability that the sum of the 2 numbers is greater than 6?
    e. What is the probability that the sum of the 2 numbers is not 5?

42. Suppose that you flip one coin and roll one die (which is six-sided).

   a. List the 12 equally likely outcomes. Each outcome has either tails or heads, and a number from 1 to 6.

   b. What is the probability that the outcome is heads, and the die is 5?

   c. What is the probability that the outcome is heads, and the die is even?

   d. What is the probability that the outcome includes tails?

   e. What is the probability that the outcome does not contain the number 4?

43. From 1900 to 1998, there were 26 years in which a major flood occurred on the Mississippi River.[49]

   a. Determine the probability that the Mississippi River will have a major flood in any given year.

   b. What is the probability that the river will *not* have a major flood in any given year?

44. A 100-year flood corresponds to the size of a flood which, on average, has occurred once every 100 years.

   a. How many 100-year floods occur, on average, every 300 years?

   b. How many 100-year floods occur, on average, every 50 years?

   c. How many 100-year floods occur, on average, every 1 year?

   d. Use part (c) to estimate the probability that a 100-year flood will occur this year.

   e. Which is likely to cause more damage, a 100-year flood or a 500-year flood? Explain.

Recurrence Interval

45. Off the coast of Oregon, Washington, and British Columbia, the Juan de Fuca crustal plate is moving under the North American plate, along the gigantic Cascadia fault. There is geological evidence that the Cascadia fault has produced great earthquakes (magnitude 8+) in the past. There have been 7 of these great quakes from 3,500 years ago to 300 years ago.[50]

   a. Determine the mean recurrence interval of these great earthquakes along the Cascadia fault.

   b. The last great Cascadia earthquake has been dated to the evening of January 26, 1700 (magnitude 9+). Should we expect another great earthquake 200 years from now? Explain briefly.

46. The North Anatolian fault in northern Turkey produced numerous large earthquakes during the twentieth century, exceeding the activity of even the San Andreas fault in California. The 1999 Izmit earthquake, with magnitude of $M = 7.4$, caused the most property destruction in Turkey's recorded history, with economic impact likely to exceed $10 billion. North Anatolian fault earthquake data from 1900–1999, listing all events with size larger than $M > 6.7$, are given in Table 2-16.[51,52]

TABLE 2-16

Year	Magnitude	Year	Magnitude
1912	7.4	1957	7.0
1939	7.8	1964	6.9
1942	7.1	1966	6.8
1943	7.3	1967	7.1
1944	7.3	1992	6.8
1949	6.9	1999	7.4
1951	6.9	1999	7.1
1953	7.2		

Source: *The Geology of Earthquakes* and *GSA Today*

   a. Determine the mean recurrence interval of earthquakes along the Anatolian fault with size $M > 6.7$.

   b. There were two large earthquakes in the same year. Explain your approach to this situation in calculating the mean recurrence interval.

47. The Office of Emergency Management for the State of Colorado has compiled information on wildfires for 2002, the worst wildfire season in Colorado's modern history,

with 3,072 fires and over half a million acres burned. There were 36 large wildfires over 180 acres in size, with the first large fire starting on April 23 and the last starting on September 5.[53] Calculate the recurrence interval of the start of large wildfires for the 2002 fire season.

48. Discharge of the Purgatoire River near Madrid, Colorado, has been measured continuously by the U.S. Geological Survey since 1972. Table 2-17 shows the peak daily discharge for years in which discharge of the Purgatoire River equaled or exceeded 4,000 cubic feet per second (cfs).[54]

TABLE 2-17

Year	Discharge (cfs)
1975	4,500
1976	14,300
1978	7,100
1979	6,620
1981	11,600
1985	4,070
1992	4,850
1993	4,680
1994	4,020
1997	4,130
2004	6,230

Source: U.S. Geological Survey

a. Determine the mean recurrence interval of peak discharges greater than or equal to 4,000 cfs for the Purgatoire River.

b. Is it possible to have two discharges greater than or equal to 4,000 cubic feet per second in the same year? Explain briefly.

SCIENCE IN DEPTH

Sinkholes and Lakes

What's going on in Florida? Giant pits or **sinkholes** swallow cars, houses, and roads without warning (Figure 2-8). How do these sinkholes form? What's their relationship to human activity?

Florida's geology is dominated by the rock known as **limestone.** Limestone dissolves easily in weak acids; statues, buildings, and headstones made of limestone are often pitted and corroded from acid rainfall. Florida's rainfall is truly acidic! For example, the average pH of rainfall in Broward County, Florida is about 5.2, almost 100 times as acidic as neutral water with a pH of 7.

Acid rainfall percolates through the soil and eventually joins the **groundwater** that fills spaces in the limestone. Acid groundwater moves slowly through cracks and crevices, dissolving the limestone as groundwater flows. The result from many years of dissolution is a system of underground caves and caverns, partially or completely filled with groundwater (Figure 2-9). The roofs of these caverns can collapse, producing **sinkholes.** Withdrawal of groundwater and lowering of the **water table** (the top of the water-saturated ground) is one of the main causes of collapse, because the groundwater filling the cave partially supports the cave roof.

Florida is pockmarked with thousands of sinkhole lakes that provide a tremendous amount of freshwater habitat. Human activities in Florida are both creating and destroying lakes and wetlands. Lakes are decreased in size or completely destroyed by lowering the water table through drainage and groundwater withdrawal, and by infilling lakes with artificial debris. But new sinkhole lakes are created during excessive groundwater withdrawals for irrigation and other human needs.

Figure 2-8 Sinkhole in Winter Park, Florida in 1981. *USGS photograph.*

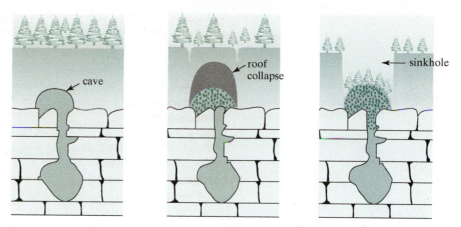

Figure 2-9 Diagrams showing underground cavern beneath a sinkhole formed by collapse of the cave's roof.

 CHAPTER PROJECT:
MEASURING HABITAT OF FLORIDA LAKES

What kinds of habitats are found at a lake? Lake-bottom muds, the water column, near-surface waters, wetlands, shorelines, and lake-shore environments provide food and shelter for a tremendous variety of plants and animals. Many organisms, such as amphibians and insects, incorporate the habitat diversity of a lake into their life cycles. The diversity of habitats and organisms at a lake also creates a variety of ecosystem functions, such as nutrient storage and water purification.

In the companion project for this chapter found at the **enviromath.com** Web site, you will quantify the amount of shoreline habitat and surface habitat associated with a group of sinkhole lakes in Florida, using ratios and percentages. You'll compare the habitat types, then evaluate two competing pieces of legislation that both claim to preserve the maximum amount of lake habitat. Which bill will you support?

NOTES

[1]Chris Niskanen, "Old Friend Found," *Birder's World*, August 2005, 20–23.

[2]Oregon State Department of Environmental Quality, Solid Waste Program, http://www.deq. state.or.us/wmc/solwaste/data/percapdisposal1998.html

[3]P. J. Meier and G. L. Kulcinski, "Energy Payback Ratio and CO_2 Emission Associated with Electricity Generation from a Natural Gas Power Plant—Preliminary Findings," University of Wisconsin–Madison, Fusion Technology Institute (2000), http://fit.neep,wisc.edu/ presentations/ pmeier_energy.pdf

[4]U.S. Dept of the Interior, National Biological Service, *Our Living Resources: A Report to the Nation on the Distribution, Abundance, and Health of U.S. Plants, Animals, and Ecosystems*, by E. T. LaRoe, G. S. Farris, C. E. Puckett, P. D. Doran, and M. J. Mac, eds. (1995), http:// biology.usgs.gov/s+t/noframe/c040.htm

[5]National Agricultural Statistics Service, http://www.nass.usda.gov

[6]U.S. Census Bureau, http://www.census.gov

[7]Chuck Clarke, "Recycling Pays Off in Many Ways," *Seattle Post-Intelligencer*, October 4, 2002.

[8]Texas Parks and Wildlife Department, http://www.tpwd.state.tx.us/wetlands/ecology/ table2_1.htm#table2_2

[9]U.S. Dept of Energy, Model Year 1999 Fuel Economy Guide, http://www.fueleconomy.gov/feg/ epadata/99feg.pdf

[10]G. Markowitz and D. Rosner, *Deceit and Denial: The Deadly Politics of Industrial Pollution* (Berkeley: University of California Press, 2004).

[11]California Dept of Fish and Game, Habitat Conservation Planning Branch, http://www.dfg.ca.gov/hcpb/species/t_e_spp/tebird/Condor%20Pop%20Stat.pdf

[12]World Lakes Network, http://www.worldlakes.org/lakeprofiles.asp?anchor=volume

[13]National Park Service, Sequoia and Kings Canyon National Parks, http://www.nps.gov/seki/shrm_pic.htm

[14]S. P. Riley and G. Skirrow (eds.), *Chemical Oceanography*, vol. 1, 2nd ed. (Academic Press, 1975). As cited in Alyn C. Duxbury and Alison B. Duxbury, *An Introduction to the World's Oceans*, 5th ed. (Dubuque, IA: Wm. C. Brown, 1997), 150.

[15]Anonymous, "Coast Guard Seizes Illegal Catch," *Blue Planet*, Summer 2004, 11.

[16]Michael Jamison, "Loon Sightings Answer Migration Questions," *Birder's World*, August 2005, 8–9.

[17]U.S. Dept of Commerce, National Hurricane Center, *The Deadliest, Costliest, and Most Intense United States Hurricanes from 1900 to 2000 (And Other Frequently Requested Hurricane Facts)*, by Jerry D. Jarrell et al., NOAA Technical Memorandum NWS TPC-1, Table 9, http://www.aoml. noaa.gov/hrd/Landsea/deadly/index.html

[18]U.S. Dept of the Interior, "Earthquake History, 1769–1989," by W. L. Ellsworth, chap. 6 of *The San Andreas Fault System, California*, by R. E. Wallace (ed.), U.S. Geological Survey Professional Paper 1515 (1990); The Southern California Earthquake Center, http://www.scec.org/

[19]International Landslide Centre, http://www.landslidecentre.org/database.htm

[20]The World Conservation Union, 2004 IUCN Red List of Threatened Species, http://www.redlist.org

[21]Elizabeth Bluemink, "Feds Take a Look at Water Case: EPA Reviewing Supervision of County Utilities Authority," *Pensacola News Journal*, September 20, 2003.

[22]Population Reference Bureau, "2004 World Population Data Sheet," http://www.prb.org/pdf04/04WorldDataSheet_Eng.pdf

[23]U.S. Environmental Protection Agency, *1997 Emissions & Generation Resource Integrated Database (EGRID 97)*, http://www.epa.gov/cleanenergy/egrid/index.htm

[24]U.S. Dept of the Interior, *Estimated Use of Water in the United States in 1990*, U.S. Geological Survey Circular no. 1081, http://water.usgs.gov/watuse/wucircular2.html

[25]San Francisco Bay Area Metropolitan Transportation Commission, *Auto Ownership in the San Francisco Bay Area: 1930–2010*, by Charles L. Purvis, 1997, http://www.mtc.ca.gov/datamart/forecast/ao/aopaper.htm

[26]Oregon Dept of Environmental Quality, Solid Waste Program, "Per Capita Solid Waste Disposed 1997–1998 by Watershed," http://www.deq.state.or.us/wmc/solwaste/data/percapdisposal1998.html

[27]UNEP–World Conservation Monitoring Centre, http://www.unep-wcmc.org/forest/

[28]World Nuclear Association, Germany Fact Sheet, http://www.world-nuclear.org/info/inf43.htm

[29]Don Thompson, Associated Press, "Health-Conscious Consumers Filling Landfills With Water Bottles," *San-Jose Mercury News*, May 30, 2003.

[30]Andrew Pollack, "Modified Seeds Found Amid Unmodified Crops," *New York Times*, February 24, 2004.

[31]U.S. National Center for Health Statistics, National Vital Statistics System, http://www.cdc.gov/nchs/nvss.htm

[32]U.S. Census Bureau, International Programs Center, http://www.census.gov/ipc/www/idbprint.html

[33]A. Johnson, "Results from Analyzing Metals in 1999 Spokane River Fish and Crayfish Samples," *Washington State Dept of Ecology Report* #00-03-017, 2000, http://www.ecy.wa.gov/biblio/0003017.html

[34]Associated Press, "Coal Plants and Mines Biggest Polluters," by John Heilprin, *Seattle Daily Journal of Commerce*, May 28, 2002.

[35]Carbon Dioxide Information Analysis Center, Trends, http://cdiac.esd.ornl.gov/trends/co2/contents.htm

[36]United Kingdom, The Environment Agency, http://www.environment-agency.gov.uk/yourenv/eff/

[37]Missouri Corn Growers Organization, Missouri Corn Online, http://www.mocorn.org/news/2003/NewsRelease11-18-03.htm

[38]University of Idaho Pest Management Center, *Use of Integrated Pest Management by Idaho Wheat Producers*, http://www.ag.uidaho.edu/ipm/ipm_reports/Wheat_ipm.htm

[39]Jorge L. Sarmiento and Nicolas Gruber, "Sinks For Anthropogenic Carbon," *Physics Today*, August 2002, 30–36.

[40]U.S. Dept of the Interior, *Comparison of Water-Quality Samples Collected by Siphon Samplers and Automatic Samplers in Wisconsin*, by David J. Graczyk et al., U.S. Geological Survey Fact Sheet FS-067-00 (2000).

[41]Susquehanna Riber Basin Commission, Information Sheet, http://www.srbc.net/docs/TMDL-Fact-Sheet.pdf

[42]Arnold Schecter et al., "Polybrominated Diphenyl Ethers Contamination of United States Food," *Environmental Science and Technology ASAP Article* 10.1021/es0490830, 2004, http://pubs.acs.org/subscribe/journals/esthag-a/38/free/es0490830.html

[43]Carlo Barbante et al., "Historical Record of European Emissions of Heavy Metals to the Atmosphere Since the 1650s from Alpine Snow/Ice Cores Drilled Near Monte Rosa," *Environmental Science and Technology* 38, 2004, 4085–4090.

[44]Arizona-Sonoma Desert Museum, "Biological Survey of Ironwood Forest National Monument: Distribution and Status of Saguaros and Trees," http://www.desertmuseum.org/programs/ifnm_saguaro.html

[45]Washington State Dept of Ecology, *Mercury in Edible Fish Tissue and Sediments from Selected Lakes and Rivers of Washington State*, by S. Fischnaller, P. Anderson, and D. Norton, Publication no. 03-03-026 (2003).

[46]U.S. Geological Survey, Patuxent Wildlife Research Center, "Dyeing Tadpoles: Protocol for Big Bend National Park Amphibian Monitoring Field Crews," http://www.pwrc.usgs.gov/amphib/primenet/dye.html

[47]C. M. Nixon, W. R. Edwards, and L. L. Eberhardt, "Estimating Squirrel Abundance from Live-Trapping Data," *Journal of Wildlife Management* 31, 1967, 96–101.

[48]Samprit Chatterjee, "Estimating Wildlife Population by the Capture Recapture Method," in F. Mosteller et al. (eds.), *Statistics by Example*, vol. 4 (Reading, MA: Addison-Wesley, 1973).

[49]U.S. Dept of Commerce, *Floods on the Lower Mississippi: An Historical Economic Overview*, by Paul S. Trotter et al., U.S. National Weather Service Technical Attachment SR/SSD 98-9 (1998).

[50]B. F. Atwater and E. Hemphill-Haley, "Recurrence Intervals for Great Earthquakes of the Past 3,500 Years at Northeastern Willapa Bay, Washington," *U.S. Geological Survey Professional Paper* 1576 (1997), 108.

[51]R. S. Yeats, K. E. Sieh, and C. R. Allen, *The Geology of Earthquakes* (Oxford: Oxford University Press, 1997), 178.

[52]R. Reilinger et al., "1999 Izmit, Turkey Earthquake Was No Surprise," *GSA Today* 10, no. 1, 2000, http://www.geosociety.org/pubs/gsatoday/gsat0001.htm

[53]Colorado Div of Emergency Management, "Colorado Wildfires 2002," http://www.dola.state.co.us/oem/PublicInformation/firebans/CO_Fires1.pdf

[54]U.S. Geological Survey, National Water Information System, http://waterdata.usgs.gov/nwis

3

Charts and Graphs

The **Rule of 4** says that quantitative information (data, numbers, values) can be represented four different ways: a written or oral description, numbers in a table, an equation, or a graph or diagram. Throughout this textbook, information is presented in all four versions as much as possible, to help you understand the material. For example, Figure 3-1 illustrates the energy use in quadrillions of British thermal units (Quads) by various regions of the world for the year 2000.[1] The energy use in this **pictograph** is approximately scaled to the size of the light bulb for a quick visual assessment.

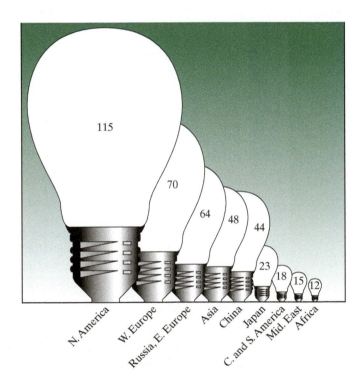

Figure 3-1 World energy use, 2000, in Quads

In this chapter, we introduce some common graphs and diagrams that visually represent numbers or values. There are two basic kinds of quantitative data that can be graphed. **Univariate data** are associated with one variable or changeable quantity. The amount of lead measured in drinking water and the energy use in a region of the world are examples of univariate data. Univariate data are often shown graphically by pie charts, bar charts, and histograms.

Bivariate data are associated with two quantities. In each household, the lead content of the tap water (one quantity) and the age of the household pipes (a second quantity) could be measured or determined. Bivariate data come in pairs (lead content and pipe age for each household) and can be represented graphically by bar charts or scatterplots.

PIE CHARTS

Pie charts (circle diagrams or circle charts) illustrate the fractions or percentages that make up a total or a sum. Each "slice" of the pie is sized according to its percentage of the total. Pie charts can be constructed by hand using a **compass** (to make circles) and a **protractor** (to measure angles). Pie charts can also be easily constructed using computer graphics programs.

EXAMPLE 3-1   Surface Water Use, Oregon, 1990

Table 3-1 and Figure 3-2 show the use of surface water from the Willamette basin in western Oregon in 1990.[2]

TABLE 3-1

Use of Water	Amount	% of Total
Public water supply	300 acre-feet per day	43%
Industrial/mining	211 acre-feet per day	30%
Agriculture	189 acre-feet per day	27%
Total	700 acre-feet per day	100%

Source: *Atlas of the Pacific Northwest*

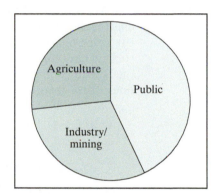

Figure 3-2 Pie chart of daily water use

Glancing at Figure 3-2, it is apparent that industrial/mining and agricultural water use are about the same, and both are less than municipal (public) water use, which is less than half of the total.

Pie charts are not as useful when some categories of data have very small percentages. For example, Table 3-2 and Figure 3-3 show the distribution of sources of fresh water worldwide.[3]

TABLE 3-2

Source	Volume	% of Total
Glaciers	27.5 million km^3	76.56%
Groundwater	8.2 million km^3	22.83%
Lakes and seas	205 thousand km^3	0.57%
Atmosphere	13 thousand km^3	0.04%
Rivers	1.7 thousand km^3	0.00005%
Total	35.9 million km^3	$\approx$ 100%

Source: U.S. Geological Survey

Three of the five categories of fresh water sources cannot be distinguished in Figure 3-3, because the values are too small. There are big differences among lakes, the atmosphere, and rivers, and those differences are not apparent on the chart. To make this pie chart more useful, the actual values or percentages could be written next to the category labels (Figure 3-4).

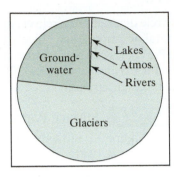

Figure 3-3 Pie chart of Earth's fresh water

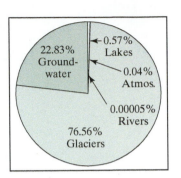

Figure 3-4 Modified pie chart with percentages labeled

BAR CHARTS

Bar charts use bars or other graphical devices to represent data that are arranged by category or rank. The lengths of the bars in a bar chart are proportional to the value for each category. Categories that are nonnumeric are often referred to as **nominal**.

An example of **categorical data** is given in Table 3-3, and its corresponding bar chart is shown in Figure 3-5.[4] The nominal categories are types of fish, and the quantity being measured is the average lead concentration. It's clear from the figure that suckers have three to four times the concentration of lead as trout and whitefish. Note the labeling of the bar chart, the use of spacing between the bars, and horizontal graph lines. These features make the bar chart more informative and easier to read.

TABLE 3-3

Type of Fish	Trout	Whitefish	Sucker
Lead Concentration	1.015 ppm	0.605 ppm	2.810 ppm

Source: Washington State Dept of Ecology

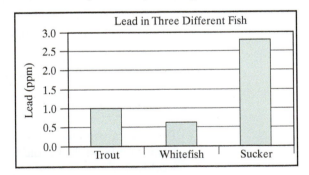

Figure 3-5 Bar chart showing lead in fish.

EXAMPLE 3-2 Sources of Mercury

About 149 tons of mercury (symbol Hg) are released into the environment each year in the United States.[5] The amount of mercury emitted by different sources is given in Table 3-4.

The chart representing these data has the sources of mercury arranged such that the corresponding emission values generally decrease from left to right (Figure 3-6). This arrangement may give the false impression that there is an inherent decrease in mercury emissions. However, the arrangement is arbitrary. We could have arranged the categories *alphabetically*, for example.

TABLE 3-4

Source of Emissions	Tons of Hg/yr
Coal-fired power plants	49
Burning municipal waste	28
Oil-fired boilers	27
Burning medical waste	15
Chlorine production	7
Cement production	4
Other (incl. manufacturing)	19

Source: U.S. Environmental Protection Agency

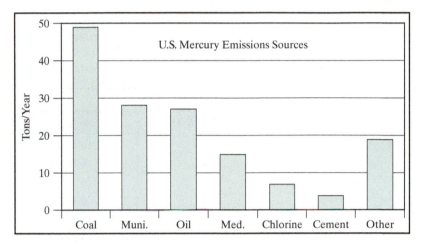

Figure 3-6 Bar chart of mercury emissions. The arrangement of the categories is arbitrary.

Some categories of data can be ranked or sorted into an ordered sequence. Data with this property are referred to as **ordinal**. Ordinal data can be arranged or sorted in a nonnumeric fashion; for example, the three stages of growth of a Magellanic penguin (chick, fledgling, and mature). Other ordinal data can be arranged numerically.

EXAMPLE 3-3 Environmental Rankings of Cities

Suppose an environmental organization ranks 170 cities in the United States according to their environmental quality. This hypothetical survey has five categories, from best cities (#1) to worst cities (#5). A bar chart (Figure 3-7) shows the number of cities in each category.

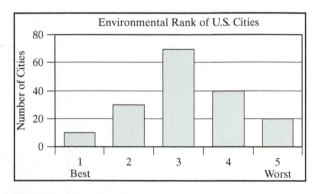

Figure 3-7 Environmental ranking is an example of ordinal data that are numerical.

In this example, the information along the horizontal axis is numerical instead of names or categories, but there is no mathematical significance to the numbers. A #2 city need not be twice as environmentally bad as a #1 city.

FREQUENCY HISTOGRAMS

Environmental science is filled with examples of repeated measurements of a single quantity. A chemist measures the ozone level along city streets in Tokyo every day for six months. A marine biologist counts the number of eggs in 105 loggerhead turtle nests on a beach in Madagascar. The number of tires dropped off every week are recorded by a recycler in Chicago. To represent large univariate data sets such as these, a special diagram called a **histogram** is used. Our first example involves measurements of the water temperatures of streams.

We start with 178 measurements of the temperature of small **closed-canopy** streams in Washington State, taken in June 1999.[6] Canopied streams have an umbrella of vegetation shading the water from the sun, keeping the water cool and suitable for fish such as salmon. In Table 3-5, the temperatures have been grouped into **bins** with a 1° Celsius temperature interval. In the table, the **bin widths** are constant; that is, the difference between the lower and upper boundaries of each bin is 1° Celsius. There are 19 bins containing all 178 temperature measurements. The number of measurements that fall in each bin is referred to as the **frequency**.

TABLE 3-5

Bin	Frequency	Bin	Frequency
9.5°C to 10.5°C	1	19.5°C to 20.5°C	5
10.5°C to 11.5°C	1	20.5°C to 21.5°C	6
11.5°C to 12.5°C	1	21.5°C to 22.5°C	5
12.5°C to 13.5°C	10	22.5°C to 23.5°C	2
13.5°C to 14.5°C	16	23.5°C to 24.5°C	1
14.5°C to 15.5°C	36	24.5°C to 25.5°C	3
15.5°C to 16.5°C	29	25.5°C to 26.5°C	1
16.5°C to 17.5°C	26	26.5°C to 27.5°C	0
17.5°C to 18.5°C	23	27.5°C to 28.5°C	1
18.5°C to 19.5°C	11	**Total:**	178

Source: Center for Water and Watershed Studies

A histogram displaying the binned data from Table 3-5 is shown in Figure 3-8. The height of each bar is scaled to the frequency. For example, 10 stream temperatures fall into the bin ranging from 12.5°C to 13.5°C and therefore the height of the bar is 10. One bin (26.5°–27.5°C) has a frequency of zero, because no temperature was measured in that interval. The 14.5–15.5°C bin has the highest frequency, with 36 temperature measurements, and hence the tallest bar. The most frequent bin is called the **modal bin**. Notice that there are no gaps between the bars, unlike many bar charts.

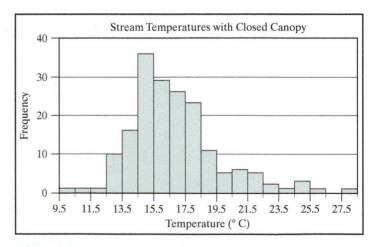

Figure 3-8 Frequency histogram of stream temperatures. The heights of the bars correspond to the frequencies of temperature measurements.

The histogram summarizes a large amount of data in a compact graphical form. From the shape of the graph, it's apparent that stream temperatures around 15–18°C were the most common in June of 1999. A large number of warmer stream temperatures can be seen in the high temperature "tail" in the diagram. The long "tail" of warm streams is not surprising, as there are many ways to overheat a stream, such as thermal pollution from waste discharge. Besides being unsuitable for certain types of fish, warm streams usually have less dissolved oxygen and a lower biodiversity.

Constructing a Frequency Histogram

How is a frequency histogram created? Let's work with a smaller data set from the same study of stream temperatures as before (Table 3-6). These temperature measurements were taken at **open-canopy** localities, where the sun was able to shine directly on the streams. The 67 temperature measurements have units of degrees Celsius.

TABLE 3-6

17.8	18.0	13.0	16.0	13.0	14.0	16.0	15.6	20.0	15.5
23.3	18.5	24.0	14.4	15.0	15.0	19.0	15.0	20.8	11.0
14.4	27.5	26.5	15.6	16.0	22.0	18.5	18.0	18.0	18.5
14.7	23.0	17.0	14.4	17.0	20.0	17.0	11.7	19.0	22.0
18.0	23.0	15.0	21.0	17.0	22.5	18.0	13.0	17.8	
18.0	17.5	15.9	21.5	20.0	23.0	15.0	13.0	13.0	
18.0	14.0	13.0	13.0	16.5	16.0	24.5	11.5	16.5	

Source: Center for Water and Watershed Studies

The first task is to sort the data, to make the binning process easier. Sometimes you can sort quickly by hand, but sorting with technology can save time and cut down on mistakes (see TI-83/84 calculator instructions in the next section). The sorted data are shown in Table 3-7.

TABLE 3-7

11.0	13.0	14.4	15.5	16.0	17.5	18.0	19.0	22.0	24.0
11.5	13.0	14.7	15.6	16.5	17.8	18.0	20.0	22.0	24.5
11.7	13.0	15.0	15.6	16.5	17.8	18.0	20.0	22.5	26.5
13.0	14.0	15.0	15.9	17.0	18.0	18.5	20.0	23.0	27.5
13.0	14.0	15.0	16.0	17.0	18.0	18.5	20.8	23.0	
13.0	14.4	15.0	16.0	17.0	18.0	18.5	21.0	23.0	
13.0	14.4	15.0	16.0	17.0	18.0	19.0	21.5	23.3	

The 67 measurements of open-canopy temperatures vary from a low of 11.0°C to a high of 27.5°C. We could pick 10°C as the lower boundary of the first bin, and choose nine bins with each bin 2°C wide (bin width = 2°C). The upper boundary of the highest bin will therefore equal 28°C (10° + 18° = 28°) and all of the temperatures will fall in the total bin interval (Table 3-8).

TABLE 3-8

Bin Interval	Frequency
10°C–12°C	3
12°C–14°C	7
14°C–16°C	15
16°C–18°C	13
18°C–20°C	12
20°C–22°C	6
22°C–24°C	7
24°C–26°C	2
26°C–28°C	2
Total:	67

In Table 3-7 there are two measurements equal to 14.0°C. These values fall exactly on a bin boundary. Which bin should these measurements be placed in? The most common practice is to "bin up" by placing a value that falls on a bin boundary in the *higher* bin. We will follow this common practice. After placing all values in the appropriate bins, it's good to add up the frequencies to ensure that they sum to the total number of measurements. A histogram of the frequency distribution in Table 3-8 is shown in Figure 3-9.

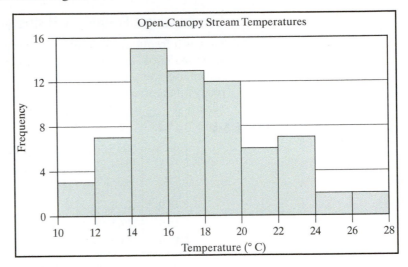

Figure 3-9 Histogram of open-canopy stream temperatures

The histogram is a concise summary of 67 measurements of temperature for streams that are not shaded by vegetation. The most common bin or modal bin is 14°–16°C. Temperatures from 16°C to 20°C are also quite frequent. There is a "tail" to the right of higher temperatures.

The choice of bin width can have a dramatic effect on the appearance of the histogram. In Figure 3-10 we plot two histograms of the same 67 measurements of open-canopy stream temperatures. The new histogram on the right has a bin width of 1°C rather than 2°C. Notice the more jagged character, the presence of empty bins, and the multiple peaks in the histogram with small bins.

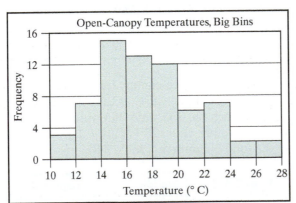 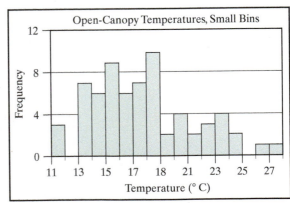

Figure 3-10 Comparison of large and small bins for the same data

Which diagram is "best"? There is no simple answer to this question. The histogram should be a fair and accurate graphical representation of the numerical data. The histogram should be readily understood and should not be misleading. The graphing calculator or computer software makes the construction of histograms much easier and faster and allows us to explore how bin width affects the resulting histogram shape.

EXAMPLE 3-4   Lengths of Bonnethead Sharks

Bonnethead sharks (*Sphyrna tiburo*) are the smallest member of the hammerhead family of sharks. Bonnetheads are very specific feeders, preying mainly on blue crabs. To better understand the ecology of bonnetheads, biologists measured the lengths of 402 bonnethead sharks in Tampa Bay, Florida. The binned lengths are given in Table 3-9.[7] Create a histogram.

TABLE 3-9

Bin	Frequency	Bin	Frequency	Bin	Frequency
40–45 cm	6	65–70 cm	36	90–95 cm	37
45–50 cm	5	70–75 cm	43	95–100 cm	46
50–55 cm	3	75–80 cm	47	100–105 cm	23
55–60 cm	11	80–85 cm	56	105–110 cm	2
60–65 cm	25	85–90 cm	61	110–115 cm	1

Source: *Canadian Journal of Fisheries and Aquatic Sciences*

Solution The data in the table are already binned, which sets the number of bins and bin width for the histogram (Figure 3-11).

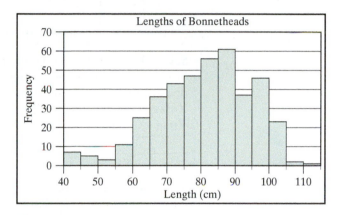

Figure 3-11 Histogram of bonnethead shark lengths

The modal bin for shark lengths is 85–90 cm. Most bonnetheads are 60–105 cm in length. There are few sharks that exceed 1 m in length.

USING TECHNOLOGY: HISTOGRAMS

Constructing a histogram often involves manipulating large data sets; technology can greatly reduce the workload. And it's easy to change the appearance of histograms created with technology until you get a satisfying result. Instructions for creating a histogram on the TI-83/84 graphing calculator are given below. Visit the text Web site for instructions on how to use online computer applications that perform the same task: (**enviromath.com**).

Entering and Sorting Data on the TI-83/84

Select **STAT>EDIT**, enter the 67 open canopy stream temperatures (Table 3-6) in L1, and return to the homescreen (**QUIT** or **2nd MODE**). Store your original data into L2 by entering **L1 STO L2** (see top of next screen).

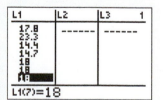

Under **STAT** select **SortA(** to paste it onto the homescreen, then type **L1** and press **ENTER**. Press **STAT>EDIT** to return to L1 to view the sorted data.

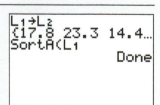

Plotting a Histogram on the TI-83/84

Clear out or turn off equations under **Y=**. Under **STATPLOT** (**2nd Y=**), choose **Plot 1**.

Scroll to **On** and select **ENTER**. In Plot 1, choose the histogram symbol under **Type**, and enter **L1** under **Xlist**. Set **Freq** to **1**.

Select **WINDOW** to set the size of the graph and the bin width. Use the same bins and bin width for the open-canopy histogram in Figure 3-9. Set **Xmin** = 10, **Xmax** = 28, and **Xscl** = 2 (bin width of 2°C). Set **Ymin** = 0, **Ymax** = 16, and **Yscl** = 1. Leave **Xres** = 1.

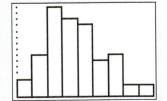

Select **GRAPH** and compare screen at right to the left side of Figure 3-10. Return to **WINDOW**, change **Xscl** to 1 (bin width of 1° C), **Ymax** to 12, and press **GRAPH**. Compare to right side of Figure 3-10.

RELATIVE FREQUENCY HISTOGRAMS

The examples of histograms seen so far have displayed the frequencies of bin values. The frequencies are the number of measurements that fall into each bin. If the total number of measurements is very high, then the frequencies for most bins will be high as well, whereas if the data set is small, then most of the frequencies will be low. It can be difficult to compare two histograms with different total measurements, especially if the two histograms are plotted on the same scale. Therefore, the frequencies are often **normalized** to the total number of measurements to create **relative frequencies**.

TABLE 3-10

Bin Interval	Actual Frequency	Relative Frequency
10°C–12°C	3	$\frac{3}{67} = 0.045 = 4.5\%$
12°C–14°C	7	$\frac{7}{67} = 0.104 = 10.4\%$
14°C–16°C	15	$\frac{15}{67} = 0.224 = 22.4\%$
16°C–18°C	13	$\frac{13}{67} = 0.194 = 19.4\%$
18°C–20°C	12	$\frac{12}{67} = 0.179 = 17.9\%$
20°C–22°C	6	$\frac{6}{67} = 0.090 = 9.0\%$
22°C–24°C	7	$\frac{7}{67} = 0.104 = 10.4\%$
24°C–26°C	2	$\frac{2}{67} = 0.030 = 3.0\%$
26°C–28°C	2	$\frac{2}{67} = 0.030 = 3.0\%$
Total:	67	$\frac{67}{67} = 1.000 = 100.0\%$

Relative frequencies are ratios of the actual frequency to the total number of measurements. Other examples of ratios and normalization are discussed in Chapter 2.

We return once more to the temperatures of streams with an open canopy of vegetation. To determine the relative frequencies, each frequency is divided by the total number of measurements, which is 67 (Table 3-10).

The relative frequencies represent the fractions of the total measurements that fall into each bin, therefore the sum of the relative frequencies is equal to 1 or 100%. The relative frequency histogram (Figure 3-12) has the same overall shape as the histogram of actual frequencies, but the vertical scale has changed.

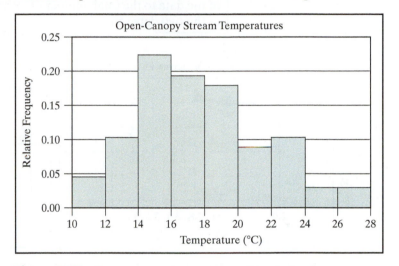

Figure 3-12 Open-canopy temperatures with relative frequency

To illustrate the utility of relative frequencies in comparing data sets, let's look at an example of two groups of lakes in California.

EXAMPLE 3-5 Elevations of Lakes with and without Fish

Biologists have divided lakes in the central Sierra Nevada of California into two groups: those with fish and those without.[8] The scientists are concerned about the impact of invasive fish on the ecology of mountain lakes and are also concerned about acid deposition in the lakes from smog. In the study area, lakes are found from about 2 km above sea level to as high as about 4 km above sea level (Figure 3-13). The numbers of lakes found in certain intervals of elevation (100-meter in-

Figure 3-13 A lake at high elevation in the Sierra Nevada.
Photo courtesy of USGS.

crements) were tallied and the presence and absence of fish was recorded. There are 554 lakes with fish in the study area but only 46 lakes without fish. How can we compare the elevations of these two types of lakes graphically?

Solution When plotted with the same vertical scale, the frequency histogram for the lakes without fish is barely readable (Figure 3-14). Relative frequency histograms of the same data at the same scale (Figure 3-15) allow the two histograms to be compared directly and support the assertion by the biologists that there is little difference in the elevation distributions of lakes with and without fish. Normalization of the data to the total number of measurements permits a more direct comparison of the two data sets.

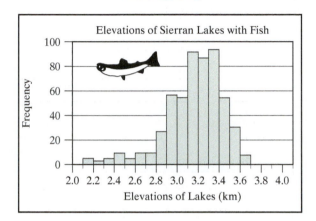

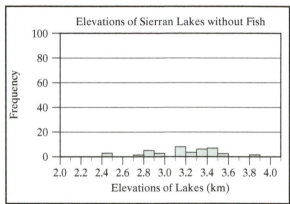

Figure 3-14 Frequency histograms of elevations of lakes with and without fish, plotted at the same vertical scale.

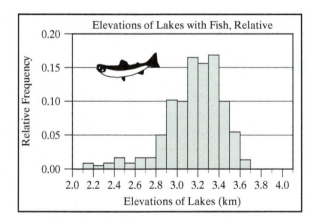

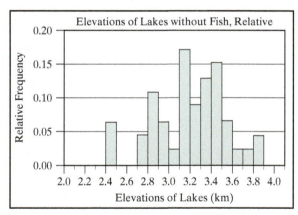

Figure 3-15 Relative frequency histograms of elevations of lakes with and without fish, plotted at the same vertical scale.

SCATTERPLOTS

A **scatterplot** is a common way of displaying "x-y" or **bivariate** data that consist of pairs of numerical values. For example, the concentrations of two heavy metals, lead and zinc, in 10 different fish from the Spokane River in Idaho and Washington are given in Table 3-11.[9] The concentrations are given in parts per million (ppm). These heavy metals are the legacy of gold and silver mining in northern Idaho; runoff from mines and mine tailings have carried these metals into rivers and streams. The metals have entered the food chain and can be concentrated in higher-order predators such as fish.

In the scatterplot (Figure 3-16), lead concentration is shown along the horizontal axis and zinc concentration is shown along the vertical axis. Each point on the scatterplot represents one pair of lead-zinc values corresponding to one of the analyzed fish. The pairs of values are scattered or distributed across the graph; hence the name *scatterplot*. The data show a distinct trend of increasing zinc content with increasing lead content, but scatterplots can show other kinds of trends or no trend at all!

	Fish Type	Lead (ppm)	Zinc (ppm)
TABLE 3-11	trout #1	0.73	45.3
	trout #2	1.14	50.8
	trout #3	0.60	40.2
	trout #4	1.59	64.0
	sucker #1	4.34	150.0
	sucker #2	1.98	106.0
	sucker #3	3.12	90.8
	sucker #4	1.80	58.8
	whitefish #1	0.65	35.4
	whitefish #2	0.56	28.4

Source: Washington State Dept of Ecology

Figure 3-16 Scatterplot of zinc versus lead in fish

When constructing scatterplots by hand or with technology, it's important to choose appropriate scales for the horizontal and vertical axes. Even though the lead and zinc concentrations have the same units (parts per million), we chose very different scales for the two axes, because the lead concentrations are much smaller than the zinc concentrations. The lead values vary between 0.56 ppm and 4.34 ppm. Therefore, we chose a horizontal scale that extended from 0.5 ppm to 4.5 ppm, to encompass all of the lead values, but not to enclose too much empty space. A similar rationale motivated our choice for the vertical scale for zinc. Choosing the right scales means that the data fill the diagram rather than being bunched to one side or crammed into one corner.

USING TECHNOLOGY: SCATTERPLOTS

Creating scatterplots by hand is time-consuming, especially with large data sets. Instructions for creating a scatterplot on the TI-83/84 graphing calculator are given next. Visit the text Web site for instructions on how to use online computer applications that perform the same task: (**enviromath.com**).

Scatterplots on the TI-83/84

Select **STAT>EDIT** and enter the lead-zinc pairs from Table 3-11 into L1 and L2. If you do not enter the same number of values in each column, you will eventually get the error message **DIM MISMATCH**. Return to the homescreen (**QUIT** or **2nd MODE**).

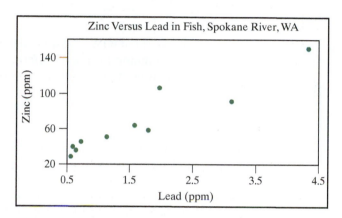

Clear or turn off equations under **Y=**. Go to **STATPLOT** (**2ⁿᵈ Y=**). In **Plot1**, select **On**, then choose the scatterplot icon in **Type**. Assign **L1** (lead) to **Xlist** and **L2** (zinc) to **Ylist**.

Press **ZOOM>9:ZoomStat** to automatically scale and graph your scatterplot. Compare this scatterplot with Figure 3-16.

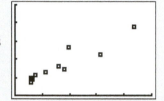

Press **WINDOW** to display the window dimensions. It's common to choose different window parameters so that both axes and both sets of tick marks are visible.

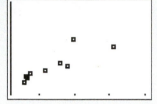

When constructing scatterplots by hand or with software, label both horizontal and vertical axes (if possible), and include the units! An informative title at the top of the graph is also desirable. Note that three different types of fish were listed in Table 3-11, but that extra information on fish type did not make it onto the graphing calculator scatterplot. A scatterplot produced with advanced software shows the different species of fish with a different symbol (Figure 3-17).

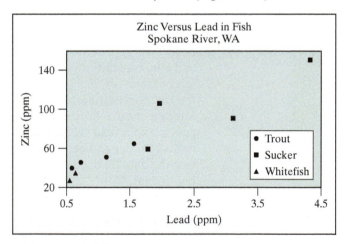

Figure 3-17 Revised scatterplot with fish identification

This small graphical modification illustrates an important feature in the data. The suckers, which prey on trout and whitefish, have higher concentrations of heavy metals. That's partially because the heavy metals become concentrated up the food chain, an example of **bioamplification**.

LINE GRAPHS

A **line graph** is a type of scatterplot with the data points connected by line segments. Connecting the data points only makes sense in certain situations, such as when time is involved (a **time series**). We would not draw line segments between lead-zinc data points (Figure 3-17), because there is no sequence to the data. Line graphs of environmental time series are quite common.

EXAMPLE 3-6   Manatee Deaths Due to Watercraft, Florida

The manatee (*Trichechus senegalensis*) is a marine mammal found in shallow waters along the west coast of Florida (Figure 3-19). Manatees are particularly vulnerable to injury and death due to high-speed collisions with power boats, as manatees often float just beneath the surface. The Florida Marine Research Institute has compiled data on manatee mortality from 1976 to 1999 (Table 3-12).[10] Create a line graph for these data.

TABLE 3-12

Year	Deaths	Year	Deaths	Year	Deaths	Year	Deaths
1976	10	1982	19	1988	43	1994	50
1977	13	1983	15	1989	50	1995	43
1978	21	1984	33	1990	49	1996	59
1979	22	1985	35	1991	52	1997	52
1980	15	1986	31	1992	38	1998	66
1981	23	1987	37	1993	34	1999	83

Source: Florida Fish and Wildlife Conservation Commission

Solution This bivariate data set includes time, which we graph on the horizontal axis (Figure 3-18). The data show an overall increasing trend of watercraft-caused mortality among manatee as time increases. One possible explanation is that there are increasing numbers of watercraft as Florida's human population grows and therefore more fatal encounters, but there could be other reasons as well.

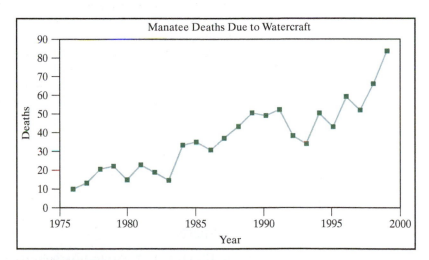

Figure 3-18 Manatee mortality caused by watercraft

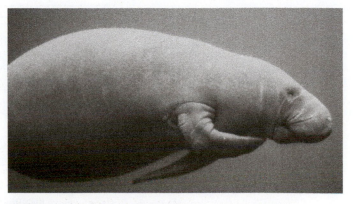

Figure 3-19 Manatee scratching. *U.S. Fish and Wildlife Service.*

CHAPTER SUMMARY

Univariate data are associated with one variable or changeable quantity. The ozone concentration of the street-level atmosphere in a city is an example of univariate data. Univariate data that are separated or arranged by category are often referred to as **categorical data**. For example, the ozone concentration could be measured in four different categories: low, mild, high, extreme.

Categorical data can be shown graphically by a **pie chart** (also called a circle graph). For example, total recyclables might be composed of 30% paper, 30% glass, 25% glass, and 15% plastics (by weight). Each percentage would be represented by its appropriately sized "slice of pie" on the circle graph.

Data categories are typically nonnumeric or **nominal**: species of plants in a wetlands, types of insulation in a home, and varieties of pesticides. Some categories can also be ranked or sorted into an ordered sequence: grade level of elementary schoolchildren or ranking of countries by their environmental legislation. Ranked or ordered categorical data are called **ordinal** data. Both nominal and ordinal data can be displayed in a **bar chart**, with one axis displaying the categories and the other axis showing the quantitative information for each of those categories.

Histograms are often used to display a collection of repeated measurements of a single quantity, such as the density of beetles on each pine tree in a Texas forest or the thickness of soil in the upland forests of Costa Rica. The data are grouped into **bins** of equal width and arranged on the horizontal axis. The vertical axis is the **frequency** or number of values in each bin. The height of each bar corresponds to the frequency. Frequencies can be normalized to the total number of measurements to produce **relative frequencies**.

Bivariate data are associated with two quantities. For example, the amounts of fertilizer in pounds per acre (one quantity) and the corresponding average heights of corn in feet (second quantity) could be measured for 50 farms in Iowa. The pairs of values (pounds per acre, average heights) for each farm are represented by a point on a **scatterplot**. A **line graph** is a scatterplot with the individual points on the diagram connected by line segments. Line graphs are typically associated with a **time series** of data.

END *of* CHAPTER EXERCISES

Pie Charts

1. The pie chart in Figure 3-20 shows the contribution to global warming of three greenhouse gases: carbon dioxide (CO_2), methane (CH_4), and nitrous oxide (N_2O).[11] Estimate the percentage contribution of each.

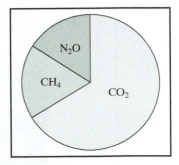

Figure 3-20 Pie chart of greenhouse gases

2. The pie chart in Figure 3-21 shows oil imports to the United States and domestic production (in millions of barrels per day) for June 2001.[12] What's wrong with this pie chart?

Oil (In Millions of Barrels per Day)

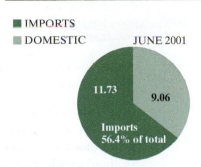

Figure 3-21 Pie chart of oil imports and domestic production

3. The U.S. Environmental Protection Agency's Toxic Release Inventory for the year 2000 gives the amount of toxic waste released to the environment by state.[13] The data for five states are listed in Table 3-13.

TABLE 3-13

State	Toxics (millions of lb)
Nevada	1,000
Utah	956
Arizona	744
Alaska	535
Texas	302

Source: U.S. Environmental Protection Agency

a. Construct a pie chart that represents the data given in Table 3-13.
b. Examine the pie chart. Which state or states is approximately twice as toxic as Alaska?
c. Compare your answer to (b) to the values in the table.

4. Values for the amount of electricity (in trillions of watt-hours) generated by sector for the United States in 1999 are given in Table 3-14.[14]

TABLE 3-14

Fuel Source	Electricity (TWh)
Coal	1,882
Natural gas	565
Petroleum	119
Nuclear	728
Hydropower	307
Other renewables	89

Source: U.S. Dept of Energy

a. Construct a pie chart for these data.
b. Fossil fuels are what percentage of the total? Shade in the fossil fuels on the pie chart.
c. Renewable energy sources are what percentage of the total? Show the renewables with cross-hatching

5. Table 3-15 shows tons of recycled materials by category for 1986 and 1998 in Washington State.[15]

TABLE 3-15

Category	1986	1998
Papers	391,994	821,994
Metals	9,528	318,710
Organics	0	815,809
Plastics	349	9,871
Glass	48,013	113,338
Others	352	87,657

Source: Washington State Dept of Ecology

a. Construct a pie chart for each year.

b. Are all categories visible on each pie chart?

c. Based on the pie charts, briefly describe two differences between the two years.

6. Lead emissions by source for Europe are given in Table 3-16 for three different years, 1955, 1975, and 1995.[16] Percentages may not sum to 100% because of rounding.

TABLE 3-16

Source	1955	1975	1995
Cars and trucks on roads	49.5%	74.9%	68.7%
Noniron metals manufacturing	21.1%	10.1%	11.8%
Stationary fuel combustion	8.7%	3.8%	9.5%
Iron and steel manufacturing	11.2%	7.0%	7.8%
Waste disposal	0.2%	0.4%	0.9%
Cement production	1.2%	1.1%	0.0%
Other	8.1%	2.8%	1.2%
Total emissions (tonnes)	**62,532**	**159,233**	**28,390**

Source: American Geophysical Union

a. Construct a pie chart for each of the three years.

b. The total amount of lead and its distribution among sources changed from 1955 to 1975. What was the most likely cause of these simultaneous changes?

c. The total amount of lead emitted decreased from 1975 to 1995 but the distribution remained almost the same. What was the most likely cause of this pattern?

Bar Charts

7. The ecological footprint of a country is the area of land and sea necessary to produce the renewable resources for that country's people. Renewable resources can grow back or replenish themselves naturally. The bar chart (Figure 3-22) shows the ecological footprint for six countries in 1999.[17]

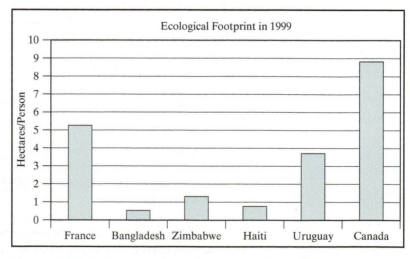

Figure 3-22 Ecological footprint for six countries in 1999

a. What are the units for the ecological footprint?

b. Name a few renewable resources consumed by people in these countries.

c. What is the ratio of the Canadian footprint to the Haitian footprint?

8. The bar chart in Figure 3-23 shows the average monthly discharge measured at the Altamira measuring station on the Xingu River, Amazon basin, Brazil.[18] Discharge is the volume of water flowing by the measuring station in a certain amount of time. In this chart, discharge is given in cubic meters per second.

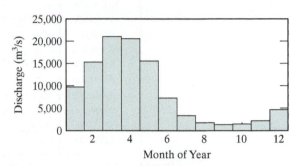

Figure 3-23 Bar chart, mean monthly discharge, Xingu River

a. Which month of the year has the highest discharge (on average)? What is that discharge?

b. Which month has the lowest discharge (on average)? What is that discharge?

c. What is the usual discharge during the month of May?

9. U.S. energy consumption (in quadrillions of BTUs) by energy source in April 2002 is given in Table 3-17.[19]

TABLE 3-17

Source of Energy	Consumption (Quads)
Nuclear	0.61
Petroleum	3.11
Natural gas	1.84
Coal	1.69
Hydroelectric	0.31

Source: U.S. Dept of Energy

a. Create a bar chart that represents these data.

b. Of the five sources of energy, which (if any) are renewable?

c. Which sources of energy are missing from the bar chart?

10. Italians are the third largest consumers of meat on planet Earth, averaging about 82 kilograms of meat per person per year.[20] Table 3-18 gives a breakdown of the categories of meat consumption by Italians.

TABLE 3-18

Category	Annual Consumption (kg/person)
Beef	26
Pork	35
Poultry	19
Mutton	2

Source: Worldwatch Institute

a. Construct a bar chart of the consumption data, in the order given in the table.

b. Should pork have been plotted as the first bar, because it has the highest consumption? Explain.

Frequency Histograms

11. Frequency histograms are often used to display population data. The age distribution of females in Bangladesh in 2000 is shown in the frequency histogram in Figure 3-24.[21]

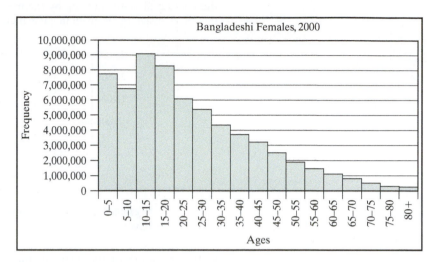

Figure 3-24 Frequency histogram of Bangladeshi females by age

  a. What is the bin width?
  b. All of the bins have the same width except one. Which bin is that?
  c. Estimate the total female population in 2000.

12. The age distribution of females in Belgium in 2000 is shown in Figure 3-25.[22]

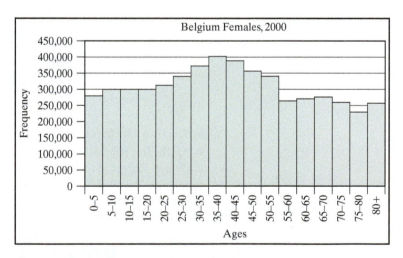

Figure 3-25 Frequency histogram of Belgium females by age

  a. What is the bin width?
  b. What is the most common bin (the modal bin)?
  c. Estimate the number of fertile women in 2000. Fertile women are defined as those of child-bearing age from about 15 to 45.

13. A **population pyramid** (Figure 3-26) for Mexico in 2000 shows the number of men and women in various age intervals.[23] The population pyramid is made of two histograms, one for the male population and another for the female population.

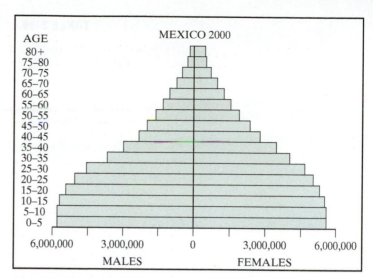

Figure 3-26 Population pyramid, Mexico, 2000

    a. What is the bin width?

    b. All of the bins have the same width except which bin?

    c. At what age do females become more populous than males?

    d. What is the total number of people in their sixties in the year 2000 (approximately)?

14. A population pyramid for Brazil in the year 2000 is given in Figure 3-27.[24]

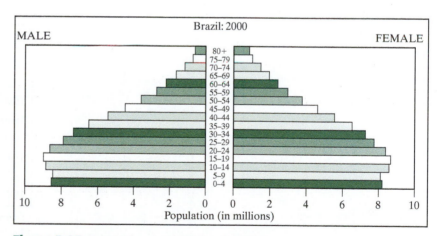

Figure 3-27 Population pyramid, Brazil, 2000

    a. The frequency is given in which units?

    b. What is the most frequent bin for men? For women?

    c. Women of child-bearing age vary from 15 to 45 years old (approximately). Estimate the number of women of child-bearing age that lived in Brazil in the year 2000.

    d. At what age do women begin to outnumber men?

15. Forty-four years of annual rainfall measurements at Union City, Tennessee, are given in Table 3-19.[25] Yearly rainfall values have been rounded off to the nearest inch.

TABLE 3-19

47	53	68	61
38	26	53	52
36	46	68	47
40	48	41	43
43	47	48	54
37	44	48	57
73	42	63	60
43	52	39	49
41	55	42	44
38	43	55	49
53	54	48	51

Source: U.S. Historical Climate Network

a. Sort and bin the data, then construct a histogram. Choose appropriate bin widths, scales, and labels.

b. Are there any empty bins in your histogram? If so, identify the bin intervals.

c. What is the most frequent or modal bin?

d. How many values fall above the modal bin and how many values fall below the modal bin?

16. Between 1900 and 2000, 65,000 dams were constructed in the United States.[26] The numbers of dams built in each decade are given in Table 3-20.

TABLE 3-20

Years	Number of Dams
1900–1910	2,127
1910–1920	1,907
1920–1930	2,252
1930–1940	3,716
1940–1950	4,053
1950–1960	11,388
1960–1970	19,310
1970–1980	13,076
1980–1990	5,017
1990–2000	2,154

Source: U.S. Army Corps of Engineers

a. Construct a frequency histogram for these data.

b. During which decade were the most dams constructed? In other words, what is the modal bin?

c. Do more dams fall above the modal bin or below the modal bin?

d. Briefly describe the history of dam construction from 1900 to 2000.

Relative Frequency Histograms

17. Annual precipitation for two stations in Pennsylvania is presented in Table 3-21.[27] The first station at Reading has 132 years of recent continuous measurement, whereas the second station at Montrose has only 35 years of recent continuous measurement. The summary data for both stations are given in the table.

TABLE 3-21

	Reading		Montrose	
Inches/Year	**Frequency**		**Inches/Year**	**Frequency**
25–30	2		25–30	0
30–35	19		30–35	6
35–40	27		35–40	6
40–45	39		40–45	9
45–50	33		45–50	10
50–55	10		50–55	2
55–60	1		55–60	2
60–65	0		60–65	0
65–70	1		65–70	0

Source: U.S. Historical Climate Network

a. Construct a relative frequency histogram for each data set, using the same bins and the same vertical scale for both.

b. What is the modal bin for each station?

c. Are the relative frequencies for the modal bins similar or different?

d. Suppose a meteorologist says, "The longer the period of measurement, the more extreme the measurements." Is this assertion supported by the data and histograms? Explain briefly.

18. For the dam construction data in Exercise 16, construct a relative frequency histogram.

Scatterplots

19. The number of gray wolves (*Canis lupis*) in Wisconsin between 1980 and 2002 is shown in a scatterplot (Figure 3-28; see also Figure 3-29).[28]

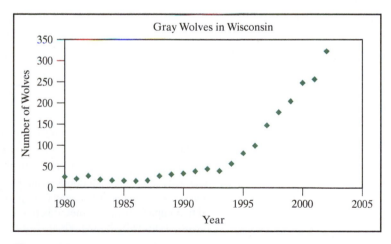

Figure 3-28 Gray wolves in Wisconsin

Figure 3-29 Gray wolf. *Photo by Gary Kramer for the U.S. Fish and Wildlife Service.*

a. The data are bivariate. What are the two variables?

b. What were the (approximately) minimum and maximum numbers of wolves in Wisconsin during the time period shown?

c. Approximately when did the Wisconsin wolf population begin to increase dramatically? Give two possible explanations for this dramatic increase.

20. The scatterplot in Figure 3-30 shows the average number of acres burned per fire from 1960 to 2002 in the United States.[29]

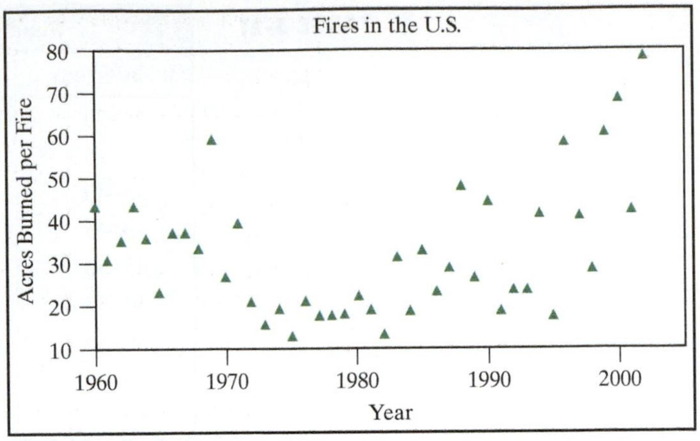

Figure 3-30 Scatterplot of U.S. fires

a. What was the lowest number of average acres per fire in this time period? What was the highest value?

b. Briefly describe what happened to the average number of acres per fire from 1960 to 2002.

c. Some fires burned thousands of acres. Why are the values in the scatterplot so low?

21. Data on the production and recycling of aluminum cans in the United Kingdom are given in Table 3-22.[30] All values represent billions of cans.

TABLE 3-22

Year	Produced	Recycled
1989	2.79	0.072
1990	4.13	0.224
1991	4.23	0.462
1992	4.62	0.694
1993	5.74	1.160
1994	6.54	1.524
1995	7.03	1.900
1996	5.31	1.500
1997	4.58	1.506
1998	4.47	1.547

Source: U.K. Dept for Environment, Food and Rural Affairs

a. By hand, construct a scatterplot of cans recycled (vertical axis) versus cans produced (horizontal axis). Choose appropriate scales and labels.

b. Briefly describe the resulting pattern of the scatterplot.

22. Scientists studying small ponds in the prairie regions of Saskatchewan, Canada, counted the ponds both from the ground and from photographs taken from airplanes.[31] The scientists wanted to compare these two methods, to see whether they gave the same results. They measured the ponds using both methods at 12 different localities. The data are given in Table 3-23.

TABLE 3-23

Locality #	Ponds from Ground	Ponds from Air	Locality #	Ponds from Ground	Ponds from Air
1	36	24	7	84	105
2	39	53	8	92	135
3	51	42	9	99	77
4	54	73	10	119	111
5	55	54	11	209	172
6	76	85	12	261	232

Source: *Journal of Wildlife Management*

a. Construct a scatterplot, by hand, of ponds from the ground (horizontal axis) versus ponds from the air (vertical axis). Use the same horizontal and vertical scale. Label appropriately.

b. Draw a straight line on the diagram that passes through the origin and the point (250, 250).

c. The data point from the fifth locality falls approximately on this line. What does this mean?

d. Which localities lie above the line? How do ground-based and aerial surveys compare for these localities?

e. Which localities lie below the line? How do ground-based and aerial surveys compare for these localities?

Scatterplots Using Technology

23. Using technology, generate a scatterplot using the same graph parameters (Xmin, Xmax, etc.) as your hand-drawn scatterplot in Exercise 21.

24. Using, technology, generate a scatterplot using the same graph parameters (Xmin, Xmax, etc.) as your hand-drawn scatterplot in Exercise 22.

SCIENCE IN DEPTH

Energy Demand and the Arctic National Wildlife Refuge

The United States is the single largest consumer of energy on Earth.[32] The total energy demand by all countries each year is approximately 400×10^{15} British thermal units or 400 quadrillion BTUs. The United States represents 25% of that total or 100 "Quads" per year. Demand for energy in the United States is growing at about 1.5% per year, about the same as the worldwide increase.

The United States consumes about 15 Quads of energy from domestically produced petroleum each year and about 25 Quads of imports. U.S. domestic petroleum production peaked in the 1970s and has declined since.[33] Known reserves of U.S. petroleum are estimated to be about 22 billion barrels of crude oil with an energy equivalent of approximately 130 Quads. The U.S. reserves of petroleum would barely last a decade at the current rate of domestic consumption. The only way to sustain this consumption without relying on more imports is to find new reserves of U.S. petroleum.

The Arctic National Wildlife Refuge (ANWR, pronounced an-whar) was established in 1980 by the Alaska National Interest Lands Conservation Act.[34] The creation of the refuge by the U.S. Congress was in response to the discovery and development of petroleum fields to the west of ANWR around Prudhoe Bay and adjacent areas on the North Slope of Alaska. Congress was concerned about the environmental impact of both exploration and extraction in northernmost Alaska and preserved ANWR to balance the damage done to the west. ANWR protects an important arctic ecosystem, which sustains the vast Porcupine caribou herd (see Chapter 10). The ANWR contains about 19 million acres of land between the Trans-Alaska Pipeline on the west and the Canadian border on the east (Figure 3-31). ANWR contains two subsections, the Wilderness Area (8 million acres) and the Coastal Plain (1.5 million acres).

The Alaskan North Slope to the west of ANWR has produced about 13 billion barrels of petroleum, equivalent to about 75 Quads of energy. Another 10 billion barrels or 60 Quads of oil remain in North Slope fields, but much of this oil is thick and gooey and

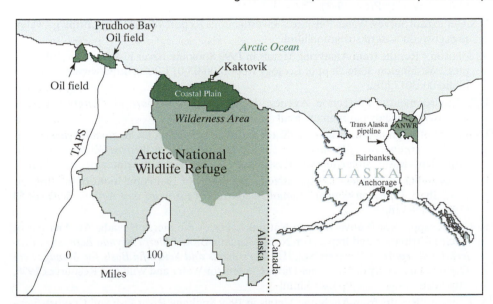

Figure 3-31 Maps showing ANWR at two scales. *Modified from the Alaska Dept. of Natural Resources.*

therefore difficult to recover economically. At its peak in the late 1980s, the Prudhoe Bay area produced about 2 million barrels of oil per day but production has declined ever since.[35] Less than a million barrels of oil now flow through the Trans-Alaska Pipeline System each day. When that flow declines to about a half a million barrels per day, the pipeline will no longer be economical and will be shut down and dismantled.

The oil companies that own the pipeline need new sources of high-quality oil, and therefore attention has shifted to the coastal region of ANWR. This subsection of ANWR was delineated by Congress because of its importance as wildlife habitat and because of its potential petroleum resources. In a 1998 assessment, the United States Geological Survey estimated that there is a moderate probability that 10 billion barrels of oil (about 60 Quads of energy) could be recovered from the coastal area of ANWR. However, the geological information about the coastal area is minimal, as exploration for oil in this region is currently forbidden. As a result, there is a lot of uncertainty in the estimates of petroleum in ANWR.

CHAPTER PROJECT: U.S. ENERGY FLOWS

The United States is the largest single consumer of energy on Earth. At the current rate of growth, energy demand in the U.S. will double in the next 50 years. Besides petroleum, what are the sources of energy consumed in the United States? How much energy comes from renewable sources? What parts of U.S. society consume the most energy?

In the companion project for this chapter found at the Quantitative Reasoning and the Environment Web site (**enviromath.com**), you will be introduced to a very rich and informative graph that illustrates U.S. energy production and consumption for the year 2001. You'll answer a number of questions about the sources and sinks of energy in the United States and look at the role of alternative and renewable energy sources in the current energy picture. Visit the Web site!

NOTES

[1] U.S. Dept of Energy, Energy Information Agency, *Annual Energy Outlook 1996*.

[2] Phillip L. Jackson and A. Jon Kimerling, *Atlas of the Pacific Northwest* (Corvallis, OR: Oregon State University Press, 1993), 77.

[3] U.S. Geological Survey, Water-Resources Investigations Report 98-4086, 1998, http://ga.water.usgs.gov/edu/waterdistribution.html

[4] A. Johnson, "Results from Analyzing Metals in 1999 Spokane River Fish and Crayfish Samples," Washington State Dept of Ecology Report 00-03-017, 2000, http://www.ecy.wa.gov/biblio/0003017.html

[5] U.S. Environmental Protection Agency, *Mercury Study Report to Congress* (1997), http://www.epa.gov/oar/mercury.html

[6] University of Washington, Center for Water and Watershed Studies, *Regional, Synchronous Field Determination of Summertime Stream Temperatures in Western Washington*, 2001, http://depts.washington.edu/cwws/Research/Projects/regionalstreamtemperature.html

[7] E. Cortés and G. R. Parsons, "Comparative Demography of Two Populations of the Bonnethead Shark (*Sphyrna tiburo*)," *Canadian Journal of Fisheries and Aquatic Sciences* 53 (1996): 709–718.

[8] Roland Knapp, "Non-Native Trout in Natural Lakes of the Sierra Nevada: An Analysis of Their Distribution and Impacts on Native Aquatic Biota," *Sierra Nevada Ecosystem Project: Final report to Congress*, vol. III, *Assessments and Scientific Basis for Management Options*. University of California-Davis, Centers for Water and Wildland Resources, 1996, http://ceres.ca.gov/snep/pubs/v3.html

[9] A. Johnson, "Results from Analyzing Metals in 1999 Spokane River Fish and Crayfish Samples."

[10]Florida Fish and Wildlife Conservation Commission, Fish and Wildlife Research Institute, http://www.floridamarine.org/

[11]U.S. Environmental Protection Agency, *Greenhouse Gases and Global Warming Potential Values—Excerpt from the Inventory of U.S. Greenhouse Gas Emissions and Sinks: 1990–2000,* http://yosemite.epa.gov/OAR/globalwarming.nsf/content/Emissions.html

[12]Lisa M. Pinsker, "Energy Notes," *Geotimes,* September 2002, 31.

[13]Associated Press, "Coal Plants and Mines Biggest Polluters," by John Heilprin, *Seattle Daily Journal of Commerce,* May 28, 2002.

[14]U.S. Dept of Energy, Energy Information Administration, *Electricity Power Annual 1999,* http://www.eia.doe.gov/cneaf/electricity/epa/backissues.html

[15]Washington State Dept of Ecology, Solid Waste and Financial Assistance Program, http://www.ecy.wa.gov/programs/swfa/solidwastedata/recyclin.asp

[16]Hans von Storch et al., "Reassessing Past European Gasoline Lead Policies," *EOS (Transactions of the AGU),* September 3, 2002, 393 and 399.

[17]World Wildlife Fund, Living Planet Report 2002, http://www.panda.org/news_facts/publications/general/livingplanet/index.cfm

[18]University of New Hampshire, Global Runoff Data Center, http://www.grdc.sr.unh.edu/html/Polygons/P3630050.html

[19]U.S. Dept of Energy, Energy Information Administration, *Monthly Energy Review: April 2002,* http://www.eia.doe.gov/cneaf/electricity/epm/epm_sum.html

[20]Worldwatch Institute, "United States Leads World Meat Stampede," http://www.worldwatch.org/press/news/1998/07/02/

[21]U.S. Census Bureau, World Population Information, http://www.census.gov/ipc/www/world.html

[22]Ibid.

[23]Ibid.

[24]Ibid.

[25]U.S. Historical Climate Network, Carbon Dioxide Information Analysis Center, http://cdiac.ornl.gov/epubs/ndp019/state/TN/tennessee.html

[26]U.S. Army Corps of Engineers, National Inventory of Dams, http://crunch.tec.army.mil/nid/webpages/nid.cfm

[27]U.S. Historical Climate Network, Carbon Dioxide Information Analysis Center, http://cdiac.esd.ornl.gov/r3d/ushcn/statepcp.html

[28]Wisconsin Dept of Natural Resources, *Status of the Timber Wolf in Wisconsin,* by Adrian P. Wydeven and Jane E. Wiedenhoeft, Endangered Resources Report #121 (2002), http://www.dnr.state.wi.us/org/land/er/publications/reports/reports.htm

[29]U.S. National Interagency Fire Center, "Wildland Fire Statistics," http://www.nifc.gov/stats/wildlandfirestats.html

[30]U.K. Dept for Environment, Food and Rural Affairs, *Digest of Environmental Statistics,* http://www.defra.gov.uk/environment/statistics/des/index.htm

[31]L. L. Strong and L. M. Cowardin, "Improving Prairie Pond Counts," *Journal of Wildlife Management* 59 (1995): 708–719

[32]Paul Weisz, "Basic Choices And Constraints On Long-Term Energy Supplies," *Physics Today,* July 2004, 47–52

[33]Jon Spencer and Steven Rauzi, "Crude Oil Supply and Demand: Long-Term Trends," *Arizona Geology* 31 (2001), http://www.azgs.az.gov/winter2001.htm

[34]U.S. Geological Survey, "Arctic National Wildlife Refuge, 1002 Area, Petroleum Assessment, 1998, Including Economic Analysis," USGS Fact Sheet FS-028-01. http://pubs.usgs.gov/fs/fs-0028-01/fs-0028-01.pdf

[35]Laura Wright, "ANWR in Black and White," *Geotimes,* May 2001, http://www.geotimes.org/may01/anwr.html

PART TWO

Function
Modeling

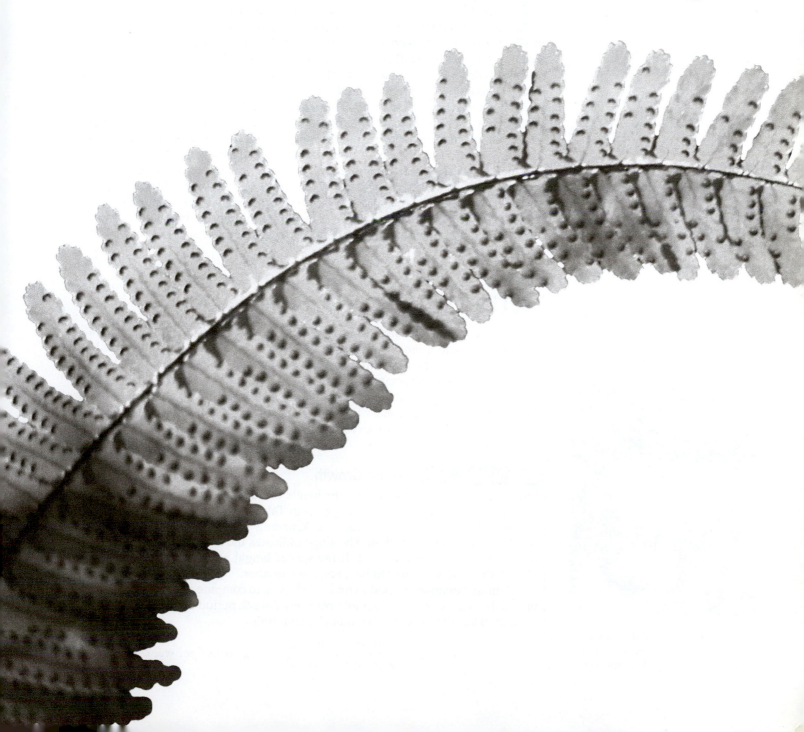

4

Linear Functions and Regression

The Glen Canyon Dam on the mighty Colorado River was built in 1963 to provide recreation, drinking water, and electricity for the arid western United States (Figure 4-1). The waters pooling behind the dam formed Lake Powell, which eventually backed up for almost 200 miles. Starting in 1999, a severe drought in the western United States caused Lake Powell to shrink. By 2004, the level of Lake Powell was dropping at a steady rate of 3 inches per day.[1] Imagine camping along the shore of Lake Powell for a week and seeing the lake level drop 21 inches! Over a 2-week stay the waters would drop 42 inches or $3\frac{1}{2}$ feet!

When a quantity changes by a constant amount each time period, or changes by a constant amount with respect to another factor, we say that the quantity **changes linearly**. The surface level of Lake Powell, dropping by a constant amount of 3 inches per day, is an example of a quantity that decreases linearly. In this chapter, we learn how to recognize and describe linear change using equations, tables, and graphs. We also examine how units of measure balance in linear equations. We then investigate techniques for approximating almost-linear data, and conclude with a discussion on measuring the strength of a linear trend.

MODELING WITH LINEAR FUNCTIONS

Scientists studying environmental problems often begin by observing qualitative (nonnumeric) and quantitative (numeric) characteristics of the subject at hand. For example, a scientist comparing old-growth and plantation forests might record the names of understory plants (qualitative) and also record soil temperature and pH (quantitative). Both qualitative and quantitative characteristics are referred to as **data**. The primary focus of this text is on quantitative data. In the next examples we look at quantitative data that are linear and model the data with linear functions.

Figure 4-1 Lake Powell behind Glen Canyon Dam. *Photo courtesy of the U. S. Bureau of Reclamation.*

EXAMPLE 4-1   Spruce Growth

Suppose that a botanist measures the height of a spruce tree each week during the summer. She records the tree's height as in Table 4-1. Notice that the height is increasing by a constant amount each week, namely 1.5 inches. This constant amount is called the **slope** or **gradient**. The slope indicates the amount that y will increase or decrease as x increases by 1. If the spruce height were decreasing (highly unlikely), then the slope would be a negative number.

A more common method to find the slope is to compute the ratio of the change in y to the change in x. For example, over any 2-week period the height increases by 3 inches. Thus the slope can be found by computing

$$\text{slope} = \frac{\text{change in } y}{\text{change in } x} = \frac{3 \text{ inches}}{2 \text{ weeks}} = 1.5 \text{ inches per week}$$

TABLE 4-1	x = time in weeks	y = height in week x (inches)
	0	70.0
	1	71.5
	2	73.0
	3	74.5
	4	76.0
	5	77.5

To some, slope makes more sense when looking at a graph, where slope can be interpreted as the ratio of rise to run between any two points on the line (Figure 4-2). Going left to right, this line is increasing, so the slope is positive. A line that is decreasing, going left to right, has a negative slope.

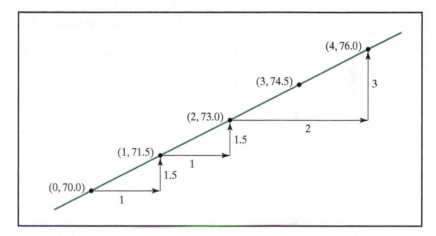

Figure 4-2 Slope can be interpreted as the amount that y increases as x increases by 1. An alternative interpretation is that slope is the ratio of the rise over run between any two points on the line.

In Table 4-2 we write each height value as the sum of the initial height (70.0) and a multiple of the slope (1.5). In the last line of the table, the height is given for any week x. This yields the general formula that describes the height of the spruce tree as a function of time: $y = 70.0 + 1.5x$.

TABLE 4-2	x = time (weeks)	y = height in week x (inches)
	0	$70.0 + (1.5)(0) = 70.0$
	1	$70.0 + (1.5)(1) = 71.5$
	2	$70.0 + (1.5)(2) = 73.0$
	3	$70.0 + (1.5)(3) = 74.5$
	4	$70.0 + (1.5)(4) = 76.0$
	5	$70.0 + (1.5)(5) = 77.5$
	$\vdots$	$\vdots$
	x	$70.0 + 1.5x$

Recall from algebra that all linear functions can be written in the form $y = mx + b$, where m represents the slope and b represents the y-intercept (the y value when $x = 0$). In this and future chapters, we'll often rearrange terms so that the equation has the form $y = b + mx$. Either form is correct.

When writing equations, it helps to use symbols or letters that are more suggestive of the quantities being measured, rather than always using x and y. Also, the variables should be defined as clearly as possible and should include the units of measure. In Example 4-1 we could use t for time and H for height and define them as follows:

Defining Variables

- Let t represent the time in weeks, with $t = 0$ referring to the time when the tree's height was first measured.
- Let H represent the height of the spruce tree in week t, measured in inches.

Using these variables, the equation for the tree's height is $H = 70.0 + 1.5t$.

EXAMPLE 4-2 Global Forest Loss

In 1990, the forests of the world covered 3,510 Mha (million hectares); in 1995 world forest area decreased to 3,454 Mha.[2] Forest loss has many negative impacts on the environment, including decreased forest habitat and carbon storage and increased soil erosion, desertification, and flooding (Figure 4-3). Assuming that the decrease in forest area is linear (i.e., losing the same amount each year) and continues into the future, how much forest will the world have in the year 2010?

Solution Let F represent the global area of forests in Mha, and let t represent time in years since 1990. The forest area data for 1990 and 1995 are presented as ordered pairs in Table 4-3.

To get the equation of the line through the data, we first compute slope using the **slope formula:**

$$m = \frac{y_2 - y_1}{x_2 - x_1}$$

The slope formula measures the change in y over the change in x between any two points (x_1, y_1) and (x_2, y_2) on the line. When using the formula, it does not matter which point is labeled (x_1, y_1) or (x_2, y_2); just be sure to subtract in the correct order in the formula.

In this example on global forest loss, we use the variables t and F instead of x and y. So the slope equation becomes

$$m = \frac{F_2 - F_1}{t_2 - t_1}$$

Letting $(t_1, F_1) = (0, 3510)$ and $(t_2, F_2) = (5, 3454)$, the slope is

$$m = \frac{3{,}454 - 3{,}510 \,\text{Mha}}{5 - 0 \,\text{yr}} = -11.2 \,\text{Mha/yr}$$

The negative sign of the slope agrees with the fact that global forest area is decreasing. Notice that by including units in both the numerator and denominator of the slope ratio, we can easily determine that the units of measure for the slope are "millions of hectares per year" or Mha/yr.

The linear equation describing forest loss will have the form $y = b + mx$, or, equivalently, $F = b + mt$. The variable b is the value of F when $t = 0$, so $b = 3,510$. Substituting these values for b and m, the equation for forest area as a function of time is

$$F = b + mt$$
$$F = 3{,}510 - 11.2t$$

We predict that the global forest area in the year 2010 (when $t = 20$) will equal

$$F = 3{,}510 - 11.2(20) = 3{,}286 \,\text{Mha}$$

It is important to express results of computations with the appropriate units, or to interpret results in a meaningful sentence. We should write something like, "In the year 2010, there will be 3,286 Mha of forest in the world." Of course, we made a bold assumption with this calculation, namely that the rate of deforestation will continue at the same pace. A better concluding sentence might be, "Assuming that forest area decreases by 11.2 Mha per year, there will be 3,286 Mha of forest in the world in the year 2010."

Figure 4-3 Deforestation leads to increased soil erosion and degradation of plant and animal habitat. *Photo: Langkamp/Hull.*

TABLE 4-3

t	F
0	3,510
1	
2	
3	
4	
5	3,454

Alternate Method to Find Slope

The slope formula is the common tool for finding the slope between two points, but the formula "hides" an important fact mentioned earlier:

Slope is the amount that y increases or decreases as x increases by 1.

Let's see how we could use this fact to find the slope for the previous example. The two original data points are presented in Table 4-4, with the unknown slope (the increase in F as t increases by 1) indicated by "$+m$."

TABLE 4-4

t	F	
0	3,510	$+ m$
1		$+ m$
2		$+ m$
3		$+ m$
4		$+ m$
5	3,454	$+ m$

To determine the value of the slope m, inspect the right-hand column of the table and note that

$$3{,}510 + 5m = 3{,}454$$

By subtracting 3,510 from each side of the equation and then dividing each side by 5, we get $m = -11.2$. This is the same result that was obtained using the slope formula.

UNITS OF MEASURE IN LINEAR EQUATIONS

Let's now look at the units in the equation from Example 4-1 on spruce growth. Note that each variable and number has units of measure attached to it.

$$H = 70.0 + 1.5\,t$$

$$in. = in. + \left(\frac{in.}{week}\right) week$$

Recall that we can combine like units and cancel units across a division sign, just as with numbers. When we do this, the units in the equation balance out.

$$in. = in. + \left(\frac{in.}{week}\right) week$$
$$in. = in. + in.$$
$$in. = in.$$

By including units, we can often decide if an equation makes sense or if the meaning of one of the variables agrees with the problem. Unfortunately, units in nonlinear equations are more difficult to balance. Let's look at one more example of a linear function.

EXAMPLE 4-3 Reducing Plastic in the Waste Stream

In 2004, the following statement was made by the U.S. Environmental Protection Agency Office of Solid Waste: "*Since 1977, the weight of 2-liter plastic soft drink bottles has been reduced from 68 grams each to 51 grams [in 2004].*"[3] Improvements in the efficiency of manufacturing goods are one focus of a growing field called **industrial ecology**.

If plastic bottles continue to decrease steadily in weight, how much will they weigh in the year 2025? To answer this question, we create a linear function to describe the bottle weight as a function of time. To get started, it's helpful to transfer the given information into a table (Table 4-5). Note that we have defined variables by letting W represent bottle weight in grams and t the time in years since 1977.

TABLE 4-5	t (years after 1977)	W (grams)
	0	68
	⋮	⋮
	27	51

Again we start by finding the slope using the slope formula:

$$m = \frac{W_2 - W_1}{t_2 - t_1}$$

$$m = \frac{51\,\text{g} - 68\,\text{g}}{27\,\text{yr} - 0\,\text{yr}}$$

$$m = -0.63\,\frac{\text{g}}{\text{yr}}$$

The value of the slope is −0.63 and the units of measure are "grams per year."

With an initial bottle weight of $b = 68$ g, the equation for bottle weight as a function of time is

$$W = b + mt$$

$$W = 68 - 0.63t$$

To predict the weight of a plastic bottle in the year 2025, substitute $t = 48$ years into the formula. Notice how we can keep track of units to make sure that the calculation is done correctly:

$$W = 68\,\text{g} - 0.63\,\frac{\text{g}}{\text{yr}} \cdot 48\,\text{yr}$$

$$= 68\,\text{g} - 30.24\,\text{g}$$

$$\approx 38\,\text{g}$$

Based on our assumption of a linear decrease in plastic bottle weight with time, we predict that 2-liter bottles will weigh approximately 38 grams in the year 2025. We must be careful extrapolating so many years into the future because the underlying assumption (linear decrease) might be wrong.

DEPENDENT VERSUS INDEPENDENT VARIABLES

In a linear equation $y = b + mx$, we call y the **dependent variable** and x the **independent variable**. In the equation $W = 68 - 0.63t$ from Example 4-3, W is the dependent variable and t is the independent variable. We often say that the dependent variable is written "in terms of" the independent variable.

In science, an independent variable is one that *drives* or *explains* the change in another variable. On the flip side, the dependent variable is one that *depends on* the independent variable. Let's look again at Example 4-3: Does bottle weight explain change in time or does time explain change in bottle weight? We'd probably answer that time (associated with technological change) explains the change in bottle weight. So bottle weight W is dependent upon time, which agrees with how we've written the equation.

If the independent variable explains the change in the dependent variable, that does not mean that it *causes* that change. To better understand this point, consider an equation that we'll find later in this chapter (see Example 4-6):

$$\text{zinc} = 28.73\,\text{lead} + 19.54$$

This equation is based upon lead and zinc concentrations that were measured in a collection of fish. With a positive slope ($m = 28.73$), this equation predicts that as lead concentration increases, the zinc concentration will also increase. But this *does not imply* that the increase in lead *causes* the increase in zinc—that's assuming too much. All we know is that lead and zinc concentrations are related by a linear formula.

GRAPHING LINEAR FUNCTIONS

You have graphed equations of lines in past algebra courses. Let's review how that is done by looking at some additional examples.

EXAMPLE 4-4   Graphing the Spruce Growth Function

To graph the spruce growth function $H = 70.0 + 1.5t$, we first need to think about which t-values we want to graph. In this function, t is the number of weeks elapsed during the summer so we might stipulate that t should fall in the interval $0 \leq t \leq 13$, reflecting the fact that there are 13 weeks of summer. Recall that $0 \leq t \leq 13$ means that t can take on all values between 0 and 13, inclusive. This interval is called the **domain** of the function. The domain is the set of values for the independent variable that make sense in the equation, from both a mathematical and practical point of view.

To graph a line accurately by hand, it is best to plot one point from the left side of the domain, and one from the right. We note that the vertical intercept (i.e., the y-intercept) is 70.0 inches, which means that the far left point is $(0, 70.0)$. When $t = 13$ weeks, $H = 89.5$ inches, and so the far right point is $(13, 89.5)$. Now plot the two points and connect them with a line. See Figure 4-4.

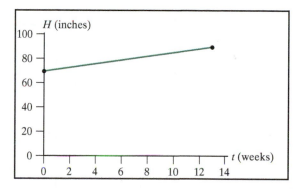

Figure 4-4 The graph of the spruce function over a 13-week period. To graph a line, plot one point on the far left, and one on the far right; then connect with a line.

Functions are sometimes plotted using a **graph break**, in which the horizontal or vertical scale begins at a number different than 0. For example, Figure 4-5 shows the spruce graph with the height axis starting at 65 inches. A graph break can make the diagram easier to read. Graph breaks on the vertical axis tend to exaggerate the slope of a line and can be employed to "mislead" the reader into thinking that a quantity is growing faster or slower than is really the case.

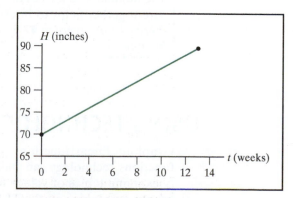

Figure 4-5 The growth of the spruce is exaggerated using a graph break, with height values starting at 65 inches rather than 0 inches.

EXAMPLE 4-5 Water Use Versus Income

In 1981, the city of Concord, New Hampshire collected data from 496 households on water use and income. The city found that, in general, as household income increased, so did water use, according to the linear function $W = 1{,}201 + 47.5I$.[4] In the formula, W is summer household water use in cubic feet and I is income in thousands of dollars. Again we see that this equation is in the form $y = b + mx$. The vertical intercept has the value $b = 1{,}201$, and the slope has the value $m = 47.5$.

To construct the graph of the water use function, we must first decide on which incomes to include. We arbitrarily allow incomes to range from \$0 to \$100,000, which means that the variable I will take on values between 0 and 100 (i.e., $0 \leq I \leq 100$).

The equation predicts that when income is $I = 0$ thousand dollars, summer water use is $W = 1{,}201$ cubic feet. The equation also tells us that if income is $I = 100$ thousand dollars, water use is $W = 1{,}201 + 47.5(100) = 5{,}951$ cubic feet. We now have two points which lie on the far left and far right sides of the graph: (0, 1201) and (100, 5951). Plotting and connecting these two points with a line gives us the graph of the equation (Figure 4-6).

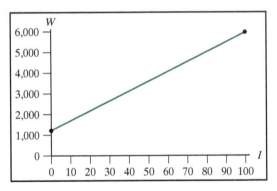

Figure 4-6 The water use function for incomes between \$0 and \$100,000

In the equation $W = 1{,}201 + 47.5I$, the vertical intercept $b = 1{,}201$ has the same units of measure as the variable W, namely cubic feet. Now, what are the units of measure for the slope $m = 47.5$? As mentioned previously, slope is the ratio of the change in y over the change in x. In this example, slope is the ratio of the change in water use over the change in income. The units of slope must match the units used in this ratio, so the units of slope are "cubic feet per thousand dollars." The number 47.5 tells us the *value* of the slope. Putting the value and units of measure together helps us interpret the meaning of the slope: *Water use in Concord increases on average 47.5 cubic feet for every one thousand dollar increase in income.*

USING TECHNOLOGY: GRAPHS AND TABLES

Graphing linear functions is not too difficult, but for graphing more complex functions, technology can make life easier. Similarly, technology can help us quickly examine tables of values for any function. We list the procedures for creating graphs and tables on the TI-83/84 calculator. Visit the text Web site for instructions on how to use online computer applications that perform the same tasks: **enviromath.com**.

Graphing Functions on the TI-83/84

To graph the water use function $W = 1,201 + 47.5I$, enter the equation in **Y=**. Use the **X,T,θ,n** key to access the variable **X**.

```
Plot1 Plot2 Plot3
\Y1∎1201+47.5X
\Y2=
\Y3=
\Y4=
\Y5=
\Y6=
\Y7=
```

Press **WINDOW** to set the dimensions of the viewing screen. To make the graphing calculator plot the graph in the same window as shown in Figure 4-6, use the settings displayed to the right.

```
WINDOW
 Xmin=0
 Xmax=100
 Xscl=10
 Ymin=0
 Ymax=6000
 Yscl=1000
 Xres=1∎
```

Press **GRAPH** to view the function. To find coordinates of points, press **TRACE** and use the left and right arrows to move along the line.

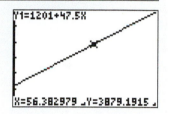

```
Y1=1201+47.5X


X=56.382979  Y=3879.1915
```

Creating Tables on the TI-83/84

To create a table for the spruce function $H = 70.0 + 1.5t$, enter the equation in **Y=**. Use the **X,T,θ,n** key to access the variable **X**.

```
Plot1 Plot2 Plot3
\Y1∎70.0+1.5X
\Y2=
\Y3=
\Y4=
\Y5=
\Y6=
\Y7=
```

Press **TBLSET** (**2nd WINDOW**) and set the parameters as displayed to the right.

```
TABLE SETUP
 TblStart=0
 ΔTbl=1
Indpnt: Auto Ask
Depend: Auto Ask
```

Press **TABLE** (**2nd GRAPH**). Use the up and down arrow keys to scroll through the table.

```
 X  │ Y1
 0  │ 70
 1  │ 71.5
 2  │ 73
 3  │ 74.5
 4  │ 76
 5  │ 77.5
 6  │ 79
Y1=79
```

To make X jump by a fixed amount (4 weeks, for example), return to the table set-up editor (**TBLSET**) and set **ΔTbl = 4**. Press **TABLE** to view the updated table.

```
 X  │ Y1
 0  │ 70
 4  │ 76
 8  │ 82
 12 │ 88
 16 │ 94
 20 │ 100
 24 │ 106
Y1=106
```

APPROXIMATING ALMOST-LINEAR DATA SETS

The bivariate (x, y) data that we've analyzed in the previous examples have all been perfectly linear, meaning that the data fall exactly along a single straight line. Most real-world data sets rarely have such perfect behavior, and some can be downright messy. In many cases, though, we can approximate the data with a line. For example, measurements of female cabezon (a marine fish, *Scorpaenichthys marmoratus*) show that, in general, as fish weight (w) increases, so does the number of eggs (E) produced. See Table 4-6 and Figure 4-7.[5] The data are a bit scattered, but there is an almost-linear trend that can be approximated with a linear function.

TABLE 4-6

Fish weight (× 100 grams)	Eggs (thousands)	Fish weight (× 100 grams)	Eggs (thousands)
14	61	34	87
17	37	37	89
24	65	40	100
25	69	41	90
27	54	42	97
33	93		

Source: California Div of Fish and Game

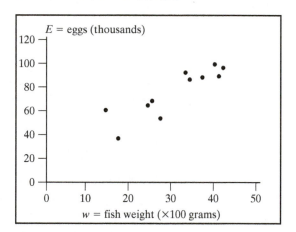

Figure 4-7 The relationship between fish weight and number of eggs is not perfectly linear but can be approximated with a linear function.

We use two basic methods for finding the equation of a line through an almost-linear data set: **straightedge method** and **least squares regression**.

Straightedge Method

In this method, we first lay a straightedge or ruler on the graph and draw a line that seems to "best fit" the data. Generally, there will be some points above and some points below the line, although the numbers of each don't have to match. The position of the line will vary depending on how one "eyeballs" the data, but usually results are fairly consistent. It's best to have gridlines on the graph when using this method. See Figure 4-8.

Once the line has been drawn, pick two points *on the line* to determine its equation. *The points do not need to be part of the original data set!* When choosing the two points, make sure that they are spread out, and (if possible) use points that fall on gridlines so that coordinates can easily be estimated. For the line drawn in Figure 4-8, we pick points $(5, 20)$ and $(45, 110)$ and compute the slope:

$$m = \frac{110 - 20}{45 - 5} = 2.25$$

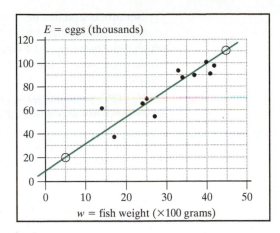

Figure 4-8 In the straightedge method, one first "eyeballs" a line through the data. Two points are chosen on the line (open circles) to determine the equation.

The equation of the line has the form $E = 2.25w + b$. Rather than read off the value of b from the graph (which would introduce more error into the equation), we use the point $(5, 20)$ and substitute to determine b. We get $20 = 2.25(5) + b$. Solving this equation results in $b = 8.75$. Therefore, the linear approximation using the straightedge method is

$$E = 2.25w + 8.75$$

Note that when the value of b is not known in advance, the equation of the line can also be determined using the **point-slope formula:** $y - y_1 = m(x - x_1)$. If you prefer to use this formula, that's perfectly OK, but we will not be demonstrating its use in this text.

Least Squares Regression

In the straightedge method, we draw a line that seems to "best fit" the data. That naturally raises the question "how do you define *best*?" Statisticians have devised several methods to answer this question; their most common definition for the best line is "the line which minimizes the sum of the squared vertical distances between the points and the line."

Understanding this strange phrase can be difficult; as always, a picture is worth a thousand words. Figure 4-9 shows six data points and an approximating line drawn through them. The vertical distances are marked d_1 through d_6. The best line will minimize the sum of the squared distances: $d_1^2 + d_2^2 + d_3^2 + d_4^2 + d_5^2 + d_6^2$. The line that makes this sum the smallest is called the **least squares regression line**. We'll often refer to this line as simply "the regression line" or the "linear regression."

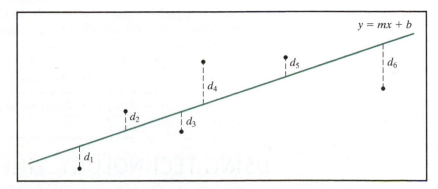

Figure 4-9 The vertical distances between the points and the line are indicated as d_1 through d_6. The "best" line will have a slope (m) and a y-intercept (b) that minimize the sum of the squared distances. That line is called the regression line.

Scientists and mathematicians don't often compute and minimize the sum of the squared distances by hand for a data set. Instead, they use formulas (based on calculus) which determine m and b directly. Because these formulas are rather complicated, the whole process to obtain the regression line is programmed into many calculators and computer software packages. For example, the TI-83/84 produces the following output screen for the cabezon data (directions on finding the regression equation using technology are given later in this chapter).

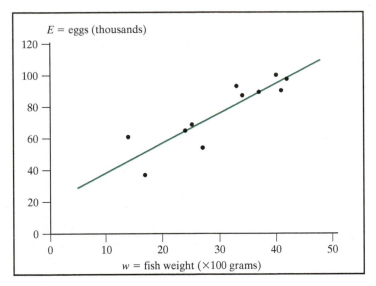

```
LinReg
 y=ax+b
 a=1.869955157
 b=19.76681614
 r²=.7784851077
 r=.8823180309
```

The regression equation is displayed using a for slope and b for the vertical intercept. The approximate values for the slope and vertical intercept are $a \approx 1.87$ and $b \approx 19.8$. Thus the least squares regression equation is approximately $y = 1.87x + 19.8$ or

$$E = 1.87w + 19.8$$

The graph of the regression line and the cabezon data are displayed in Figure 4-10.

Figure 4-10 The least squares regression line through the cabezon data

We've now explored two methods to find a line through an almost-linear data set. Table 4-7 summarizes the results of each method. We generally consider the least squares regression line "the best" fit to the data, although a line obtained by the straightedge method will often have a similar slope and y-intercept.

TABLE 4-7 Results of the two methods for approximating the cabezon data set

Method	Straightedge	Least Squares Regression
Equation	$E = 2.25w + 8.75$	$E = 1.87w + 19.8$

USING TECHNOLOGY: LINEAR REGRESSION

The following box shows the procedures required to find and graph a regression line on the TI-83/84 calculator. The instructions are based on the cabezon data set. Visit the text Web site for instructions on how to use online computer applications that perform the same tasks: **enviromath.com**.

Entering Bivariate Data on the TI-83/84

Select **STAT** > **1: Edit** to view the list editor. You should see the window displayed to the right showing lists L1, L2, and L3.

*Note 1: To clear a list, use the arrow keys to highlight the list name (e.g., L1); then press **CLEAR** > **ENTER**. Do not press **DEL** or you will delete the list from the list editor.*

*Note 2: If any of the lists L1 through L6 are not present in the list editor, select **STAT** > **5:SetUpEditor**; then press **ENTER**.*

Enter the 11 data points from the cabezon data (Table 4-6) under two lists, such as L1 and L2. Scan the lists to ensure you've entered the data correctly and that each list contains an equal number of values.

Linear Regression on the TI-83/84

Press **STAT** > **CALC**. Scroll down to **4:LinReg(ax+b)** and press **ENTER**.

The command **LinReg(ax+b)** will be pasted onto the homescreen of your calculator. Then enter **L1, L2**. *Note: The comma key is found directly above the number 7 key.*

Press **ENTER** to obtain the regression equation.

Plotting Data and the Regression Line on the TI-83/84

Select **STATPLOT** (**2nd Y=**) and then select **1:Plot1** to set up the plot features.

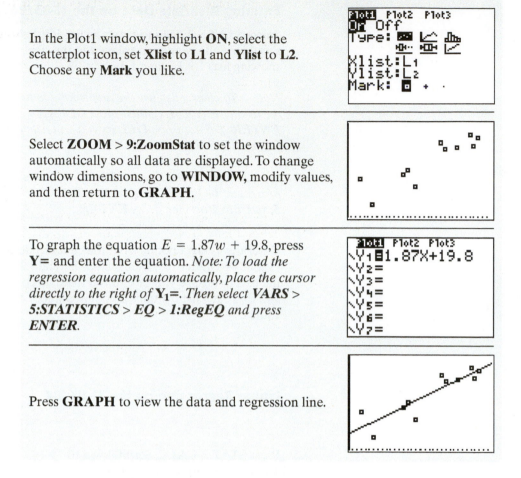

In the Plot1 window, highlight **ON**, select the scatterplot icon, set **Xlist** to **L1** and **Ylist** to **L2**. Choose any **Mark** you like.

Select **ZOOM > 9:ZoomStat** to set the window automatically so all data are displayed. To change window dimensions, go to **WINDOW,** modify values, and then return to **GRAPH.**

To graph the equation $E = 1.87w + 19.8$, press **Y=** and enter the equation. *Note: To load the regression equation automatically, place the cursor directly to the right of* **Y₁=**. *Then select* **VARS > 5:STATISTICS > EQ > 1:RegEQ** *and press* **ENTER**.

Press **GRAPH** to view the data and regression line.

THE CORRELATION COEFFICIENT: *r*

The cabezon data are somewhat scattered, although there is a noticeable linear trend. Many real-world data sets have similar patterns, although some are more scattered, and some tend to lie more in a straight line. Statisticians use the **correlation coefficient** (denoted by the letter r) to measure the strength of the linear pattern in data.

The value of the correlation coefficient r will always lie in the interval $-1 \le r \le 1$. A linear trend with a positive slope (as x increases, y increases) will have a correlation coefficient that is positive. Points that line up perfectly along a straight line with positive slope have a correlation coefficient of $r = 1$. For a linear trend with negative slope (as x increases, y decreases) the correlation coefficient will be negative, and if the points fall perfectly on a straight line, the correlation coefficient will equal $r = -1$. The more scattered the points, the closer r gets to zero. See data sets in Figure 4-11.

A simple interpretation of the correlation coefficient is that it measures how close the data points fit a straight line. Another interpretation is that it measures how

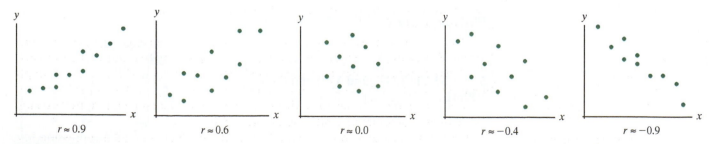

$r \approx 0.9$ $r \approx 0.6$ $r \approx 0.0$ $r \approx -0.4$ $r \approx -0.9$

Figure 4-11 Examples of data sets and their correlation coefficients

well the independent variable (x) can be used to predict the value of the dependent variable (y). In a data set with correlation coefficient close to $r = 1$ or $r = -1$, the value of x can be used to accurately predict the value of y. When $r \approx 0$ (third diagram in Figure 4-11), we don't know whether the y-values of the data will increase, decrease, or stay the same as we move in the positive x-direction. Thus x has no ability to predict y.

USING TECHNOLOGY: THE CORRELATION COEFFICIENT

The value of r can be computed manually using a rather complex formula, but most people just let their calculator or computer handle the computation. Often the correlation coefficient is displayed along with the least squares regression line. In the section that follows, we explain how to display the correlation coefficient on the TI-83/84 graphing calculator (if it is not displayed already).

The Correlation Coefficient on the TI-83/84

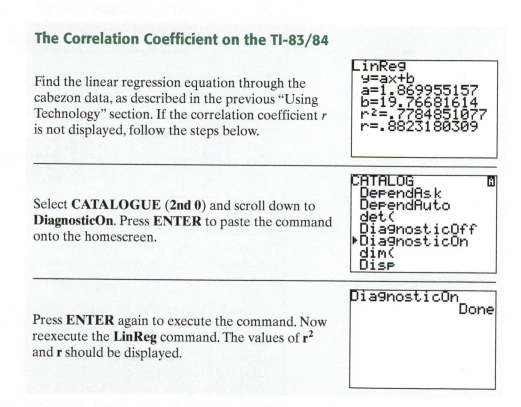

Find the linear regression equation through the cabezon data, as described in the previous "Using Technology" section. If the correlation coefficient r is not displayed, follow the steps below.

Select **CATALOGUE (2nd 0)** and scroll down to **DiagnosticOn**. Press **ENTER** to paste the command onto the homescreen.

Press **ENTER** again to execute the command. Now reexecute the **LinReg** command. The values of $\mathbf{r^2}$ and $\mathbf{r}$ should be displayed.

CORRELATION FALLACIES

A data set with a correlation coefficient of $r = 0.8$ will be more linear than one with a value of $r = 0.7$. But how much more? That is a difficult question. In fact, we must be careful not to read too much into the value of r. One common fallacy is to assume that a value of $r = 0.8$ is twice as good as a value of $r = 0.4$. *This is not necessarily true.*

Another fallacy is to assume correlation implies cause and effect. *This is not always the case.* For example, suppose we sample several cities and find a strong correlation between pollution clean-up money and hamburger sales (Figure 4-12). Does this imply that more pollution clean-up money causes higher hamburger sales? Most likely not! A better explanation is that cities with large amounts of clean-up money are also larger cities—and larger cities tend to sell more hamburgers.

Let's now look at a more practical example in which we find a regression line and compute the correlation coefficient.

Figure 4-12 Correlation does not imply causation

EXAMPLE 4-6 Heavy Metals in Fish

Heavy metals can enter a watershed from many sources. Once in the watershed's streams and rivers, the metals are taken in by fish and other aquatic animals. Table 4-8 gives metal concentrations in a sample of 10 fish from the Spokane River.[6]

TABLE 4-8

Fish	Lead Concentration (ppm)	Zinc Concentration (ppm)
Rainbow trout	0.73	45.3
Rainbow trout	1.14	50.8
Rainbow trout	0.60	40.2
Rainbow trout	1.59	64.0
Large-scale sucker	4.34	150.0
Large-scale sucker	1.98	106.0
Large-scale sucker	3.12	90.8
Large-scale sucker	1.80	58.8
Mountain whitefish	0.65	35.4
Mountain whitefish	0.56	28.4

Source: Washington State Dept of Ecology

The following are graphing calculator screens showing the entered data, the linear regression equation, and a plot of the data and the regression line.

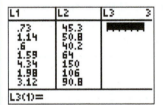 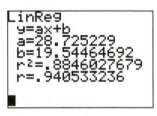 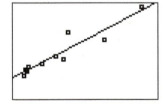

We note that the data have a positive slope and a strong linear pattern. This is reflected in the high r-value. With x corresponding to lead concentration and y the zinc concentration, the regression equation can be written as

$$\text{zinc} = 28.73 \text{ lead} + 19.54$$

With such a high r-value, we would say that the concentration of lead (x) is a good predictor of the concentration of zinc (y). In fact, if we had the task of measuring both lead and zinc concentrations in a fish, we could simply measure the lead concentration and use the equation to predict the zinc concentration. This is one advantage of having a high correlation coefficient along with a regression equation!

The correlation between lead and zinc (however strong it might be) *does not imply causation*. We cannot say that the presence of lead in fish causes the presence of zinc in fish, or vice versa. We can say that fish with low amounts of lead have low amounts of zinc, and fish with high amounts of lead have high amounts of zinc. We can say that the pattern between the lead and zinc amounts tends to be quite linear.

One explanation for the positive correlation between lead and zinc is that heavy metals **bioaccumulate** in fish over time. As fish live longer, more zinc and more lead accumulate in their bones and tissue. So the real connection between zinc and lead may be *exposure time* to these metals.

CHAPTER SUMMARY

A quantity that changes by a *constant amount* each time period, or with respect to another variable, is said to undergo **linear change**. If y changes linearly with respect to x, then the relationship between these variables can be written in the form $y = mx + b$, which is called a **linear function**. This type of function can also be written as $y = b + mx$. In this function, x is called the **independent variable** and y the

dependent variable. The values that can be substituted for the independent variable *x* make up the **domain** of the function.

The value of *b* in a linear function, the **y-intercept**, tells us where the graph of the line crosses the *y*-axis. If *x* represents time, the *y*-intercept can be interpreted as the "starting value" of the function (at time = 0). The value of *m* in a linear function, the **slope**, indicates the amount that *y* increases or decreases as *x* increases by one unit. The slope can be determined from any two points given in a table or read from a graph, by using the slope formula:

$$m = \frac{y_2 - y_1}{x_2 - x_1}$$

If the slope is positive, *y* will increase as *x* increases, and the line will slant up and to the right. If the slope is negative, *y* will decrease as *x* increases, and the line will slant down and to the right (Figure 4-13).

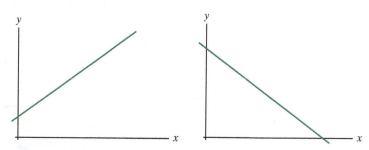

Figure 4-13 Linear graphs with positive slope (left) and negative slope (right)

An almost-linear trend in data can be approximated with a line. The **straightedge method** involves "eyeballing" a straight line through the data manually. Two points on the line can then be used to find the approximating linear equation. Almost-linear data can also be approximated with **least squares regression**. The least squares regression line is typically found using technology. Of all lines drawn through the data, the regression line is the line of "best fit," in the sense that it will have the minimum total squared vertical distance between itself and the data.

If bivariate (*x,y*) data exhibit a linear or almost-linear trend, the strength of that trend can be measured with the **correlation coefficient**. The correlation coefficient, *r*, tells us "how close" the data fall on a straight line. The value of *r* will lie in the interval from −1 to 1. If the data are increasing, the value of *r* will be positive; if the data are decreasing, *r* will be negative. Data with a strong linear trend have a correlation coefficient close to *r* = 1 or *r* = −1. For data with a weak linear trend, the correlation coefficient will lie close to *r* = 0. An alternative interpretation of the correlation coefficient is that it measures the ability of *x* to predict *y*.

END *of* CHAPTER EXERCISES

Writing Linear Functions

1. Write a linear function that models each of the following scenarios. Be sure to explain (define) clearly what your variables represent.
   a. Global temperatures are predicted to increase at an average rate of 0.03°F each year. In 2001, the global average temperature was 58.2°F.[7]
   b. The total fertility rate (TFR) is the average number of children women will bear through all childbearing years. In 2001, the United States TFR was 2.034 and was decreasing by approximately 0.022 each year.[8]

2. Define variables; then write a linear function (i.e. linear model) for each scenario.
   a. Between 1992 and 1998, the municipal solid waste disposal rate in Texas stayed fairly steady, averaging 6.5 pounds per person per day.[9]

b. Aquaculture, the farming of aquatic organisms, totaled 45.7 million metric tonnes (Mmt) worldwide in year 2000. The amount is increasing by approximately 2.7 Mmt each year.[10]

In Exercises 3–6, assume that the dependent variable changes linearly with respect to the independent variable; then do the following:

a. *Find the value of the slope, m, and give the units for the slope.*

b. *Write an equation for the linear function.*

c. *Fill in the remainder of the table.*

3.

Example A		Example B	
t = time (years)	P = population (people)	d = distance (km)	P = population density (people/km^2)
0	2,000	0	5,000
1	2,400	1	4,450
2		2	
3		3	
4		4	

4.

Example A		Example B	
t = time (years)	P = population ($\times 10^6$ people)	d = depth (ft)	V = stream discharge (ft^3/sec)
0	14.5	0	0
1	14.0	1	4.5
2		2	
3		3	
4		4	

5.

Example A		Example B	
I = income ($\times$ \$1000)	E = annual home energy use (kWh)	t = time in years since 2000	C = water consumption ($\times 10^6$ gallons)
0		0	
2	12,000	5	2,650
4		10	
6	15,000	15	2,100
8		20	

6.

Example A		Example B	
w = vehicle weight (lb)	G = gasoline mileage (mi/gal)	P = pesticide concentration (ppm)	D = cancer deaths per 1,000 people
2,000	45	300	
3,000		400	0.065
4,000		500	
5,000		600	0.095
6,000	15	700	

Balancing Units in Equations

7. Review the global forest loss equation in Example 4-2. Explain how all units balance in the equation.

8. Review the equation in Example 4-3 that gives the weight of 2-liter plastic bottles as a function of years. Explain how all units balance in the equation.

Applications

9. According to the U.S. Census Bureau, the population of the United States increased from 281.4 million people in April 1, 2000, to 284.8 million people as of July 1, 2001.[11] Assume that population grew by the same amount each month.

   a. Define variables for population and time. Include units.

   b. Compute slope and interpret its meaning.

   c. Write an equation that expresses U.S. population as a function of time.

   d. Use your equation to predict when the U.S. population will reach 300 million people.

   e. Graph your equation over the domain April 1, 2000, to April 1, 2020. On your graph, show the location of the original two data values. Is extrapolating out to the year 2020 reasonable or not?

10. **Roundwood** is wood used for either fuel or construction purposes. Between 1950 and 1995, world production of roundwood grew at an overall steady pace, from 1,421 million cubic meters to 3,461 million cubic meters.[12]

    a. Define variables for roundwood production and time. Include units.

    b. Compute the slope and interpret its meaning.

    c. Write a linear equation relating world roundwood production to time.

    d. Identify a domain for your equation and explain why you chose the interval that you did.

    e. Graph the roundwood production function over the domain that you determined in part (d).

11. Toward the end of the last ice age about 10,000 years ago, glaciers retreated from the northern Midwest states, leaving virtually no soil in place. Since that time, soils have slowly formed, with soil thickness today averaging 2 to 3 feet.[13]

    a. Assume that soil thickness today in northern Midwest states is 2.5 feet, or 30 inches. Write a linear equation which relates inches of soil thickness, S, to time in years, t. Let $t = 0$ correspond to 10,000 years ago.

    b. Interpret the meaning of the slope.

    c. Make a table that exhibits soil thickness every 1,000 years, from 10,000 years ago until today.

    d. Poor agricultural practices, such as farming on steep slopes and not planting ground cover, often lead to soil erosion. Suppose that a heavy storm washes 1/8 inch of topsoil from an unplanted field. According to your equation in part (a), how many years would it take for nature to replace this lost soil?

12. Country-by-country statistics show a general decline in fertility rates as contraceptive prevalence increases. Table 4-9 gives contraceptive use and total fertility rate estimates for Pakistan and Poland.[14] (For more details, see the Chapter 4 project.)

TABLE 4-9

	Percent of Married, Fertile Women Using Contraceptives	Total Fertility Rate (TFR)
Pakistan	18	5.6
Poland	75	1.5

Source: Johns Hopkins INFO Project

   a. Assume that the relationship between percent contraceptive use and fertility rate is linear. Using the Pakistan and Poland values, create a linear model showing how TFR depends on percent contraceptive use.

   b. What is the domain for your equation? Use the domain to graph your equation. Label points identifying Pakistan and Poland on the graph.

   c. If a third country reports that its fertility rate is 4.3, predict the percentage of married, fertile women using contraceptives in that country.

13. Data from many of the world's countries show a linear relationship between wealth and energy consumption. In general, as a country's wealth increases, so does its energy consumption. Table 4-10 shows wealth and energy consumption data for Spain and the United States for the year 2001.[15]

TABLE 4-10

	Wealth Measured as Gross Domestic Product (billion U.S. dollars)	Total Energy Consumption (quadrillion BTU)
Spain	723	5.7
United States	9,040	97.1

Source: U.S. Dept of Energy

a. Identify variables and write a linear equation that models energy consumption as a function of wealth.

b. Graph your equation. Label points for Spain and the United States.

c. Determine the units of measure for the slope. What is the meaning of the slope?

d. Which country uses more energy to produce the equivalent of $1 in wealth?

14. Oil demand in Asia reached 21 million barrels a day in 2004. The demand is projected to increase to 38 million barrels a day by 2025.[16]

a. Define variables for time and oil demand; then write a linear equation modeling Asia's demand for oil.

b. Graph your equation on graph paper for years 2004–2025. Label and scale properly.

c. Suggest two likely reasons for the increase in Asia's demand for oil.

15. In 2002, the estimated total amount of recoverable coal on planet Earth was about 1,081 billion tons. The world consumption rate of coal energy in that year was 9.83×10^{16} BTU per year (Figure 4-14).[17]

a. Convert the consumption rate into units of billions of tons per year. Note that the energy content of coal is about 2.068×10^7 BTU/ton.

b. Let A = the amount of coal in billions of tons and t = time in years since 2002. Assume that the coal consumption rate remains constant and that no additional coal is found. Write a linear equation expressing A as a function of t.

c. Use your equation to predict the amount of coal remaining in the year 2052.

d. According to your equation, in which year will coal reserves equal 200 billion tons?

e. Give the most important reason why your answers to the previous two questions are likely false.

16. A vehicle with a fuel efficiency of 27.5 miles per gallon (mpg) will emit 54 tons of CO_2 over its lifetime. A less efficient vehicle getting 14 mpg will emit 100 tons of CO_2 over its lifetime. These estimates are based on year 2002 U.S. emissions standards.[18]

a. Plot the data on graph paper, with fuel efficiency on the horizontal axis and lifetime CO_2 emissions on the vertical axis. Connect the two points with a line. Then extend the line so that it intersects both the horizontal and vertical axes.

b. What is the domain of your graph? Does the domain seem realistic?

c. Set up variables and write an equation for the line.

d. For each additional increase of 3 mpg in fuel efficiency, how many tons of CO_2 will be saved over the lifetime of an average car?

Figure 4-14 An underground coal mine near Edri, Pennsylvania, in 1904. Many of today's coal mines are large-scale, surface operations. *Historical photo courtesy of USGS/R. W. Stone.*

Approximating Linear Data, Regression, and Correlation

17. Chicken farms produce a lot of waste—approximately 1 pound of litter (manure, feathers, uneaten food, and bedding) per pound of chicken. Treatment and disposal of this waste is an important environmental issue, as the waste can contaminate both groundwater and surface water near the farms. Table 4-11 shows water quality data collected from several wells at a Florida chicken farm.[19]

a. Plot the data on graph paper with potassium concentration on the horizontal axis and nitrate concentration on the vertical axis. Use at least a half sheet of graph paper.

b. How linear are the data? Are there any **outliers** (i.e., data points that appear unusual)? Explain.

c. Use the straightedge method to approximate the data with a line. Before you find the equation of your "eyeballed" line, you must pick two points that lie on the line.

TABLE 4-11

Potassium Concentration (mg/L)	Nitrate Concentration (mg/L)	Potassium Concentration (mg/L)	Nitrate Concentration (mg/L)
1.5	18	0.6	11
1.8	19	0.6	26
1.6	20	0.4	5.2
1.3	21	0.4	5.4
0.7	9.4	0.4	5.0
0.7	10	0.1	4.9

Source: U.S. Geological Survey

d. Using technology, find the least squares regression line through the data. What is the value of the correlation coefficient, and what does it tell you?

e. If a new well were drilled and the potassium concentration were 1.1 mg/L, what nitrate concentration would the regression equation predict?

18. In the Pacific Northwest, glaciers store water that is used in irrigation, power production, and salmon hatcheries (Figure 4-15). Data on recent changes in length of the Rainbow Glacier on Mt. Baker in Washington State are shown in Table 4-12.[20] Position refers to the location of the glacier's leading edge, as compared to its 1985 position. Negative position values indicate that the glacier has decreased in length, with the leading edge farther up-hill than in 1985.

Figure 4-15 What are the consequences of global warming for Pacific Northwest economies that rely on glacier water? *Photo: Langkamp/Hull.*

TABLE 4-12

Year	Position (meters)	Year	Position (meters)
1985	0	1993	−96
1986	−11	1994	−116
1987	−22	1995	−137
1988	−33	1996	−161
1989	−44	1997	n/a
1990	−55	1998	−201
1991	−60	1999	−241
1992	−75	2000	−246

Source: North Cascades Glacier Climate Project

a. Renumber the years so that 1985 refers to time $t = 0$. Using at least a half sheet of graph paper, plot glacier position P as a function of time t. Label accurately.

b. How linear are the data? Are there any outliers?

c. Use the straightedge method to find a line that approximates the data.

d. Using technology, find the least squares regression line through the data. Discuss the meaning of the slope of the regression line.

e. What is the correlation coefficient r for the data? How do you know that the sign of the correlation coefficient is correct?

f. The Rainbow Glacier was 3,750 meters long in 1950. Assuming that it has been decreasing in length at the same rate each year, how long was the glacier in 1985?

19. The Hawaii-Emperor chain of volcanoes stretches from its active end on the Big Island of Hawaii northwest toward the Aleutian trench near the Kamchatka Peninsula (see Figure 4-16). The Hawaii-Emperor chain is a classic example of a hot spot track—a geologic track on the Earth's surface formed when a crustal plate moves over a hot magma source. The Hawaii-Emperor chain has a prominent bend near the underwater volcano Daikakuji; it is thought that the crustal plate changed directions at the time Daikakuji was formed. Table 4-13 displays data on the age and distance of the volcanoes; age is measured in millions of years (mega-annum or Ma) and distance is measured in kilometers along the chain from the youngest volcano, Kilauea. Kilauea lies on the southern end of the Big Island of Hawaii.[21]

a. Plot the data on graph paper with age on the horizontal axis and distance on the vertical axis. Use at least a half sheet of graph paper.

b. How linear are the data? Are there any outliers?

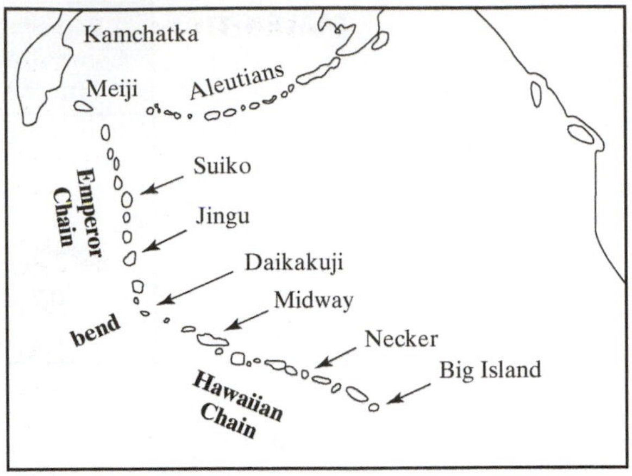

Figure 4-16 The Hawaii-Emperor chain, consisting of about 110 volcanoes, stretches almost 5,000 km across the Pacific Ocean. Today the hot spot is located near the southern end of the Big Island.

c. Use the straightedge method to find a line that approximates the data. Be sure to identify your variables.

d. Using technology, find the best-fitting regression line through the data.

e. What is the value of the correlation coefficient? What does the *size* of its value tell you? What does the *sign* of its value tell you?

f. One volcano in the Hawaii-Emperor chain, Northampton, is located 1,841 km from Kilauea. Estimate the age of this volcano.

TABLE 4-13

Volcano Name	Estimated Age (Ma)	Distance (km)
Kilauea (Big Island of Hawaii)	0.20	0
West Maui	1.32	221
Lanai	1.28	226
West Molokai	1.90	280
Kauai	5.10	519
Necker	10.30	1,058
Laysan	19.90	1,818
Midway	27.70	2,432
Daikakuji	42.40	3,493
Kimmei	39.90	3,668
Koko	48.10	3,758
Jingu	55.40	4,175
Suiko	64.70	4,860

Source: Geological Society of America

20. Cerro Negro is an active volcano in the northwest part of Nicaragua, one of a series of active volcanoes in the "Ring of Fire," which rims the Pacific Ocean (Figure 4-17). Table 4-14 shows the cumulative volume of ash and cinders from all Cerro Negro eruptions dating back to 1899.[22]

a. Renumber the time column so that the year 1899 corresponds to year $t = 0$. Then, using a half sheet of graph paper, plot cumulative volume V (in millions of cubic meters) as a function of time t. Label and scale accurately.

b. Would you consider the data to be almost linear? Were there any large periods of time without an eruption?

Figure 4-17 Aerial view of Cerro Negro Volcano erupting on July 24, 1947. *Source: USGS/R. E. Wilcox*

TABLE 4-14

Year	Cumulative Volume ($\times 10^6$ cubic meters)	Year	Cumulative Volume ($\times 10^6$ cubic meters)
1899	1	1960	85
1914	2	1961	85
1919	2	1962	86
1923	29	1963	86
1929	29	1968	105
1947	41	1969	105
1948	41	1971	132
1949	41	1992	142
1950	58	1995	142
1954	58	1995	150
1957	64		

Source: Geological Society of America

c. Use the straightedge method to find a linear function that approximates the data.

d. Determine the least squares regression line using technology.

e. Interpret the meaning of the slope of the regression line.

21. The data in Table 4-15 were acquired at a station along the Columbia River below Grand Coulee Dam at a distance of 13 feet from the edge of the river. The depth of the Columbia River at this spot was about 12 feet at the time of measurement. Velocities were measured at various depths in the water column.[23]

TABLE 4-15

Depth below Surface (feet)	Velocity (ft/sec)	Depth below Surface (feet)	Velocity (ft/sec)
0.66	1.55	7.26	0.91
1.98	1.11	8.58	0.59
2.64	1.42	9.90	0.59
3.30	1.39	10.56	0.41
4.62	1.39	11.22	0.22
5.94	1.14		

Source: U.S. Geological Survey

a. What is the equation of the best-fitting regression line?

b. How well can depth be used to predict velocity? How do you know?

c. Give an interpretation of the slope of the regression equation.

d. Find the horizontal intercept for the regression line. What does this point represent?

e. As you move from the surface downward toward the river bottom, the velocity of the water decreases. Why would the velocity of river water be slower when the depth is greater?

22. Vehicles that use ethanol, electricity, or natural gas instead of petroleum fuels, such as gasoline and diesel, are referred to as **alternative fueled vehicles** or AFVs (Figure 4-18).

Figure 4-18 A road rally in California featuring prototype AFVs powered by fuel cells. Fuel cells burn hydrogen gas, emitting nearly zero smog-forming pollutants. Hydrogen gas, though, must be produced using some form of energy, such as that from fossil fuel, nuclear, or hydropower sources. *Photo with attribution to the California Fuel Cell Partnership.*

Growing concerns over the environmental impacts of burning fossil fuels and the U.S. dependence on foreign oil resulted in an increase in AFVs on the road since the mid-1990s (Table 4-16).[24] These numbers do not include gasoline or diesel-electric hybrid vehicles.

TABLE 4-16

Year	Number of AFVs
1995	246,855
1996	265,006
1997	280,205
1998	295,030
1999	322,302
2000	394,664
2001	425,457
2002	471,098
2003	510,805
2004	547,904

Source: U.S. Dept of Energy

a. Construct a new data table with years expressed as "years after 1995" and the number of AFVs rounded to the nearest 1,000. In the table record the number of AFVs in units of thousands. For example, the first row of the table should be (0, 247).

b. Use technology to plot the data and the best-fitting regression line. What is the equation of the regression line?

c. According to the regression equation, what number of AFVs were on the road in 1997? How does that number compare with the actual value?

d. What is the correlation coefficient for the data set? How linear are the data?

SCIENCE IN DEPTH

Population Growth

World population doubled from 1960 to 1999, and is projected to reach 7 billion people by the year 2013. How many people can we cram onto this small blue planet? When will population growth level out, and at how many people?

Births, deaths, emigration, and immigration all play an important role in population dynamics. The **birth rate** is defined as the number of live births each year for every 1,000 individuals in the population. The 2002 global birth rate was approximately 20 births per 1,000 people; individual countries range from 10 to 40 live births per thousand.[25]

Men don't give birth, so let's refine the birth rate without them. The **total fertility rate** is the average number of live births for each woman during her childbearing years (ages 15 to 45). Worldwide, the total fertility rate was about 2.8 live births per woman in 2002, with individual countries ranging from 1 to 8.[26] The replacement rate is 2.1, which is the number of children needed to reach reproductive age so as to replace their parents. The replacement rate of 2.1 is slightly higher than 2 to account for child mortality before reproduction. A fertility rate that exceeds the replacement rate means that the population is growing.

Figure 4-19 shows four stages of a country or a region undergoing a change in birth and death rates known as the **demographic transition**.[27] In Stage 1, both birth and death rates are quite high and approximately equal. The balance between births and deaths results in a stable population overall. Countries or regions in Stage 1 typically have large segments of the population living in poverty. As living conditions improve in Stage 2, there is less malnutrition, less disease, and greater access to health care, and infant mortality begins to decline. However, the fertility rate remains high, and therefore the population tends to grow quickly with the addition of many children.

The third stage of the demographic transition is marked by a decline in the birth rate due to the lower infant mortality, better access to contraceptives, family planning, and education. The population still grows but at a slower rate. Ultimately, a population will enter Stage 4, where low birth rates and low death rates are balanced, once again giving rise to a stable population size.

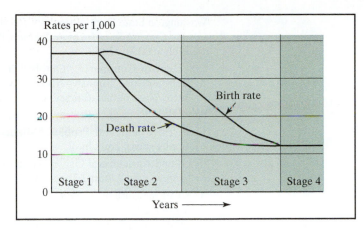

Figure 4-19 Demographic transition in birth and death rates with increased industrialization and development

Demographers and sociologists have traditionally linked the demographic transition with a change from a "less-developed" economy to a "more-developed" economy (Stage 1 to Stage 4). Populations in the early, high birth–high death phase (Stage 1) are often dominated by subsistence agriculture, where families produce just enough food to survive, or unskilled menial labor, where wages are marginal. Advancements in agricultural practices or an increase in wages results in fewer childhood deaths and subsequently a lower birth rate (Stage 2 to 3) as family life improves.

New evidence suggests that women in some populations are taking a shortcut around the demographic transition, moving from Stage 1 to Stage 3 directly. In countries such as Brazil, Mexico, Egypt, and India, the fertility rates have fallen rapidly. Average fertility rates in developing countries dropped from six children per mother in the 1970s to four children per mother in the 1990s.[28] During the same time period, average contraceptive use rose from 27% to 40% in developing countries, reflecting changing social norms and access to family planning.

CHAPTER PROJECT:
FERTILITY RATES IN DEVELOPING COUNTRIES

Approximately 90% of the world's countries now support some sort of family planning and distribution of contraceptives either directly through their governments or indirectly through nongovernmental organizations. Family planning is one factor that directly impacts fertility rates. Can income or education levels be correlated with fertility rates in developing countries?

In the companion project for Chapter 4 found at the text Web site (**enviromath.com**), you will analyze the factors that influence fertility rates for a group of less-industrialized countries. And you'll draw conclusions about which social policies will have the biggest impact on lowering fertility rates and stabilizing population growth. Visit the Web site!

NOTES

[1]Sandra Blakeslee, "Drought Unearths a Buried Treasure," *New York Times*, November 2, 2004

[2]Food and Agricultural Organization of the United Nations, *FactFile: Forest Areas by Region in 1995*, http://www.fao.org/

[3]U.S. Environmental Protection Agency, "Source Reduction and Reuse," http://www.epa.gov/epaoswer/non-hw/muncpl/sourcred.htm

[4]Lawrence C. Hamilton, *Regression with Graphics: A Second Course in Applied Statistics* (Belmont, CA: Duxbury Press, 1992), 35.

[5]C. P. O'Connell, "The Life History of the Cabezon *Scorpaenichthys marmoratus* (Ayres)," *California Div of Fish and Game, Fish Bulletin* 93 (1953). As cited in Robert R. Sokal and F. James Rohlf, *Biometry* (San Francisco: W. H. Freeman, 1969), 483.

[6]Washington State Dept of Ecology, *Results from Analyzing Metals in 1999 Spokane River Fish and Crayfish Samples*, by A. Johnson, Publication no. 00-03-017 (2000). http://www.ecy.wa.gov/biblio/0003017.html

[7]U.S. Dept of Commerce, "October 2001 Global Temperature Warmest On Record: U.S. Temperature Slightly Above Average," NOAA News Releases, November 2001, http://www.publicaffairs.noaa.gov/releases2001/nov01/noaa01117.html

[8]Population Reference Bureau, *2003 World Population Data Sheet*, http://www.prb.org/

[9]Texas Environmental Profiles, "Municipal Solid Waste in Texas," http://www.texasep.org/html/wst/wst_2mtx.html

[10]Food and Agricultural Organization of the United Nations, *Aquaculture Production Trend Analysis*, by Albert J. Tacon, 2003.

[11]U.S. Census Bureau, *Table US-2001EST-01—Time Series of National Population Estimates: April 1, 2000 to July 1, 2001*, http://www.census.gov/popest/archives/2000s/vintage_2001/US-2001EST-01.html

[12]U.N. Food and Agricultural Organization, *Forest Products Yearbook* (various years). As cited in L. R. Brown, M. Renner, and C. Flavin, *Vital Signs 1997: The Environmental Trends That Are Shaping Our Future* (New York: W. W. Norton & Company, 1997), 68–69.

[13]David McConnell, University of Akron Dept of Geology, "Soils and Weathering," http://lists.uakron.edu/geology/natscigeo/Lectures/weath/weath.htm#soils

[14]The Information and Knowledge for Optimal Health Project, "Why Family Planning Matters," *Population Reports*, series J, no. 49, http://www.infoforhealth.org/pr/j49edsum.shtml

[15]U.S. Dept of Energy, Energy Information Administration, *International Energy Annual 2001*, Tables B2 & E1, http://tonto.eia.doe.gov/FTPROOT/international/021901.pdf *Note:* GDP values in 1995 dollars.

[16]Simon Romero, "As Prices Soar, Some in OPEC Balk at Cuts," *New York Times*, March 23, 2004.

[17]U.S. Dept of Energy, Energy Information Administration, *International Energy Annual 2002*, Tables 1.4 and 8.2, http://www.eia.doe.gov/emeu/iea/coal.html

[18]Union of Concerned Scientists, "Federal Fuel Economy Standards—Past, Present, and Future," http://www.ucsusa.org/clean_vehicles/archive/page.cfm?pageID=223

[19]U.S. Dept of the Interior, *Effects of Waste Disposal Practices on Ground-Water Quality at Five Poultry (Broiler) Farms in North-Central Florida, 1992–93*, by H. H. Hatzell, U.S. Geological Survey Water Resources Investigation Report 95–4064 (1995).

[20]North Cascades Glacier Climate Project, http://www.nichols.edu/departments/Glacier/

[21]D. A. Clague and B. G. Dalrymple, "Tectonics, Geomorphology and Origin of the Hawaiian-Emperor Volcanic Chain," in E. L. Winterer, D. M. Hussong, and R. W. Decker (eds.), *The Eastern Pacific Ocean and Hawaii*, volume N in *The Geology of North America* (Boulder, CO: Geological Society of America, 1989), 188–217.

[22]Brittain E. Hill et al., "1995 Eruptions of Cerro Negro Volcano, Nicaragua, and Risk Assessment for Future Eruptions," *Geological Society of America Bulletin* 110, no. 10 (1998): 1231–1241.

[23]U.S. Dept of the Interior, *Analysis of Current Meter Data at Columbia River Gaging Stations, Washington and Oregon*, by J. Savini and G. L. Bodhaine, U.S. Geological Survey Water Supply Paper No. 1869-F (1971): 59.

[24]U.S. Dept of Energy, Energy Information Administration, "Alternatives to Traditional Transportation Fuels: 2003 Estimated Data," Table 1, http://www.eia.doe.gov/cneaf/alternate/page/datatables/atf1-13_03.html

[25]United Nations Population Information Network, http://www.un.org/popin/data.html

[26]Ibid.

[27]James Rubenstein, *The Cultural Landscape: An Introduction to Human Geography,* 8th ed. (Upper Saddle River, NJ: Prentice Hall, 2005).

[28]United Nations, *World Population Monitoring 2002: Reproductive Rights and Reproductive Health*, http://www.un.org/esa/population/publications/2003monitoring/WorldPopMonitoring_2002.pdf

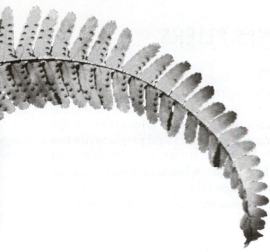

5

Exponential Functions and Regression

In 1798, an influential English economist named Thomas Malthus (Figure 5-1) published *An Essay on the Principle of Population.* In this essay, Malthus predicted that the world's food supply would eventually be insufficient to feed the growing human population, and that only "moral restraint" could prevent this looming doomsday. What was Malthus thinking? He reasoned that the food supply, growing at a linear rate, would eventually be outstripped by world's population, growing at an **exponential** rate.

Figure 5-1 Portrait of Thomas R. Malthus. *Photo courtesy of Library of Congress.*

In this chapter we investigate environmental examples in which quantities increase or decrease by a constant *percentage*. We'll learn to write exponential functions, which are the essential mathematical tools for describing this type of change. Then we discuss solving exponential equations and the important topics of doubling times and half-lives. The chapter concludes with an excursion into approximating almost-exponential data sets. We start with an example that shows just why Malthus's doomsday prediction of future starvation was wrong.

EXPONENTIAL RATES AND MULTIPLIERS

EXAMPLE 5-1   The Green Revolution

The 1900s saw several advances that greatly improved farmland productivity worldwide—the so-called Green Revolution. These technological advances included sophisticated seeds, pesticides and fertilizers, and advanced crop management strategies (Figure 5-2). To understand how dramatically farmland productivity has increased, we can model the growth in the production of barley. In 1960, an acre of barley in the United States yielded 31.0 bushels of grain. (A bushel is a bit more than 1 cubic foot, or about the size of a wastebasket.) Since 1960, the yield has grown on average by 1.7% each year.[1] What were the approximate barley yields for years after 1960?

Figure 5-2 The Green Revolution was driven by modern fertilizers, pesticides, and farm machinery. What were the environmental costs? *Photo courtesy of USDA Agricultural Research Service.*

Solution   To estimate the 1961 barley yield, we add 1.7% of the 1960 yield to the 1960 yield:

$$1961 \text{ yield} = 31.0 + 0.017(31.0)$$
$$= 31.0 + 0.527$$
$$= 31.527$$
$$\approx 31.5 \text{ bushels}$$

Let's inspect the first line of the above computation. If we factor out 31.0 bushels, we can compute the 1961 yield in a slightly different way:

$$1961 \text{ yield} = 31.0 + 0.017(31.0)$$
$$= 31.0(1 + 0.017)$$
$$= 31.0(1.017)$$
$$= 31.527$$
$$\approx 31.5 \text{ bushels}$$

Using this factoring approach, we find that the 1961 yield can be obtained by multiplying the 1960 yield (31.0 bushels) by 1.017. The number 1.017, written as a percentage, is 101.7%. Think about this—instead of *adding* 1.7% of the 1960 yield to the 1960 yield, we can take 101.7% *of* the 1960 yield to get the 1961 yield.

The 1962 yield will be 1.7% larger than the 1961 yield. To find the 1962 yield, we take 101.7% of the 1961 yield:

$$1962 \text{ yield} = 31.527(101.7\%)$$
$$= 31.527(1.017)$$
$$= 32.062959$$
$$\approx 32.1 \text{ bushels}$$

To estimate barley yields for years after 1962, we could repeatedly multiply by 1.017. But repeated multiplication is laborious, so we make use of the following shortcut: The yield in any year can be found by multiplying the yield in 1960 by some *power* of 1.017. For example, the yield in 1962 is the 1960 yield multiplied by 1.017 twice:

$$1962 \text{ yield} = 31.0(1.017)^2$$
$$= 32.062959$$
$$\approx 32.1 \text{ bushels}$$

In Table 5-1, we compute the yield for years after 1960 using this shortcut, with results rounded to one decimal place. Notice that the number of years after 1960 is the same as the power of 1.017.

To estimate the yield in year 2002, or 42 years after 1960, simply calculate

$$31.0(1.017)^{42} \approx 62.9 \text{ bushels.}$$

TABLE 5-1

Year	Years after 1960	Yield (bushels/acre)
1960	0	31.0
1961	1	$31.0(1.017)^1 \approx 31.5$
1962	2	$31.0(1.017)^2 \approx 32.1$
1963	3	$31.0(1.017)^3 \approx 32.6$
1964	4	$31.0(1.017)^4 \approx 33.2$

In general, to calculate barley yield Y at a time t years after 1960, we use the function

$$Y = 31.0(1.017)^t$$

The preceding function is called an **exponential function** because the independent variable, t, is "upstairs" as an exponent.

Each year, the 1.7% growth is added onto the barley yield from the previous year. This means that the 1.7% increase will factor in all previous 1.7% increases. This cumulative process is called **compounding**. The result of compounding is that yields will increase more and more each year and the graph of the exponential function gets steeper and steeper. We see the increasing steepness in Figure 5-3, where the barley function is plotted over a 100-year domain. This graph is said to have a J-shape, which is typical for exponentially increasing functions and gives us a clue as to how the food supply has kept up with an exploding world population.

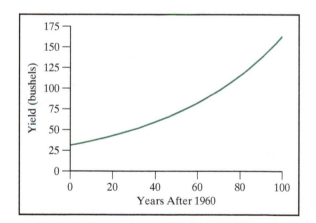

Figure 5-3 Since 1960, barley yields have increased 1.7% each year. The graph curves upward in the characteristic J-shape of exponential growth functions.

Before moving on to another example, we introduce some terminology. The percentage by which an amount grows is called the **growth rate** and is denoted by r. In the previous example, the growth rate for barley yields is $r = 1.7\% = 0.017$ each year.

Barley yields can be computed for years after 1960 by repeatedly multiplying by 1.017. In terms of the growth rate, we multiply by $1 + r = 1 + 0.017 = 1.017$. The number 1.017 is called the **growth multiplier** and is denoted by M. In the previous example, $M = 1 + r = 1.017$. Now we consider an example in which a quantity is decreasing by a fixed percentage each time period.

EXAMPLE 5-2   Sacramento River Salmon Population

Populations of chinook salmon (*Oncorhynchus tshawytscha*) along the west coast of North America have declined dramatically over the last century. Reasons for the decline include overfishing, loss of stream habitat, and pollution. In 1966, the Red Bluff Diversion Dam was built across the Sacramento River in northern California to divert water for irrigation. Fish ladders bypassing the dam (Figure 5-4) and a fish-counting facility were constructed at the same time. In 1967, when counting began, approximately 86,500 chinook salmon swam up the ladders. Since then, the chinook run has declined at an average rate of 18.1% per year.[2]

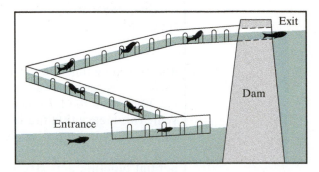

Figure 5-4 Fish ladders may allow migrating salmon to swim upstream past dams

To compute the population of salmon that swam up the ladders in 1968, we take the 1967 total and subtract off 18.1% of its value:

$$
\begin{aligned}
\text{pop. in 1968} &= 86{,}500 - 0.181(86{,}500) \\
&= 86{,}500 - 15{,}656.5 \\
&= 70{,}843.5 \\
&\approx 70{,}844 \text{ salmon}
\end{aligned}
$$

As in the previous example, we can compute the 1968 population in a slightly different way by factoring out the number 86,500:

$$
\begin{aligned}
\text{pop. in 1968} &= 86{,}500 - 0.181(86{,}500) \\
&= 86{,}500(1 - 0.181) \\
&= 86{,}500(0.819) \\
&= 70{,}843.5 \\
&\approx 70{,}844 \text{ salmon}
\end{aligned}
$$

This last computation shows that we can find the 1968 population by multiplying the 1967 population by 0.819 or 81.9%.

To compute the number of salmon in 1969, we can multiply the 1968 population by 0.819, or multiply the 1967 population by 0.819 twice:

$$
\begin{aligned}
\text{pop. in 1969} &= 70{,}843.5(0.819) \qquad\qquad \text{pop. in 1969} = 86{,}500(0.819)^2 \\
&= 58{,}020.8265 \qquad\qquad\qquad\qquad\quad\; = 58{,}020.8265 \\
&\approx 58{,}021 \text{ salmon} \qquad\qquad\qquad\qquad \approx 58{,}021 \text{ salmon}
\end{aligned}
$$

Multiplying the 1967 population by a power of 0.819, as we did in the second method, is faster and more accurate because all computations are done in one step. Using this one-step method, we estimate that by 1990 (23 years after 1967) the chinook population had dropped to $86{,}500(0.819)^{23} = 876$ salmon.

The preceding computations lead to a general exponential formula for the population of salmon, P, that pass up the fish ladders t years after 1967:

$$
P = 86{,}500(0.819)^t
$$

The annual decrease in the salmon population of 18.1% is called the **decay rate**. As with growth rates, we assign decay rates the symbol r, but we make their values negative. In this example, $r = -18.1\% = -0.181$. When we compute $1 + r$, we get the **decay multiplier** M:

$$M = 1 + r = 1 + (-0.181) = 0.819$$

Let's interpret the value 0.819 in the formula. Every year, the chinook population drops by 18.1%. This is equivalent to saying that 81.9% of the population *survives* from year to year. So the multiplier, $M = 0.819$ or 81.9%, is a yearly survival rate for the chinook population. By multiplying by 0.819 each year, we find that the population in any year is 81.9% of that in the previous year.

The salmon population graph in Figure 5-5 shows the classic shape of an exponentially decaying function. Notice that the population graph never reaches the horizontal axis, indicating that (according to our equation) the chinook salmon population never goes extinct no matter how long we wait. Of course, in reality, once the chinook population reaches some critically small number, the salmon will no longer be viable and will go extinct forever. The actual winter-run chinook salmon population in the Sacramento River in 1994–1996 was 830 fish; this salmon run has been classified as endangered by the National Marine Fisheries Service.

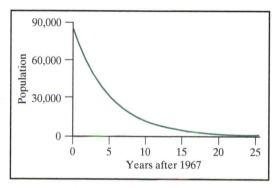

Figure 5-5 Since 1967, the chinook salmon population on the Sacramento River has decreased exponentially at an average rate of 18.1% each year. The graph of the population has the classic exponential decay shape.

THE GENERAL EXPONENTIAL MODEL

We now state a generalized formula for a quantity that grows or decays exponentially:

$$y = y_0(1 + r)^x$$

In this formula, y is the dependent variable and x is the independent variable. For many of our examples, x will represent time. The symbol y_0 stands for the initial value of y when $x = 0$, and r is the growth or decay rate. (Be sure not to confuse the growth or decay rate r with the correlation coefficient r. They are completely unrelated!) Using the substitution $M = 1 + r$, the preceding formula can be written in the form

$$y = y_0 M^x$$

A quantity can never decrease by *more than* 100%. And a quantity results in "nothing left over" if it decreases by *exactly* 100%. Thus, in an exponential function $r > -1$, and it follows that $M > 0$. The following are a few additional key concepts to remember about exponential functions:

In the formula $y = y_0(1 + r)^x$,

- if $r > 0$, then y is increasing exponentially with respect to x.
- if $r < 0$, then y is decreasing exponentially with respect to x.
- if $r = 0$, then y is a constant function.

In the equivalent formula $y = y_0 M^x$,

- if $M > 1$, then y is increasing exponentially with respect to x.
- if $M < 1$, then y is decreasing exponentially with respect to x.
- if $M = 1$, then y is a constant function.

The two basic shapes of exponential functions are displayed in Figure 5-6.

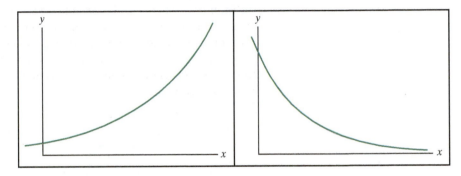

Figure 5-6 The basic shapes of graphs that are exponentially increasing (left) and decreasing (right)

Once we know the starting value (y_0) and the multiplier (M), writing the exponential function is quite easy. Determining the multiplier will often require some work. Given the growth or decay rate r, we can compute the multiplier using the formula $M = 1 + r$. The multiplier can also be computed directly from two data points, as we'll see in the next examples.

EXAMPLE 5-3   Recycling in Minnesota

Recycling efforts around the United States have generally been on the rise. For example, in 1999, citizens of Minnesota recycled 2,175,108 tons of glass, metals, food waste, paper, plastics, textiles, and hazardous materials. In the year 2000, the amount of recycled materials grew to 2,267,952 tons.[3] Assuming that Minnesota's recycling efforts continue to increase by the same percentage each year, find the growth multiplier and write the exponential function modeling the growth.

Solution We start by defining variables: Let A represent the amount of recyclables measured in tons and t represent years, with $t = 0$ referring to the year 1999. With respect to these variables, the recycling data are presented in Table 5-2.

TABLE 5-2

t (years after 1999)	A (tons)
0	2,175,108
1	2,267,952
2	
3	

$\times M$

We can find the yearly multiplier M directly from the data. Because the amount recycled in 1999 multiplied by M equals the amount recycled in 2000, we write

$$2{,}175{,}108 \text{ tons} \cdot M = 2{,}267{,}952 \text{ tons}$$

Solving for M yields

$$M = \frac{2,267,952 \text{ tons}}{2,175,108 \text{ tons}}$$

$$= 1.042684777$$

Rounded to four decimal places, the value of the multiplier is $M = 1.0427$. Notice how the units of measure cancel in the fraction above, leaving the multiplier with *no units*!

The exponential function will be of the form $y = y_0 M^x$ or $A = A_0 M^t$. Because the initial recycling value, when $t = 0$, is $A_0 = 2,175,108$ tons, it is a simple matter to write the exponential function:

$$A = 2,175,108(1.0427)^t$$

A few more comments are in order about the multiplier. In the previous example, the multiplier $M = 1.0427$ indicates that the growth rate is $r = 0.0427 = 4.27\%$. This means that the weight of recycled materials in Minnesota is increasing at an annual rate of 4.27%. A term often mentioned in the news is the **recycling rate**, which is not the same. Typically, the recycling rate is the fraction of the total amount of garbage that gets recycled (by weight). For example, in the year 2000, Minnesota's statewide recycling rate was 47.8%, indicating that 47.8 pounds out of every 100 pounds of garbage were recycled.

Note that we could have determined M by first finding the percent increase between 1999 and 2000. (See the discussion of percent change in Chapter 2.) The amount of recycled materials in these years increased by $2,267,952 - 2,175,108 = 92,844$ tons. Dividing this amount by the 1999 total gives us the percent increase r:

$$r = \frac{92,844 \text{ tons}}{2,175,108 \text{ tons}} = 0.0427 = 4.27\%$$

Using this value of r and the formula $M = 1 + r$, we find that the yearly growth multiplier is $M = 1.0427$. When computing the multiplier, use the method that makes sense to you. Hopefully, both methods make sense!

EXAMPLE 5-4   Infant Mortality

According to the U.S. Census Bureau, infant mortality is defined as the number of infant deaths (under age 1) per 1,000 registered live births. For the United States, infant mortality was 8.50 in year 1992 and 6.88 in year 2002.[4] The decline in infant mortality is generally attributed to advances in medical research and improved public health and social services. Despite the overall decrease in infant mortality, the United States still ranked 27th among 37 industrialized nations based on year 2000 data.[5] Find an exponential model to describe the decrease in infant mortality in the United States, and use the model to predict infant mortality in 2012 and 2022.

Solution The data values are given 10 years apart, so we define our time variable on a decadal (10-year) scale. Let t (the independent variable) represent the number of decades after 1992 and I (the dependent variable) represent infant mortality with units of deaths per 1,000 registered live births. Assuming that infant mortality continues to decrease by the same percentage each decade, the exponential model will be of the form

$$I = I_0 \cdot M^t$$

To determine the decay multiplier M, we again transfer the data into a table (Table 5-3).

TABLE 5-3

Years	t	I
1992	0	8.50
2002	1	6.88
2012	2	?
2022	3	?

$\times M$

We note that (infant mortality in 1992) $\cdot M =$ infant mortality in 2002 or

$$8.50M = 6.88$$

Solving for M yields

$$M = \frac{6.88}{8.50} = 0.8094117647 \approx 0.809$$

It makes sense that M is less than 1—if we multiply infant mortality by a number less than 1, infant mortality will decrease in our computations. We also note here that the decadal decay rate r is easy to determine once we know the value of M. We substitute into the formula $M = 1 + r$ to get

$$0.8094117647 = 1 + r$$

Subtracting 1 from each side of the equation produces

$$r = -0.1905882353$$

Rounded to three decimal places, the decay rate is $r = -0.191$. This value of r indicates that infant mortality decreased by about 19.1% between 1992 and 2002.

Now back to writing the exponential formula. When $t = 0$ the infant mortality is 8.50, so the exponential function is

$$I = 8.50(0.809)^t$$

In 2012, when $t = 2$, we expect that infant mortality will equal $8.50(0.809)^2 = 5.56$ deaths per 1,000 births. In 2022, we project that infant mortality will drop to $8.50(0.809)^3 = 4.50$ deaths per 1000 births. These results assume that the multiplier stays the same.

In 1982, when $t = -1$, our formula predicts that infant mortality was equal to $8.50(0.809)^{-1} = 10.50$ deaths per 1,000 births. The actual number was 11.50 deaths per 1,000 births, so this prediction is 1/10.5 or about 10% lower than the actual value.

FINDING EXPONENTIAL FUNCTIONS— THE MORE GENERAL CASE

In this section, we explore the mathematics of finding an exponential function through two points in which the x-values are not consecutive integers. This will allow us to model more general environmental data. Although the procedures are not difficult, it is good to keep in mind the role of the multiplier M in the exponential function—remember, M is the number that y is multiplied by as the independent variable x increases by 1.

TABLE 5-4 **Rules of exponents and logarithms for $a > 0$ and $b > 0$**

Property A	$a^m a^n = a^{m+n}$
Property B	$\dfrac{a^m}{a^n} = a^{m-n}$
Property C	$(a^m)^n = a^{mn}$
Property D	$(ab)^n = a^n b^n$
Property E	$\left(\dfrac{a}{b}\right)^n = \dfrac{a^n}{b^n}$
Property F	$a^0 = 1$
Property G	$a^{1/n} = \sqrt[n]{a}$
Property H	$a^{-n} = \dfrac{1}{a^n}$
Property I	$\log(a^n) = n\log(a)$
Property J	$\log(ab) = \log(a) + \log(b)$
Property K	$\log\left(\dfrac{a}{b}\right) = \log(a) - \log(b)$
Property L	$10^{\log(a)} = a$
Property M	if $a = b$, then $10^a = 10^b$
Property N	if $a = b$, then $\log(a) = \log(b)$

Many of the following examples in this chapter (and in Chapter 6) will occasionally make use of exponential and logarithmic properties. Table 5-4 lists most of the properties that we'll need to simplify expressions or solve equations involving exponents and logarithms. Don't worry—you will not need to memorize this list, nor will you have to work dozens of drill exercises to become exceptionally proficient with each property. We simply state these properties up front so that we can conveniently refer to them, and we'll explain their use as we need them. Now on to those examples!

EXAMPLE 5-5   Finding an Exponential Growth Function

Suppose we want to find the exponential function that passes through the two ordered pairs $(x, y) = (0, 20)$ and $(x, y) = (4, 100)$. As a visualization aid, we put the data in a table (Table 5-5) and label the multiplier M.

TABLE 5-5

x	y	
0	20	$\times M$
1		$\times M$
2		$\times M$
3		$\times M$
4	100	
5		

We know that the initial y-value is $y_0 = 20$, so the function is of the form $y = 20\,M^x$. Once we find the value of M, we will have enough information to write the complete exponential function. From the table we observe that if we multiply 20 by M four times, we will get 100. In equation form this is written as

$$20M^4 = 100$$

Dividing by 20 yields

$$M^4 = 5$$

Recall that equations like the preceding one, which involve *even* powers of exponents, can have both a positive and a negative solution. Because M is always greater than 0 in exponential models, we will disregard the negative solution.

To isolate M, we take the fourth root of each side of the equation. But by Property G ($a^{1/n} = \sqrt[n]{a}$), this is the same as raising each side to the 1/4 power. Because it's often easier to work with exponents, we'll try that:

$$(M^4)^{1/4} = 5^{1/4}$$

Property C ($(a^m)^n = a^{mn}$) allows us to multiply exponents, which results in M^1 or simply M on the left side of the equation. Thus we get

$$M = 5^{1/4}$$

The approximate value of M can be found with a calculator. When doing this, be sure to include parentheses around the 1/4 power.

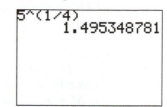

We find that the approximate value of the multiplier is $M \approx 1.495348781$. The exponential equation, with multiplier approximated to three decimal places, is

$$y = 20(1.495)^x$$

Using technology, we create a table and graph of this function for x-values from 0 to 6. Notice that when $x = 4$ the function does not give an output of exactly 100.

The function's *y*-values do not match those in the original data perfectly because of the rounded-off multiplier.

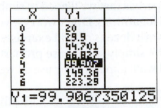
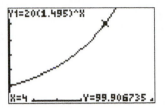

EXAMPLE 5-6 North Dakota's Population

The U.S. Census Bureau estimated that the April 1, 2000, North Dakota population was 642,200 and the July 1, 2001, population was 634,448.[6] In order to create an exponential equation that models the decreasing population, we organize the data in a table and make the first data point correspond to "time zero." See Table 5-6.

TABLE 5-6

t = Months Beginning April 1, 2000	P = Population
0	642,200
1	
2	
⋮	⋮
15	634,448

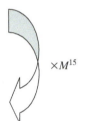

$\times M^{15}$

From the table we conclude that $642{,}200 M^{15} = 634{,}448$. Now divide each side by 642,200 and raise each side to the 1/15 power to isolate M:

$$M^{15} = \frac{634{,}448}{642{,}200}$$

$$(M^{15})^{1/15} = \left(\frac{634{,}448}{642{,}200}\right)^{1/15}$$

$$M = \left(\frac{634{,}448}{642{,}200}\right)^{1/15}$$

When evaluating M on your calculator, be sure to include parentheses around *both* the fractional base and the 1/15 power.

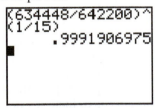

Using all the decimal places for the multiplier displayed in the preceding calculator screen, the population function is

$$P = 642{,}200(0.9991906975)^t$$

The percentage by which North Dakota's population decreases each month is given by the monthly decay rate r. To determine this value, we rearrange the formula $M = r + 1$ to read as $r = M - 1$. Substituting for M gives us

$$r = M - 1$$

$$= 0.9991906975 - 1$$

$$\approx -0.00081$$

The decay rate tells us that North Dakota's population is decreasing by about 0.081% each month.

We computed the *monthly* percent decrease in North Dakota's population. So what is the *yearly* percent decrease? Surprisingly, we do not multiply the monthly percent decrease by 12. Monthly rates of change *compound* (build upon each other), making the annual rate greater than the sum of the 12 monthly rates. How then do we equate monthly rates to yearly rates? The answer lies in the multiplier: Twelve compounded monthly multipliers are equivalent to one giant yearly multiplier, as we compute next.

$$M_{year} = (M_{month})^{12}$$
$$= (0.9991906975)^{12}$$
$$= 0.9903314817$$

Now that we have the yearly multiplier M, we can determine the value of the yearly decay rate r:

$$r_{year} = M_{year} - 1$$
$$= 0.9903314817 - 1$$
$$= -0.0096685183$$

This last result indicates that North Dakota's population is decreasing by about 0.97%, or almost 1%, each year.

SOLVING EXPONENTIAL EQUATIONS

In Example 5-3, we modeled the amount of recycled materials in Minnesota by the exponential growth function $A = 2{,}175{,}108(1.0427)^t$, where A represented tons of material and t the number of years after 1999. To predict the year when the recycling amount reaches 3 million tons, we need to solve the exponential equation

$$2{,}175{,}108(1.0427)^t = 3{,}000{,}000$$

Isolating the exponent in an exponential equation is not too difficult but requires the use of **logarithms**. Before solving the preceding equation, we review some basics on logarithms and introduce the procedures required to tackle such an equation.

Recall from Chapter 1 that the common logarithm or log of a number tells us the power of 10 equal to that number. For example, log(100) tells us the power of 10 equal to 100. The power is 2, so log(100) = 2. We can use the preceding definition of the logarithm to solve exponential equations in which the multiplier (i.e., the base of the exponent) is the number 10.

EXAMPLE 5-7   Solving Exponential Equations with Multiplier $M = 10$

a. Solve for x: $10^x = 3{,}400$.
Solution: By the definition of the logarithm, $x = \log(3{,}400) \approx 3.53$. We check this by finding that $10^{3.53} \approx 3{,}400$.

b. Solve for n: $4 \cdot 10^n = 80$.
Solution: First divide each side of the equation by 4 to get $10^n = 20$. Then by the definition of the logarithm, the solution is $n = \log(20) \approx 1.3010$.

c. Solve for t: $10^t = -5$.
Solution: We might start by writing $t = \log(-5)$. But when we try to find $\log(-5)$ on our calculator it gives us an error message! We can't take the log of a negative number because there is no power of 10 that equals a negative number. There is no solution to this equation.

How do we solve exponential equations that involve multipliers other than 10, such as $2^x = 15$? Our method will again use common logarithms, but in an entirely different manner. Let's start by investigating a simple example to "discover" a property of logarithms.

EXAMPLE 5-8 An Important Property of Logarithms

With a calculator, find the value of the following expressions:

a. $\log(2^3)$ and $3\log(2)$

Calculator output: Both are approximately equal to 0.903089987.

b. $\log(7^5)$ and $5\log(7)$

Calculator output: Both are approximately equal to 4.2254902.

c. $\log(3^{-1})$ and $-1 \cdot \log(3)$

Calculator output: Both are approximately equal to -0.4771212547.

These examples illustrate Property I, which says $\log(a^n) = n\log(a)$. Property I is the "power comes down" property. This is probably the most useful property when solving exponential equations. Before we demonstrate its use, recall that Property N says if $a = b$, then $\log(a) = \log(b)$. Property N is true as long as $a > 0$ and $b > 0$, because we can only take logs of positive numbers. Let's see how we can put these two properties to work when solving exponential equations that involve multipliers other than 10.

EXAMPLE 5-9 Solving an Exponential Equation

Solve $2^x = 15$.

Solution: We need to get x down from its "upstairs position" and isolate it on one side of the equation. Applying Property N, we log each side of the equation to get

$$\log(2^x) = \log(15).$$

Now we use Property I and bring the x down in front of $\log(2)$:

$$x\log(2) = \log(15)$$

The left side of the preceding equation is read as "x times the log of 2." The number $\log(2)$ can be approximated on a calculator, but we leave it alone for now. Our next step is to divide each side by $\log(2)$ to isolate x:

$$\frac{x\log(2)}{\log(2)} = \frac{\log(15)}{\log(2)}$$

which reduces to

$$x = \frac{\log(15)}{\log(2)}$$

A common mistake at this point is to calculate $\log(15/2)$. This is not correct! The following screen shows how to enter the expression correctly on your graphing calculator:

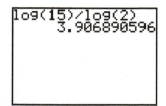

```
log(15)/log(2)
          3.906890596
```

The calculator output indicates that the solution to the equation $2^x = 15$ is approximately 3.906890596. We finish by checking that our answer is correct: $2^{3.906890596} \approx 15$.

EXAMPLE 5-10 Solving an Exponential Equation

Solve $200(0.95)^t = 45$.

Solution: We first divide each side of the equation by 200 and simplify:

$$\frac{200(0.95)^t}{200} = \frac{45}{200}$$

$$0.95^t = 0.225$$

Next we take the log of each side and bring the *t* down in front:

$$\log(0.95^t) = \log(0.225)$$

$$t\log(0.95) = \log(0.225)$$

Finally, divide each side by log(0.95) and approximate:

$$t = \frac{\log(0.225)}{\log(0.95)} \approx 29.1$$

EXAMPLE 5-11   The Minnesota Recycling Equation

Having seen a few examples of solving exponential equations, we are now ready to answer the question regarding Minnesota's recycling: In how many years after 1999 will recycling reach 3 million tons? We start by setting up the exponential equation:

$$2{,}175{,}108(1.0427)^t = 3{,}000{,}000$$

As in the previous examples, we divide by the initial amount to isolate the multiplier raised to the power *t*:

$$1.0427^t = \frac{3{,}000{,}000}{2{,}175{,}108}$$

Now take the log of each side of the equation, and bring the *t* down in front:

$$\log(1.0427^t) = \log\left(\frac{3{,}000{,}000}{2{,}175{,}108}\right)$$

$$t\log(1.0427) = \log\left(\frac{3{,}000{,}000}{2{,}175{,}108}\right)$$

Finally, divide each side by log(1.0427) to isolate *t*:

$$t = \frac{\log\left(\dfrac{3{,}000{,}000}{2{,}175{,}108}\right)}{\log(1.0427)} \approx 8 \text{ years}$$

Assuming that recycling efforts continue to increase by $r = 4.27\%$ each year, we predict that Minnesota will recycle 3 million tons of material by the year 2007.

DOUBLING TIMES AND HALF-LIVES

The values of the growth or decay rate *r* and multiplier *M* indicate how quickly a quantity will exponentially increase or decrease. Another way of communicating exponential rates of growth or decay is by stating the amount of time needed for a quantity to double, triple, get cut in half, and so on. Let's look at a few practical examples.

EXAMPLE 5-12   Lung Cancer

The amount of time it takes a quantity to double in size (increase by 100%) is called the **doubling time**. Exposure to hazardous substances such as radon gas or toxic chemicals can cause cancer (although smoking is, by far, the leading culprit. See Figure 5-7). Through a process called **mitosis**, cells divide in half, producing two daughter cells identical to one another and to the original parent cell. The two daughter cells eventually go through the mitosis process, each dividing to produce a total of four cells. The dividing process goes on and on. The amount of time it takes a malignant lung cell to divide into two malignant daughter cells, the doubling time, is approximately 3 months.

Suppose that one cell becomes malignant and starts to divide. Table 5-7 tallies the total number of malignant cells every 3 months, or every doubling period. We let *d* represent the number of doubling periods and *C* the number of malignant cells. The exponential function relating *C* to *d* is $C = 1(2)^d$, or, more simply, $C = 2^d$.

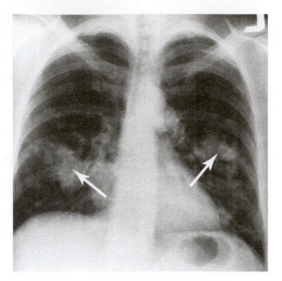

Figure 5-7 An x-ray image of a human chest shows both lungs. In the lower portion of each lung are cancerous growths (indicated by arrows). *Courtesy of the National Cancer Institute.*

TABLE 5-7

Months	d = Doubling Periods	C = Number of Cells
0	0	1
3	1	2
6	2	4
9	3	8
12	4	16
15	5	32
18	6	64

A cluster of malignant cells is visible on an x-ray when the cluster's diameter is about 1 cm, or when there are about 1 billion cells. How long will it take before the cluster is detectable by x-ray? To determine this time, we solve the equation $2^d = 1,000,000,000$ to find the number of doubling periods using the approach we've just explained:

$$2^d = 1,000,000,000$$
$$\log(2^d) = \log(1,000,000,000)$$
$$d\log(2) = 9$$
$$d = \frac{9}{\log(2)} \approx 30$$

A lung cancer victim would have to wait 30 doubling periods, or 7.5 years, before the cancer is detectible by x-ray! Cancers vary in doubling time from 30 to 500 days, averaging 100 days, so the doubling time that we used in this example is fairly typical.[7]

EXAMPLE 5-13 Radiocarbon Dating

The amount of time it takes a quantity to be reduced by half (decrease by 50%) is called the **half-life**. In the study of atomic physics we observe that unstable **isotopes** (forms) of elements naturally break down into other forms. When isotopes break down they release energy or radiation; we call these isotopes **radioactive**. One example of an unstable, radioactive isotope is carbon-14 (given the symbol C-14). Laboratory observations indicate that C-14 has a half-life of about 5,730 years;[8] let's see what this half-life really means.

Suppose we start with a sample of 100 μg of C-14. (*Note*: 1 μg = 1 microgram = 1 millionth of a gram.) The half-life information indicates that in 5,730 years the amount of C-14 will be reduced to half this amount, or 50 μg. In an additional 5,730 years, there will be half this amount again, or 25 μg.

Let's develop a model for the amount of C-14 remaining after a given number of years. Start by displaying the preceding information in a table:

t = Time (years)	A = Amount of C-14 (μg)
0	100
5,730	50
11,460	25

$\times M^{5,730}$

Now write an equation relating 100 and 50, and then solve for M, the yearly decay multiplier:

$$100M^{5730} = 50$$
$$M^{5730} = 0.5$$
$$M = 0.5^{1/5730}$$

We could approximate M at this point, but for greater accuracy we leave it in exact form. The equation for the amount of C-14 remaining after t years is given by

$$A = 100(0.5^{1/5730})^t$$

Using Property C ($(a^m)^n = a^{mn}$), we rewrite the exponential function as

$$A = 100(0.5)^{t/5730}$$

Let's check that this equation provides the values that we expect:

- When $t = 0$, $A = 100(0.5)^{0/5730} = 100(0.5)^0 = 100(1) = 100$ μg. ✔
- When $t = 5,730$, $A = 100(0.5)^{5730/5730} = 100(0.5)^1 = 50$ μg. ✔
- When $t = 11,460$, $A = 100(0.5)^{11460/5730} = 100(0.5)^2 = 100(0.25) = 25$ μg. ✔

We have created a valid function that describes how a 100 μg sample of C-14 will decay. The actual practice of estimating the age of dead organisms (called **radiocarbon** or **carbon-14 dating**) requires the measurement of the *ratio* of the C-14 to C-12 in a sample. C-12 is the stable isotope of carbon. When an organism is alive, the ratio of C-14 to C-12 within its cells is about the same as the background ratio in the atmosphere. When the organism dies, the amount of stored C-14 decays exponentially while the amount of C-12 stays constant, so the ratio of C-14 to C-12 also decays exponentially over time. By measuring the ratio of C-14 to C-12 in dead organisms, and comparing that to the background ratio, scientists can estimate the number of years since the organism has died. Scientists use carbon-14 dating on objects up to about 60,000 years old. After this amount of time, roughly 10 half-lives of C-14, the mass of C-14 present in the sample is often too small to measure accurately.

EXAMPLE 5-14   Population of Kenya

In the year 2002, the population of Kenya was about 31 million people and was growing at an annual rate of $r = 1.4\%$.[9] The exponential equation modeling Kenya's population (in millions) since the year 2002 is

$$P = 31(1.014)^t$$

Determining the doubling time of Kenya's population is easy. We simply compute the number of years until the population reaches $2(31) = 62$ million people. Next we show the steps in the solution process:

$$62 = 31(1.014)^t$$
$$2 = 1.014^t$$
$$\log(2) = \log(1.014^t)$$
$$\log(2) = t\log(1.014)$$

$$\frac{\log(2)}{\log(1.014)} = t$$

$$t \approx 50 \text{ years}$$

We found that Kenya's population will double in 50 years; consequently, it will double twice (quadruple) in 100 years! Of course, this assumes that Kenya's growth rate will remain the same.

Doubling time is useful to demonstrate the long-term effect of a growth rate but should be used with caution to project population size. Many of the more industrialized countries currently have very low growth rates and, as a result, their doubling times are in hundreds or thousands of years. During such long periods of time, who knows what may happen? Many less industrialized countries, such as Kenya, currently have high growth rates that are associated with short doubling times. The populations of these countries are expected to grow more slowly as birth rates decline. In a 1998 edition of *National Geographic* magazine, an article began with the following line: *Kenya's growth rate hovers around 4% a year, highest in the world; its population will double in about 17 years to 46 million.*[10] Clearly, *National Geographic*'s assumption that Kenya's population would continue to increase by 4% each year was much off the mark!

APPROXIMATING ALMOST-EXPONENTIAL DATA SETS

In Chapter 4, we explored two basic methods to approximate almost-linear data sets, and now we do the same for data sets that are almost-exponential. Once again, the two methods are the **straightedge method** and **least squares regression**.

Straightedge Method

If you are familiar with the straightedge method used with almost-linear data, you might be asking, "How can we use a *straight*edge to approximate exponential data?" The answer to this question is quite remarkable and makes use of the following two important facts:

- If (x, y) data are perfectly exponential, then $(x, \log(y))$ data are perfectly linear. The converse or "reverse" statement is also true: If $(x, \log(y))$ data are perfectly linear, then (x, y) data are perfectly exponential.
- If (x, y) data are *almost* exponential, then $(x, \log(y))$ data are *almost* linear. The converse statement is also true.

The latter fact is the more useful of the two, because data are not often perfectly exponential. To investigate the straightedge method, let's consider an example concerning an important topic in environmental studies—that of renewable energy.

Photovoltaic cells convert sunlight to electricity and show great promise because of their portability and dependence on a renewable resource—sunlight. Most cells today are made of atoms whose outermost electrons are easily knocked loose by incoming solar radiation. The flow of loose electrons creates an electric current, with a typical solar cell generating about 0.5 volts. Most graphing calculators take four 1.5-volt batteries, requiring a total of 6 volts. Thus, a solar powered graphing calculator (wouldn't that be nice?) would require about 12 standard photovoltaic cells.

From 1985 to 1999, the total capacity of U.S. shipments of photovoltaic cells increased dramatically, indicating a greater demand for this type of renewable energy source.[11] Figure 5-8 displays the data, with $x = 0$ referring to the year 1985. We'll use x and y as variables in this example to make the mathematics easier to follow. It appears that the capacity is increasing in an exponential manner.

Table 5-8 lists the data points from the graph—we refer to these as the **original data**. We start the straightedge method by making a new table in which we compute the logarithms of the y or capacity values (Table 5-9). We leave the x-values as before. The resulting $(x, \log(y))$ points are called the **transformed data**.

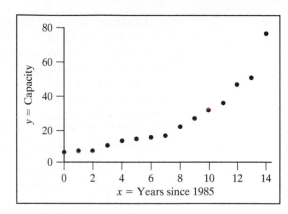

Figure 5-8 The capacity of photovoltaic cells appears to be increasing exponentially.

TABLE 5-8 Original data		**TABLE 5-9** Transformed data	
Year	**Capacity in MW**	**Year**	**log(Capacity) in log(MW)**
0	6.0	0	$\log(6.0) \approx 0.78$
1	6.6	1	$\log(6.6) \approx 0.82$
2	6.9	2	$\log(6.9) \approx 0.84$
3	10.0	3	$\log(10.0) = 1.00$
4	13.0	4	$\log(13.0) \approx 1.11$
5	13.8	5	$\log(13.8) \approx 1.14$
6	14.9	6	$\log(14.9) \approx 1.17$
7	15.6	7	$\log(15.6) \approx 1.19$
8	21.0	8	$\log(21.0) \approx 1.32$
9	26.1	9	$\log(26.1) \approx 1.42$
10	31.1	10	$\log(31.1) \approx 1.49$
11	35.5	11	$\log(35.5) \approx 1.55$
12	46.4	12	$\log(46.4) \approx 1.67$
13	50.6	13	$\log(50.6) \approx 1.70$
14	76.9	14	$\log(76.9) \approx 1.89$

Source: Statistical Review of *World Energy*

The plot of the transformed data (Figure 5-9) shows an almost-linear relationship between years and log(capacity). This is exactly what we expect from our earlier fact: If (x, y) data are *almost* exponential, then $(x, \log(y))$ data are *almost* linear.

We continue with the straightedge method by laying a straightedge or ruler on the almost-linear transformed data and "eyeballing" a best-fitting line. Then we choose

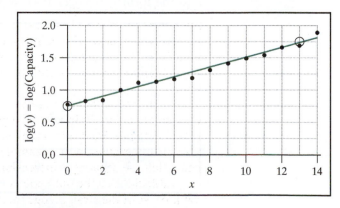

Figure 5-9 The plot of transformed data shows an almost-linear relationship between x and $\log(y)$.

two data points from the line to determine its equation. The points that we choose are $(0, 0.75)$ and $(13, 1.75)$.

Now we find the equation of the line through these points. Clearly the vertical intercept is $b = 0.75$. The slope can be found using the slope formula:

$$m = \frac{1.75 - 0.75}{13 - 0} \approx 0.0769230769$$

Note that it's important for future computations to use all the decimal places that our calculator produces for the slope.

Now that we have the slope and vertical intercept of the line through the transformed data, we can write the equation of the line. Because the transformed data are $(x, \log(y))$ points, the line through these data is of the form

$$\log(y) = mx + b$$

For this particular example, the linear equation through the transformed data is

$$\log(y) = 0.0769230769x + 0.75$$

Recall that our goal is to find an equation that approximates the almost-exponential original data. Quite remarkably, we can find that exponential equation by solving the equation above for y. All we need are a few of the exponential and logarithmic properties. Property M (if $a = b$, then $10^a = 10^b$) tells us that if we have an equation, we can raise 10 to each side of the equation and the results will be equal. We try this on our preceding linear equation:

$$10^{\log(y)} = 10^{0.0769230769x + 0.75}$$

By Property L ($10^{\log(a)} = a$), the left side of the equation reduces to y:

$$y = 10^{0.0769230769x + 0.75}$$

Now we use Property A, the addition rule for exponents, in its reverse order ($a^{m+n} = a^m a^n$) to rewrite the right side of the equation:

$$y = 10^{0.0769230769x} 10^{0.75}$$

The right side can be manipulated even further by using Property C in its reverse order, $a^{mn} = (a^m)^n$, and rearranging factors:

$$y = 10^{0.75} (10^{0.0769230769})^x$$

This last equation has the form $y = y_0 M^x$ with $y_0 = 10^{0.75}$ and $M = 10^{0.0769230769}$. Approximating the values of y_0 and M on a calculator gives us the function that fits the almost-exponential data:

$$y = 5.62(1.194)^x$$

To conclude our example, we display in Figure 5-10 the plots of the original (x, y) and transformed $(x, \log(y))$ data sets. In each graph we list the equation that approximates the data.

The work that we did in converting the equation $\log(y) = 0.0769230769x + 0.75$ into the form $y = 5.62(1.194)^x$ was a bit complex and something that we'd rather

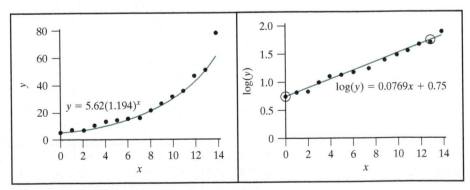

Figure 5-10 Plots of the original data (left) and transformed data (right). The equation fitting the original data can be found from the equation fitting the transformed data.

avoid each time we use the straightedge method. The last step in the conversion was to approximate y_0 and M using the values

$$y_0 = 10^{0.75} \quad \text{and} \quad M = 10^{0.0769230769}$$

But 0.75 was the vertical intercept b of the line, and 0.0769230769 was the slope m. In general, the following conversion formulas will apply:

$$\boxed{y_0 = 10^b \quad \text{and} \quad M = 10^m}$$

The following is a summary of the steps needed to find the equation through almost-exponential data using the straightedge method:

Straightedge Method Summary

1. Start with the original data consisting of points in the form (x, y). Take the logarithms of the y-values (leave the x-values alone) to form a new data set with points of the form $(x, \log(y))$. Call this new data set the transformed data.

2. Plot the transformed data and "eyeball" a line through them. Pick two points on the line to find the slope m and vertical intercept b. The line through the transformed data will be in the form $\log(y) = mx + b$.

3. Solve the equation obtained in step 2 for y, or find M and y_0 using the formulas $y_0 = 10^b$ and $M = 10^m$.

Least Squares Regression

As with linear data, we can use technology to find a best-fitting curve through exponential data. The output from the TI-83/84 graphing calculator is given next. (Instructions on how to find the exponential regression follow shortly.)

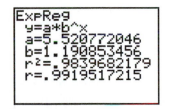

```
ExpReg
 y=a*b^x
 a=5.520772046
 b=1.190853456
 r²=.9839682179
 r=.9919517215
```

The calculator uses the variable a in place of the vertical intercept y_0, and b in place of the multiplier M. The approximate regression equation is

$$y = 5.52(1.191)^x$$

So how does the calculator find the best-fitting exponential regression equation? Essentially, the calculator is programmed to work through the same steps as the straightedge method, although it uses a least squares regression program to find the best-fitting line through the transformed data. Using the calculator's exponential regression feature makes life easy, but we encourage you to learn the straightedge method for two reasons. First, you should know some of the mathematics behind the calculator's "black box" or automated feature. Second, you should develop an intuitive understanding of the fact that "if (x, y) data are almost exponential, then $(x, \log(y))$ data are almost linear." We hope you do.

The correlation coefficient r for an exponential regression model measures how well the data fit an exponential curve and indicates the extent to which the x-variable is able to predict the y-variable. More technically, the correlation coefficient measures the strength of the linear relationship among the transformed $(x, \log(y))$ data. A correlation coefficient close to $r = 1$ indicates a very good exponential fit to the data and that the data are generally increasing ($M > 1$). When the correlation coefficient is close to $r = -1$, we know that there is a very good exponential fit to the data and that the data are generally decreasing ($M < 1$). As with linear data, a correlation coefficient close to $r = 0$ indicates a poor exponential fit to the data. It is important not to confuse the correlation coefficient r with the growth rate r for the exponential model—they are completely unrelated.

In this last example, the correlation coefficient value of $r = 0.992$ indicates that time is a very good predictor of photovoltaic cell capacity. In practical terms, the exponential regression equation will do a very good job in predicting the actual photovoltaic cell capacity for years from 1985 to 1999.

We've now completed our two methods to find an approximating equation through an almost-exponential data set. Table 5-10 lists the equations found with each method—we see that the results are quite similar.

TABLE 5-10 Results of the two methods for approximating the almost-exponential photocell data set

Method	Straightedge	Least Squares Regression
Equation	$y = 5.62(1.194)^x$	$y = 5.52(1.191)^x$

USING TECHNOLOGY: EXPONENTIAL REGRESSION

Instructions for obtaining the exponential regression equation on the TI-83/84 are given in this section. More detailed instructions on entering data, graphing the regression equation, and finding the correlation coefficient can be found in Chapter 4. Visit the text Web site for instructions on how to use online computer applications that perform the same tasks: **enviromath.com**

Exponential Regression on the TI-83/84

Press **STAT > Edit**; then enter the original (unlogged) data into two lists, such as **L1** and **L2**.

Now select **STAT > CALC** and scroll down to **0:ExpReg**.

Press **ENTER** to paste ExpReg onto the homescreen. Now enter **L1, L2**.

Press **ENTER** once more. The exponential regression equation is approximately $y = 5.52(1.191)^x$, and the correlation coefficient is about $r = 0.99$. *See instructions in Chapter 4 if the correlation coefficient is not being displayed.*

CHAPTER SUMMARY

A quantity that increases by a *constant percentage* is said to exhibit **exponential change**. The basic exponential function has the form $y = y_0(1 + r)^x$. The y-intercept, y_0, indicates where the function crosses the y-axis. If x represents time, the y-intercept can be interpreted as the "starting value" of the function (at time $= 0$). The percentage by which the function is increasing or decreasing is given by the value of r. If the function is increasing, r is positive and is called the **growth rate**; if the function is decreasing, r is negative and is called the **decay rate**.

Using the substitution $M = 1 + r$, the basic exponential function above can be written in the form $y = y_0 M^x$. If the function is increasing, M is greater than 1 and is called the **growth multiplier**; if the function is decreasing, M is between 0 and 1 and is called the **decay multiplier** (Figure 5-11). The multiplier is the value that y is multiplied by as x increases by one unit.

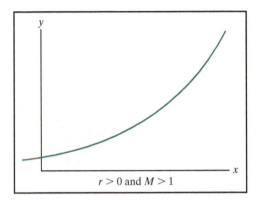
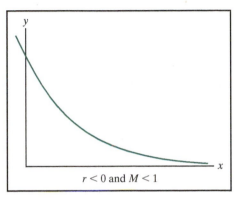

$r > 0$ and $M > 1$ $r < 0$ and $M < 1$

Figure 5-11 An exponentially increasing graph (left) and decreasing graph (right)

An exponential equation such as $200 = 10(1.25)^x$ can be solved for x by using the properties of exponents and logarithms in Table 5-4. To solve this equation, first divide each side by 10 to get $20 = 1.25^x$. Logging each side results in $\log(20) = \log(1.25^x)$. The x in this last equation can be "pulled down" using Property I, resulting in $\log(20) = x\log(1.25)$. Finally, divide each side by $\log(1.25)$ to obtain the exact solution of $x = \log(20)/\log(1.25)$. The approximate solution is $x = 13.43$.

The amount of time it takes an exponentially-growing quantity to double in size is called the **doubling time**. The **half-life** for exponentially-decaying quantities is the amount of time for a quantity to be reduced by half. To find the doubling time for a function such as $y = 80(1.06)^t$, substitute 160 for y to get $160 = 80(1.06)^t$. Solving for t yields the doubling time of $t = \log(2)/\log(1.06) \approx 11.9$. The half-life of an exponentially-decaying function can be found in a similar manner.

Almost-exponential (x, y) data will appear linear when plotted in the form $(x, \log(y))$. This is the basis of the **straightedge method**. The steps in the straightedge method are as follows: (1) Take the logarithms of the y-values of the original data. (2) Plot the $(x, \log(y))$ transformed data and "eyeball" a straight line through them. Pick two points on the line to find the slope, m, and vertical intercept, b. The equation of the line through the transformed data can be written in the form $\log(y) = mx + b$. (3) Write this last equation in the form $y = y_0 M^x$ by solving for y using the properties of exponents and logarithms in Table 5-4. Or use the conversion formulas $y_0 = 10^b$ and $M = 10^m$.

A second method to approximate almost-exponential data is **least squares regression**. Exponential regression is based on the straightedge method, but uses the minimization procedure summarized in Chapter 4 to find a line through the transformed $(x, \log(y))$ data. The **correlation coefficient**, r, can be found for exponential-like data; this number indicates "how close" the data lie to an exponential curve. A correlation coefficient close to 1 or -1 indicates a very good exponential fit to the data. A correlation coefficient close to 0 indicates a poor exponential fit to the data.

END *of* CHAPTER EXERCISES

Writing Exponential Functions

1. Write an exponential function that models each of the following scenarios. Be sure to include definitions for the variables used.

   a. In 2000, the population of Mexico was approximately 99.9 million people. The U.S. Census predicts that in the period 2000–2010, the population will grow by 1.2% yearly.[12]

   b. Chlorofluorocarbons (CFCs) have been linked to the destruction of atmospheric ozone, creating the ozone holes over the North and South Poles. In 1988, total world production of CFCs peaked at 1,074,465 metric tons. From that year until 1991, world CFC production was reduced by 26.6% annually.[13]

2. Define variables; then write an exponential function modeling each scenario.

   a. In 1991, world paper production was 243,336,270 metric tons. This amount increased at an average annual rate of 2.79% until 2001.[14]

   b. Indiana manufacturers reduced their releases of toxic chemicals from 134 million pounds in 1991 to 54 million pounds in 2000, a 9.6% annual decrease.[15]

3. For the examples in Table 5-11, assume exponential growth or decay. For each example, do the following:

   a. Find the growth or decay multiplier M.

   b. State the growth or decay rate r in decimal and percent form.

   c. Write an exponential equation and fill in the remainder of the table.

TABLE 5-11

Example A		Example B	
t	A	t	P
0	20	0	120,000
1	24	1	114,000
2		2	
3		3	
4		4	

4. For the examples in Table 5-12, assume exponential growth or decay. For each example, do the following:

   a. Find the multiplier M.

   b. State the value of r in both decimal and percent form.

   c. Write an exponential equation and fill in the remainder of the table.

TABLE 5-12

Example A		Example B	
t	A	t	P
0	1,000	0	50
1	850	1	56
2		2	
3		3	
4		4	

Solving Exponential Equations

5. Solve each exponential equation using logarithms. Round off only on the last step of your solution process.

   a. $4^x = 40$ b. $1.0625^W = 10$

   c. $52(3)^t = 150$ d. $0.6 = 10(4)^x$

6. Solve each exponential equation using logarithms. Round off only on the last step of your solution process.

   a. $5.2^t = 90$ b. $2.85^C = 1.05$

   c. $20(1.043)^x = 16$ d. $25(2.2)^T = 4,000,000$

Applications

7. One substitute for petroleum-based fuels is **biodiesel**, which is derived from oilseed crops such as flax, mustard, and canola (rapeseed). Biodiesel is nontoxic, biodegradable, and can be burned in most standard diesel engines. Compared to gasoline, biodiesel releases 78% less carbon dioxide—a greenhouse gas linked to global warming. In 1999 about 5 million gallons of biodiesel were produced and used in the United States. By 2004 that amount had increased sixfold to 30 million gallons, representing a 43.1% annual increase.[16]

   a. Set up variables; then find an exponential function that relates U.S. biodiesel production to time. Use 1999 as the base year (time = 0).

   b. Using your function, create a table of production values from 1999 to 2010. Then graph the table's values. Label and scale axes.

   c. According to your table, did biodiesel production increase sixfold between 1999 to 2004? How many years will it take until production increases sixfold again?

8. **Permeability** is a measure of how easily a fluid can pass through rock and is measured in units of speed (distance per time). A typical permeability value for fractured rock near the surface of the Earth is about 1.3 mm/hr (millimeters per hour). Permeability decreases deeper into the Earth's crust, because the weight of rock pressing down closes the cracks and fractures, sealing the rock shut. Measurements suggest that permeability decreases about 4.1% for every kilometer of depth into the Earth.[17]

   a. Set up variables; then find the exponential equation between permeability and depth.

   b. Create a table of permeability values for depths from 0 to 50 kilometers by increments of 5 kilometers. Round permeability values to two decimal places. Then graph the table's values. Label and scale axes.

   c. Would you agree with the statement *Permeability at any depth is only 95.9% of the permeability at a depth 1 km closer to the Earth's surface*? Explain.

   d. Suppose that the Deep Disposal Company would like to pump liquid hazardous waste down an injection well deep into the Earth's crust (out of sight and out of mind). The company assumes that if underground rocks have a permeability between 0.5 and 0.6 mm/hr, then there are enough cracks and empty pockets for liquids to seep in, but not spread too fast. To what range of depths can the company drill its wells?

9. Demographics for many of the world's countries can be found on the U.S. Census Bureau's International Data Base Web site. Table 5-13 shows data for two countries that had shrinking populations in the year 2000.[18]

TABLE 5-13

Country	Year 2000 Population ($\times 10^6$)	Annual Decay Information
Belarus	10.4	$r = -0.17\%$
Hungary	10.1	$M = 0.9971$

Source: U.S. Census Bureau

   a. For each country, find an exponential equation that models the statistics given. Use P to represent population in millions, and t for years since 2000.

   b. Which country's population is declining the fastest?

   c. According to your formulas, how many more people will Belarus have in year 2015 as compared to Hungary?

10. Follow-up to previous exercise. Table 5-14 features data for three countries that had growing populations in the year 2000:

TABLE 5-14

Country	Year 2000 Population ($\times 10^6$)	Annual Growth Information
Algeria	30.4	$r = 1.46\%$
Canada	31.3	$r = 0.0102$
Kenya	30.0	$M = 1.0211$

Source: U.S. Census Bureau

a. For each country, find an exponential equation that models the statistics given. Use P to represent population in millions and t for years since 2000.

b. Which country is projected to have more people in 10 years: Algeria, Canada, or Kenya?

c. Suppose that Algeria cut its growth rate to 1.26% in year 2000 and maintained this new growth rate until 2005. How many fewer people would Algeria have had in the year 2005 (as compared to maintaining the year 2000 growth rate)?

11. In 1994, residential vehicles in the United States traveled an estimated 1,793 billion miles. This amount was 282 billion miles more than in 1988.[19]

a. Assuming that total miles grew exponentially between 1988 and 1994, find the yearly growth multiplier and yearly growth rate for total vehicle miles. *Round to four decimal places*.

b. Set up variables and write an equation relating the number of vehicle miles driven to the number of years since 1988.

c. Use your equation from part (b) to predict the total vehicle mileage in 2040, assuming the same growth rate.

d. Sketch the graph of your equation on graph paper for years 1988 to 2040. Use technology for assistance.

e. Estimate the doubling time for vehicle mileage from the graph, and then find it more accurately using your equation.

f. Why might we expect total miles to grow exponentially?

12. In 1992, the total amount of irrigated land in Colorado was 2.65 million acres. By 1997 that amount grew to 2.76 million acres.[20]

a. Assume that total irrigated acreage grew exponentially between the years 1992 and 1997. Find the yearly growth multiplier and yearly growth rate. *Round to four decimal places*.

b. Write a function relating the number of irrigated acres to years since 1992.

c. Use your function to predict total irrigated acreage in the year 2020, assuming the same growth rate.

d. Sketch your function on graph paper over the domain 1992 to 2020. Use a solid curve for known years and a dotted curve for predicted years.

e. Estimate the number of years it takes for irrigated acreage to double (assuming the annual growth rate does not change).

f. Give one factor that might inhibit acreage doubling in the future.

13. In 1996, California harvested 2.27 billion board feet of timber. In 2002, the harvest decreased to 1.69 billion board feet.[21]

a. Assume that the harvest decreased exponentially from 1996 to 2002. Determine the yearly decay multiplier (rounded to four decimal places); then write the exponential equation modeling the timber harvest.

b. By what percent did the timber harvest decrease each year?

c. In 1998, the average single-family home required 14,000 board feet of lumber (Figure 5-12).[22] How many single-family homes could be built with California's timber harvest in that year?

14. The African Elephant Database project estimates that there were 225,200 African elephants (*Loxodonta africana*) in Central Africa in 1995 (Figure 5-13).[23] The estimate for the same region in 1998 was 125,500 elephants. For this project, Central Africa consisted of Cameroon, Central African Republic, Chad, Congo, Democratic Republic of Congo, Equatorial Guinea, and Gabon.

a. Assume that the elephant population was decreasing by the same rate each year. Calculate the yearly decay multiplier for Central African elephants; then write an exponential function modeling the elephant population. Be sure to identify variables that are used in the equation.

b. Without calculations (just look at the data), estimate the half-life for this elephant population. Now calculate the half-life for this population using your answer to part (a). How does your calculation compare with your rough estimate?

c. Suppose that the critical value of the Central African elephant population is 1,000 individuals; if the population falls below this level, the Central African elephants will go extinct. Assuming the same decay rate, when will this population level be reached? Show your work.

d. From 1995 to 1998, there was extensive civil strife in the Congo and the Democratic Republic of Congo. How might civil unrest affect (1) estimates for the number

Figure 5-12 Many homes today are built with manufactured lumber products. Pictured here is a carpenter nailing floor joists made of wood chips and glue. *Photo: Langkamp/Hull.*

Figure 5-13 Poaching and habitat loss have decimated the population of African elephants. *U.S. Fish and Wildlife Service/photo by Gary M. Stolz.*

of elephants in these two countries, (2) the actual number of elephants in these two countries, and (3) the actual number of elephants in the Central African region as a whole?

15. The genetic diversity of domestic livestock is diminishing rapidly and threatening the raw material used in animal breeding programs. According to Keith Hammond, a United Nations senior officer for Animal Genetic Resources, at least 1,500 of the roughly 5,000 domesticated livestock breeds worldwide are now rare. (*Rare* means being represented by less than 20 breeding males on Earth or less than 1,000 breeding females.) Hammond estimates that 5% of livestock breeds disappear each year.[24]

    a. Write an exponential equation that models the total number of living domestic livestock breeds. Be sure to define variables.

    b. Find the half-life of domestic livestock breeds using your equation. Interpret the meaning of your answer.

16. Glyphosate is the active ingredient in some herbicides (weed killers) such as Roundup®, Rodeo®, and Accord®, and typically is sprayed on agricultural fields and forests. When glyphosate lands on plants, it is quickly absorbed through the leaves; in high enough concentration, glyphosate will kill a plant in 24 hours. Glyphosate that lands on soil eventually breaks down, although it can kill microorganisms that are beneficial to the soil. The average half-life of glyphosate in soil is 40 days—sometimes more or less depending on soil conditions. Recommended application rates of glyphosate vary from 0.3 to 4.0 pounds per acre.[25]

    a. With a half-life of 40 days, determine the daily decay rate of glyphosate in soils.

    b. Suppose that 4 pounds of glyphosate are applied to a 1-acre agricultural field, with 1 pound landing on the soil. Compute the number of days until the glyphosate in the soil breaks down to 0.25 pounds. What is special about your answer?

17. The U.S. Census Bureau monitors population changes for each county in the United States. Population data for the three fastest-growing counties from April 1, 2000, to July 1, 2001, are displayed in Table 5-15.[26]

TABLE 5-15

Rank	County	State	April 1, 2000	July 1, 2001	April 1, 2000, to July 1, 2001 Percent Change
1	Douglas	CO	175,766	199,753	13.6
2	Loudoun	VA	169,599	190,903	12.6
3	Forsyth	GA	98,407	110,296	12.1

Source: U.S. Census Bureau

    a. For each county, determine the monthly percent increase from April 1, 2000, to July 1, 2001.

    b. For each county, write an exponential function modeling the population growth. Use April 1, 2000, as the initial date (time = 0).

    c. Determine the population doubling time for each county.

18. As the world seeks alternate means of generating electricity, wind power has emerged as the fastest-growing energy source. In 2004, the generating power of wind turbines was about 40,000 megawatts and was doubling every three years.[27]

    a. Determine the annual percent increase in wind power, rounded to the nearest tenth of a percent.

    b. Write an exponential function that models the growth in wind power since 2004.

    c. How many years will it take until wind power generation equals 160,000 megawatts?

Approximating Exponential Data, Regression, and Correlation

19. Mauna Loa, Hawaii, is considered one of the most favorable locations for measuring atmospheric carbon dioxide (CO_2) levels because of its geographic isolation and the high elevation of the monitoring site—about 11,000 ft above sea level (Figure 5-14). Measurements of atmospheric CO_2 concentrations at Mauna Loa constitute the longest continuous record available. The graph of the April and October CO_2 concentrations for the

Figure 5-14 The observatory on Mauna Loa, Hawaii. *Courtesy of Commander John Bortniak/NOAA Corps Collection.*

period 1959–2002 (Figure 5-15) has two important features. First, atmospheric CO_2 concentrations fluctuate regularly each year, and, second, there has been an overall exponential-like increase in atmospheric CO_2 concentrations since monitoring began.[28]

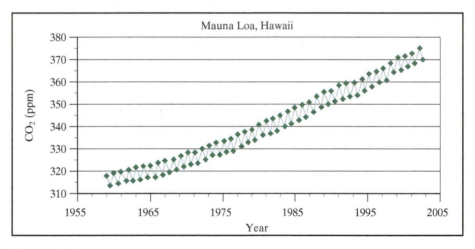

Figure 5-15 CO_2 concentrations recorded in April and October of each year, from 1959 to 2002

a. The least squares regression equation through the Mauna Loa data is $C = 312(1.004)^t$, where C represents the atmospheric CO_2 concentration in parts per million and t the number of years since 1959. What is the annual growth rate in the atmospheric CO_2 concentration?

b. If atmospheric CO_2 concentration continues to increase according to the regression equation, when will the concentration equal 400 ppm?

c. Use the regression equation to estimate the doubling time for the Earth's atmospheric CO_2 concentration.

d. During growth, plants absorb CO_2; when plants decay and rot, CO_2 is released. How might this explain the annual fluctuations in atmospheric CO_2 concentrations at Mauna Loa? *Hint: Notice the first point on the graph, which is April 1959. Each year, the April CO_2 readings are higher than the October CO_2 readings.*

20. Rainbow trout (*Oncorhynchus mykiss*) taken from four different localities along the Spokane River during July, August, and October of 1999 were analyzed for heavy metals by the Washington State Department of Ecology. As part of this study, the length (in millimeters) and weight (in grams) of each trout were measured. Table 5-16 provides a subset of the data.[29]

a. Plot the data on graph paper. How exponential do the data look?

b. Use technology to find the best-fitting exponential regression equation through the data. What is the value of the correlation coefficient? How well can length be used to predict weight?

TABLE 5-16

Length	Weight	Length	Weight
405	715	270	209
455	975	347	432
460	895	259	202
365	540	265	223
390	660	324	353
385	609	337	363
360	557	318	340
392	623	351	506
413	754	502	1,300
395	584		

Source: Washington State Dept of Ecology

c. What weight does the regression equation predict for a rainbow trout 0.6 meters in length?

21. Survival rates of individual organisms can be estimated by monitoring the number of survivors from a single cohort (group) over time. For example, researchers in the Galapagos Islands tagged and followed a cohort of 210 cactus finches (*Geospiza scandens*) for 16 years. The number of surviving finches up to age 15 is given in Table 5-17.[30]

TABLE 5-17

Age	Survivors	Age	Survivors
0	210	8	15
1	91	9	14
2	78	10	11
3	70	11	10
4	65	12	4
5	62	13	3
6	42	14	2
7	23	15	1

Source: *Ecology*

a. Plot the cactus finch data on graph paper. Do the data look exponential?

b. Transform the original data by taking the logarithms of the survivor values. Make a new table showing age and log(survivors) values.

c. Plot the transformed data on graph paper. Draw a line through the data with a straightedge; then find the equation of the line. Write your equation in the form $\log(y) = mx + b$.

d. Use the equation found in part (c) to find the exponential equation through the original data. Show your work.

e. Use technology to find the exponential regression equation through the original data. How does it compare to the equation that you found using the straightedge method?

f. According to the regression equation, what is the annual death rate for cactus finches on the Galapagos Islands?

22. Until the mid-1920s, animal manure was the primary source of nitrogen fertilizer for agricultural fields. Just before World War I, German researcher Fritz Haber discovered a commercial process to convert ammonia gas into fertilizer; this discovery led to the creation of modern-day manufactured fertilizers. The heavy use of commercial nitrogen today is a two-edged sword: Current levels of agricultural production could not be sustained without it, yet studies consistently demonstrate its negative impacts on water quality and ecosystems. Data from Sweden (Table 5-18) illustrate the growth in use of manufactured fertilizer for a typical industrialized country.[31]

TABLE 5-18

Year	Manufactured Fertilizer Input (kg nitrogen/hectare)
1925	26
1935	47
1945	87
1955	164
1965	294
1975	461
1985	476

Source: *Agriculture, Ecosystems, Environment*

a. Reconfigure the data so that the year 1925 becomes the base year (time = 0). Then plot the data on graph paper.

b. How exponential do the data appear? Are there any **outliers** (data that do not fit the trend)?

c. Use the straightedge method to find an exponential equation that approximates the data. Show all steps clearly.

d. Find the best-fitting exponential regression equation using technology.

e. What is the value of the correlation coefficient, and what does it indicate about the data?

f. By what percentage has manufactured fertilizer input increased from 1925 to 1985 in Sweden? What was the average annual percent increase?

23. Dichloro-diphenyl-trichloroethane (DDT) is a pesticide that was once hailed as a miracle chemical because it is toxic to a broad spectrum of insect pests and is inexpensive. In the 1960s an alarm was sounded about DDT and similar chemicals as they were found to be harmful or fatal to many beneficial organisms. Rachel Carson's book *Silent Spring*, published in 1962, issued dire warnings about the dangers of such chemicals. One problem with DDT is that it is **persistent**, meaning that it doesn't break down very readily. Persistent chemicals can become more and more concentrated up the food chain—this is called **biological magnification**. To illustrate this point, consider a pesticide study of a Long Island, New York, salt marsh. In the salt marsh, zooplankton are eaten by minnows, which are eaten by needlefish, which are eaten by osprey. As the feeding level in the food chain increased, so did the DDT concentrations (Table 5-19).[32]

TABLE 5-19

Organism	Consumer Level	DDT Concentration (ppm)
Zooplankton	1	0.04
Minnow	2	0.23
Needlefish	3	2.07
Osprey	4	13.8

Source: *Science*

a. Complete the following table, expressing the DDT concentration in fraction form and in scientific notation. Example: $0.6 \text{ ppm} = \dfrac{0.6}{1,000,000} = 6 \times 10^{-7}$. Then find the logarithm of the DDT concentration.

x = Consumer Level	y = DDT Concentration (fraction)	y = DDT Concentration (scientific notation)	log(y) = log(DDT Concentration)
1			
2			
3			
4			

b. On graph paper, plot the $(x, \log(y))$ data from the table. Do the $(x, \log(y))$ data look linear? How well will an exponential function approximate the original (x, y) data?

c. Find the linear function that approximates the $(x, \log(y))$ data. Write your equation in the form $\log(y) = mx + b$. Convert your equation into the corresponding exponential equation that fits the (x, y) data. Show your work.

d. Examine your exponential model found by the straightedge method. By what factor does DDT concentration change for each one-step increase in consumer level?

e. Using technology, find the best-fitting exponential regression function through the original (x, y) data.

f. What is the value of the correlation coefficient? How well does the food-chain level predict DDT concentration?

g. An osprey is a hawklike predator that consumes mostly fish. Predict the concentration of DDT in a level 5 consumer that feeds on osprey.

24. The American bald eagle population was greatly reduced in the early 1900s by hunting and loss of habitat and further decimated in the twentieth century by chemical contaminants. The pesticide DDT (see the previous exercise) proved extremely harmful to eagles because DDT concentration increases as it is passed up the food chain. The levels of DDT in eagles were high enough to disrupt their calcium absorption, and the females laid eggs whose shells were too thin. Over the last 30 years, the use of DDT and related chemicals has been prohibited in the United States and Canada and the bald eagle population has rebounded dramatically. Figure 5-16 shows the number of pairs of nesting bald eagles in the lower 48 states of the United States, from 1963 to 2000.[33]

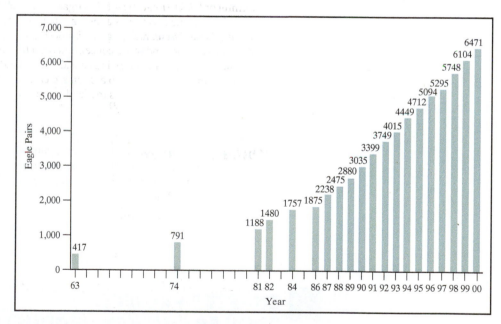

Figure 5-16 Bald eagle pairs in the lower 48 states, 1963–2000

a. In some years (for example, 1965) there are no data given in the diagram. Is it correct to assume that there were 0 breeding pairs for those years?

b. Let 1963 correspond to the base year 0. Use the straightedge method to find an exponential equation through the data. Show all steps.

c. Use technology to find the exponential regression equation. According to the equation, what was the annual percent increase in bald eagle pairs from 1963 to 2000?

d. In which year, approximately, can we expect the number of bald eagle pairs to reach 15,000?

SCIENCE IN DEPTH

Chicken Nation

Chicken production has increased dramatically in the United States over the last 40 years. About 8 billion broiler chickens are raised in the United States annually today. Broiler chickens are typically grown in large metal barns with automatic feed and water systems (Figure 5-17). The chicken feed is very rich, to speed weight gain and reduce time to market. There is concern among environmentalists that high-quality foodstuffs such as corn, wheat, and rice, which can be consumed directly by people, are being fed to chickens instead.

Chicken feed can be spiked with various antibiotics and growth hormones to decrease chicken mortality and speed weight gain. The growth rate of a modern chicken is about twice that of chickens 30 years ago.[34] Environmentalists are concerned about the long-term, cumulative effects on people, particularly young children, who ingest antibodies and hormones residing in the processed and cooked chicken. Unfortunately, and not surprisingly, there is very little concrete information about this issue. Organic (sometimes called "free-range") chicken has recently come onto the market, offering chicken without extra hormones and antibiotics; however, there is an ongoing struggle among organic farmers, agribusinesses, the United States Food and Drug Administration, and the United States Department of Agriculture over which chickens can be labeled organic.

Chickens produce approximately 1 pound of litter (manure, feathers, uneaten food, and bedding) per pound of chicken. Treatment and disposal of this waste, which is extremely rich in nitrogen and phosphorus, is an important environmental issue. Typically, chicken litter has been spread on agricultural fields as fertilizer; however, both groundwater and surface water can become polluted from rainfall runoff from these fields. High quantities of nitrogen and phosphorus will create algal blooms that can seriously alter and degrade freshwater ecosystems.

Chickens are often shipped some distance to centralized butchering and rendering facilities, whose automation and economy of scale keep processing costs low. A modern rendering plant can process up to 180 birds per minute on a single inspection line.[35] Leftover carcasses, viscera, and entrails are processed into animal feed for cows, pigs, cats, dogs, and even other chickens.

Contamination of both chicken parts for human consumption and processed animal food is very common. Salmonella, *E. coli*, listeria, and campylobacter are some of the more notable pathogens.[36] These pathogens can pose grave risks to children, the elderly, and those with compromised immune systems and can cause serious illness among even healthy adults. Contamination of raw chicken is routine. In 1985, the Centers for Disease Control and Prevention reported about 57,000 cases of salmonella infection; however, the actual number of salmonella infections is much higher, around 2 million cases per year. Proper handling and cooking of raw chicken is essential.

Figure 5-17 Poultry in an industrial-sized feed barn. *Photo by Jeff Vanuga, USDA National Resources Conservation Service.*

CHAPTER PROJECT:
BROILER CHICKEN PRODUCTION

In the 1930s, politicians' promise of "a chicken in every pot" reflected the scarcity of chicken in the U.S. diet. What sort of trends in U.S. chicken production have there been since that time? How has U.S. population growth driven those trends? In the companion project for Chapter 5 found at the text's Web site (**enviromath.com**), you will model the change in both chicken production and human population in the United States since 1960. You'll investigate alternative explanations for the dramatic increase in U.S. chicken production and consumption. And you'll explore the environmental implications of 8 billion chickens on planet Earth (and counting!). Check it out!

NOTES

[1] U.S. Dept of Agriculture, National Agricultural Statistics Service, *Historical Track Records: April 2004*, http://www.usda.gov/nass/

[2] U.S. Dept of Commerce, *Status Review of Chinook Salmon from Washington, Idaho, Oregon, and California*, by J. M. Myers et al., NOAA Tech. Memo. NMFS-NWFSC-35 (1998), 443, http://www.nwfsc.noaa.gov/publications/techmemos/tm35/index.htm#toc

[3] Minnesota Office of Environmental Assistance SCORE Database, http://www.moea.state.mn.us/lc/score.cfm

[4] U.S. Census Bureau International Database, http://www.census.gov/ipc/www/idbnew.html

[5] U.S. Dept of Health and Human Services, National Center for Health Statistics, *Health, United States, 2004*, Table 25, http://www.cdc.gov/nchs/hus.htm

[6] U.S. Census Bureau, *Table ST-2001EST-01—Time Series of State Population Estimates: April 1, 2000 to July 1, 2001*, http://www.census.gov/popest/archives/2000s/vintage_2001/ST-2001EST-01.html

[7] The British Thoracic Society, "Lung Cancer," http://www.brit-thoracic.org.uk/lung_cancer.html See also Chest X-Ray: Your Thoracic Imaging Resource, "Doubling Time," http://www.chestx-ray.com/SPN/DoublingTime.html

[8] Cecie Starr and Ralph Taggart, *Biology: The Unity And Diversity Of Life*, 6th ed. (Belmont, CA: Wadsworth Publishing Co., 1992), 21.

[9] U.S. Census Bureau International Database.

[10] Robert Caputo, "A Population Exploding," *National Geographic*, December 1988, 918.

[11] BP Amoco, *Statistical Review of World Energy 2000*, 39.

[12] U.S. Census Bureau International Database.

[13] Alternative Fluorocarbons Environmental Acceptability Study, "Production and Sales of Fluorocarbons," http://www.afeas.org/

[14] Based on data from Food and Agricultural Organization of the United Nations, *FAOSTAT Forestry Database 2004*, http://www.fao.org/forestry/site/6768/en

[15] Indiana Dept of Environmental Management, *Indiana State of the Environment Report 2002*, http://www.in.gov/idem/soe2002/

[16] Kristin Dizon, "Pumped Up About Cleaner Fuel," *Seattle Post-Intelligencer*, February 18, 2005.

[17] C. E. Manning and S. E. Ingebritsen, "Permeability of the Continental Crust: Constraints from Heat Flow Models and Metamorphic Systems," *Reviews of Geophysics* 37 (1999): 127–150.

[18] U.S. Census Bureau International Database.

[19] U.S. Dept of Energy, Energy Information Administration, "Chapter 3: Vehicle-Miles Traveled," http://www.eia.doe.gov/emeu/rtecs/chapter3.html

[20] U.S. Dept of Agriculture, Economic Research Service Database, "Colorado State Fact Sheet," http://www.ers.usda.gov/StateFacts/CO.htm

[21] California Dept of Commerce, *California Statistical Abstract 2003*, Table G-27, http://www.dof.ca.gov/html/fs%5Fdata/stat%2Dabs/sa%5Fhome.htm

[22] George Couch, "Shaping Tomorrow's Building Blocks," *Wisconsin Natural Resources Magazine*, December 2003, http://www.wnrmag.com/stories/2003/dec03/forprod.htm

[23] International Union for Conservation of Nature and Natural Resources, *African Elephant Database 1998*, http://www.iucn.org/themes/ssc/sgs/afesg/aed/aed98.html

[24] Janet Raloff, "Dying Breeds: Livestock are Developing a Largely Unrecognized Biodiversity Crisis," *Science News Online*, October 4, 1997, http://www.sciencenews.org/pages/sn_arc97/10_4_97/bob1.htm

[25] U.S. Dept of Agriculture, U.S. Forest Service, *Glyphosate Pesticide Fact Sheet*, prepared by Information Ventures, Inc., http://infoventures.com/e-hlth/pestcide/glyphos.html

[26] U.S. Census Bureau, *Table CO-EST2001-11—100 Fastest Growing Counties by Percent Change: April 1, 2000 to July 1, 2001*, http://www.census.gov/popest/archives/2000s/vintage_2001/CO-EST2001-11.html

[27] Jim Motavalli, "Catching the Wind: The World's Fastest-Growing Renewable Energy Source Is Coming of Age," *E: The Environmental Magazine*, January/February 2005, 27–39.

[28] C. D. Keeling and T. P. Whorf, "Atmospheric CO_2 Records from Sites in the SIO Air Sampling Network," *Online TRENDS: A Compendium of Data on Global Change*, http://cdiac.ornl.gov/trends/trends.htm; *Note*: The April 1964 value was estimated from the May value of that same year.

[29]Washington State Dept of Ecology, *Results from Analyzing Metals in 1999 Spokane River Fish and Crayfish Samples*, by A. Johnson, Publication no. 00-03-017 (2000). http://www.ecy.wa.gov/biblio/0003017.html

[30]Peter R. Grant and B. Rosemary Grant, "Demography and the Genetically Effective Sizes of Two Populations of Darwin's Finches," *Ecology* 73: 766–784.

[31]M. Hoffman et al., "Leaching of Nitrogen in Swedish Agriculture—A Historical Perspective," *Agriculture, Ecosystems & Environment* 80, no. 3, 277–290.

[32]George M. Woodwell et al., "DDT Residues in an East Coast Estuary: A Case of Biological Concentration of a Persistent Insecticide," *Science* 156: 821–824.

[33]U.S. Fish and Wildlife Service Region 3, "How Many Bald Eagles Are There?," http://midwest.fws.gov/eagle/population/2000chtofprs.html

[34]E. Dransfield and A. A. Sosnicki, "Relationship Between Muscle Growth And Poultry Meat Quality," *Poultry Science* 78 (1999): 743–746.

[35]K. Chao and Y. Chen, "High-Speed Poultry Inspection Using Visible/Near-Infrared Spectrophotometer," *SPIE Proceedings: Monitoring Food, Safety, Agriculture, And Plant Health* 5271 (2003), 51–61.

[36]U.S. Centers For Disease Control and Prevention, Division of Bacterial and Mycotic Diseases; http://www.cdc.gov/ncidod/dbmd/diseaseinfo/default.htm

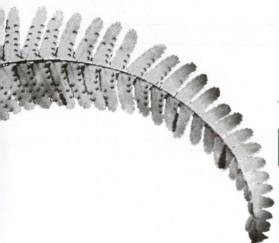

6

Power Functions

The number of bird species on different islands near Papua New Guinea in the southwest Pacific Ocean has been compared to the areas of those islands.[1] As you might expect, the bigger the island, the more bird species. But the relationship between area and number of species is neither linear nor exponential. The observed relationship between the number of bird species S and island area A (in square kilometers) is given by a **power function**:

$$S = 21.83A^{0.18}$$

According to this power function, an island with an area of 100 km² should have approximately 50 species of birds. The relationship between the number of bird species and the area of islands is useful for determining the impact of habitat destruction by invasive species or human activity. If we know an island's area, we can calculate the *expected* number of birds using the power function, and then compare our calculation to the *actual* number of species, to quantify the impact of habitat destruction and other factors (Figure 6-1).

In Chapters 4 and 5, you learned about two common functions, the linear and exponential functions. In this chapter, we introduce a third function called the **power function** or **power law**. We start with some simple power functions and then apply the power function model to real data. We then focus on a special type of frequency histogram with power law behavior and explore its relationship to the topic of fractals in environmental science.

Figure 6-1 Tropical birds are at risk from loss of habitat and introduced predators. *Courtesy of PD Photo.*

BASIC POWER FUNCTIONS

Recall the general form of the exponential function:

$$y = y_0M^x$$

The variable x is found as an exponent; hence the name, exponential function. Exponential functions have M (the multiplier) raised to a variable exponent. In the exponential function, if $x = 0$, then $y = y_0$, the initial value.

Power laws or power functions, on the other hand, have the variable x raised to a power c:

$$y = kx^c$$

In the power function, x and y are the independent and dependent variables, and k and c are constants. The constant c is the power of the variable x. Let's explore the shape of a power function when c is positive. Three examples of simple power functions that have $k = 1$ and $c > 0$ are given in Figure 6-2.

We observe that all three functions are increasing, which is always the case when $c > 0$. However, the shape of each of the functions is different. Power functions with

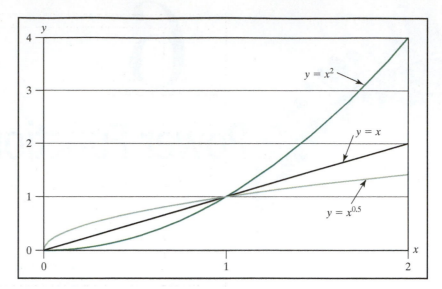

Figure 6-2 Three simple increasing power functions

$c > 1$, such as $y = x^2$, curve upward. Those functions with $0 < c < 1$, such as $y = x^{0.5}$, curve downward. When $c = 1$, the function is linear ($y = x$). All power functions with $c > 0$ pass through the origin $(0, 0)$, for if $x = 0$, then $y = 0$:

$$y = k \times 0^c = k \times 0 = 0$$

EXAMPLE 6-1 Shrimp Ponds

A growing source of food is **aquaculture**, the artificial rearing of marine and freshwater organisms. Suppose there are several shrimp ponds circular in shape. Each pond has a different radius (5 meters, 10 meters, and 20 meters). What are the areas of these shrimp ponds?

Solution The function relating area A to radius r for a circle is

$$A = \pi r^2$$

The equation for the area of a circle is a power function because the radius r (the independent variable) is raised to the power 2. The value of k in this equation is π (approximately 3.14159). The radii and corresponding areas for the three shrimp ponds are given in Table 6-1.

TABLE 6-1

Radius	Area
5 m	$\approx 79\,\text{m}^2$
10 m	$\approx 314\,\text{m}^2$
20 m	$\approx 1{,}257\,\text{m}^2$

Notice that as the radius doubles, the area increases by a factor of 4 (check this on your calculator). The relationship between radius and area is graphed in Figure 6-3. The graph of the function passes through $(0, 0)$, is increasing, and curves upward.

What happens if the power c is negative in a power function? Three examples of simple power functions with $c < 0$ and $k = 1$ are shown in Figure 6-4. These examples illustrate that power functions with negative powers are both decreasing and curved upward. These functions also get closer and closer to the x- and y-axes but never touch them.

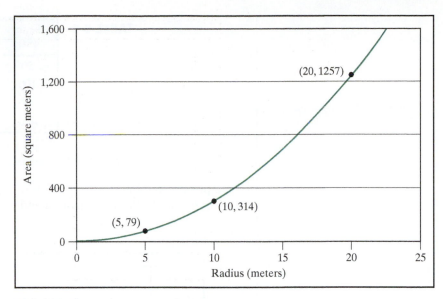

Figure 6-3 Pond area versus radius

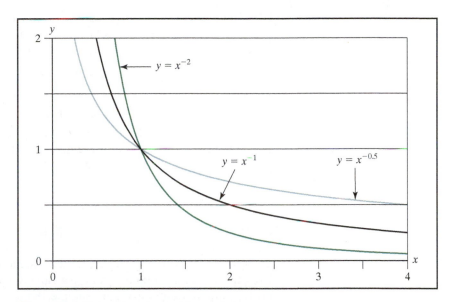

Figure 6-4 Power functions with $c < 0$

Figure 6-5 Transmission lines produce electromagnetic fields that decrease in strength with increasing distance to the lines. *U.S. Dept. of Energy.*

EXAMPLE 6-2   Magnetic Fields and Distance

Magnetic fields are produced by currents passing through electrical transmission lines (Figure 6-5). There is concern about a possible relationship between human health and strong electromagnetic fields associated with large electric lines. Engineers with the National Grid of the United Kingdom measured magnetic fields for one year at 43 different localities, and averaged those results.[2] They discovered that the strength of the magnetic field *MF* (in microteslas or μT) is related to the distance *d* (in meters) from the electric lines according to the power law:

$$MF = 1{,}217d^{-1.98}$$

What are the magnetic field strengths at 50, 100, 200, and 400 meters from the electrical lines? What happens to the field strength as the distance is doubled?

Solution   The field strengths at the given distances are listed in Table 6-2. Examination of the values in the table reveals that the field strength decreases by a factor of about 4 as the distance from the transmission lines doubles. The graph

TABLE 6-2	**Distance (m)**	**Field Strength (μT)**
	50	0.526
	100	0.133
	200	0.034
	400	0.009

Source: U.K. National Grid

of the power function is given in Figure 6-6. Note that the function is decreasing and approaches both the horizontal and vertical axes of the graph.

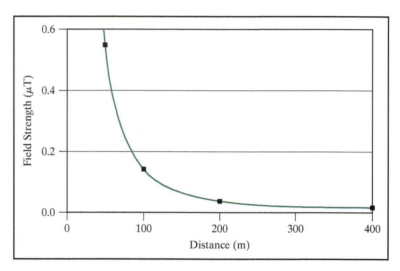

Figure 6-6 Magnetic field strength versus distance

SOLVING POWER EQUATIONS

In the previous examples, we substituted a numerical value for *x* and calculated *y*. However, there may be situations where a value for *y* is given and we must determine the corresponding value for *x*. For example, the diameters and heights of about 300 Douglas firs (*Pseudotsuga menziesii*) in southwest Oregon were measured (Figure 6-7).[3] The relationship between diameter *D* and height *H* (both in meters) of those trees can be described by the power function:

$$D = 0.003H^{1.4567}$$

Suppose we encounter a large Douglas fir on a forest hike and measure its diameter of 0.6 meters. How tall is this tree? To answer this question, we substitute the diameter into the formula:

$$0.6 = 0.003H^{1.4567}$$

To solve this equation, first divide both sides by 0.003:

$$200 = H^{1.4567}$$

To isolate the height *H*, raise both sides to the 1/1.4567 power and simplify:

$$(200)^{1/1.4567} = (H^{1.4567})^{1/1.4567}$$

$$38 \text{ meters} \approx H$$

A Douglas fir that is 0.6 m in diameter should be approximately 38 meters tall.

Figure 6-7 Student measuring the height of a Douglas fir. *Photo: Langkamp/Hull.*

EXAMPLE 6-3   Drainage Area and Stream Length

Streams and rivers form a branching network within a **drainage basin** (Figure 6-8). All of the rain that falls within a drainage basin eventually works its way into the master stream, which is the longest stream in the drainage basin. The area of the drainage basin A and the length L of the largest stream were measured for about 75 small drainage basins in the eastern United States.[4] Basin area (in square miles) and stream length (in miles) are related by the power law formula:

$$L = 1.40A^{0.568}$$

If the largest stream is 55 miles long, what is the area of the corresponding drainage basin?

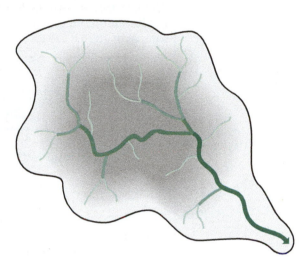

Figure 6-8 Idealized stream network in a drainage basin

Solution Substitute 55 miles for L and then divide each side by 1.40:

$$55 = 1.40A^{0.568}$$

$$\frac{55}{1.40} = A^{0.568}$$

$$39.286 \approx A^{0.568}$$

Isolate A by taking the 1/0.568 power of both sides:

$$(39.286)^{1/0.568} \approx (A^{0.568})^{1/0.568}$$

$$641 \text{ mi}^2 \approx A$$

A drainage basin with a master stream 55 miles long is approximately 641 square miles in area. The power law relationship between drainage area and stream length could be very useful in evaluating different habitats using topographic maps of drainage basins.

APPROXIMATING POWER-LIKE DATA

In Chapters 4 and 5 we developed two basic methods to approximate almost-linear and almost-exponential data sets. The two methods are the **straightedge method** and the **least squares regression**. We now apply those same methods to model or approximate a data set by a power function.

Straightedge Method

Modeling exponential-like data using the straightedge method in Chapter 5 started with a transformation of the y-values by taking their logarithms. To model a data set that shows a power-like behavior, *both* variables are transformed by taking the logarithm of x and the logarithm of y. We note the following about these transformed data:

- If (x, y) data fit a power model perfectly, then $(\log(x), \log(y))$ data are perfectly linear. The converse of this statement is also true; if $(\log(x), \log(y))$ data are perfectly linear, then (x, y) data are "perfectly power."
- If (x, y) data *almost* fit a power function, then $(\log(x), \log(y))$ data are *almost* linear. The converse statement is true as well.

EXAMPLE 6-4 Automobile Engine Size and Fuel Economy

The U.S. Department of Energy has compiled data on engine size and fuel economy for automobiles manufactured in the 1999 model year.[5] Data for an arbitrary sample of 18 cars are given in Table 6-3, with engine size in liters and fuel economy in miles per gallon.

TABLE 6-3

Engine Size (liters)	Fuel Economy (mpg)	Engine Size (liters)	Fuel Economy (mpg)
3.2	26	2.3	30
2.8	26	3.5	23
2.5	27	2.5	26
2.8	26	3.4	25
5.7	28	1.5	41
5.7	13	2.3	31
1.8	29	4.3	25
5.0	23	6.8	16
6.0	19	1.0	47

Source: U.S. Dept of Energy

The scatterplot of the data (Figure 6-9) shows a trend of decreasing fuel economy as engine size increases, suggesting that a power function model with a negative exponent might be a reasonable match.

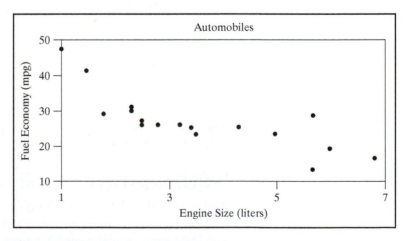

Figure 6-9 Engine size and fuel economy

To find the power function that models these data using the straightedge method, first transform the data using logarithms, and then find a straight line that fits the logged data. Then work "backward" to find the original power function. The logarithms of engine size and fuel economy are given in Table 6-4.

TABLE 6-4	log (liters)	log (mpg)	log (liters)	log (mpg)
	0.505	1.415	0.362	1.477
	0.447	1.415	0.544	1.362
	0.398	1.431	0.398	1.415
	0.447	1.415	0.531	1.398
	0.756	1.447	0.176	1.613
	0.756	1.114	0.362	1.491
	0.255	1.462	0.633	1.398
	0.699	1.362	0.833	1.204
	0.778	1.279	0.000	1.672

The plot of the transformed data (Figure 6-10) shows a reasonably linear trend, though there is some scatter in the data at large engine sizes. Now draw a line using a straightedge through the almost-linear transformed data and pick two points that lie on that line (shown as open circles). The two points need not be part of the data set. In this example, we chose two points that fall on grid intersections, to make life easier! The two points have coordinates of (0.2, 1.6) and (0.8, 1.2).

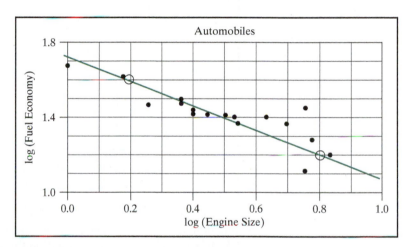

Figure 6-10 Scatterplot of the logarithm of the fuel economy versus the logarithm of the engine size

Using the two points, calculate the slope m and vertical intercept b of the straightedge line. The slope is

$$m = \frac{1.2 - 1.6}{0.8 - 0.2} = -0.67$$

Let ES stand for engine size (the independent variable) and FE stand for fuel economy. The equation of the straight line through the transformed data is

$$\log(FE) = b + m\log(ES)$$

To determine the vertical intercept b, substitute the slope and one of the points into this equation. Choosing the second point (0.8, 1.2), we find that

$$1.2 = b + (-0.67)(0.8)$$
$$1.74 = b$$

The equation for the line passing through the transformed automobile data is therefore

$$\log(FE) = 1.74 - 0.67\log(ES)$$

We now take this linear equation through the transformed data and determine the power function for the original, untransformed data. In simple terms, our goal in the next series of steps is to "undo the logarithms." Start by raising 10 to each side of the linear equation using Property M in Chapter 5 (Table 5-4):

$$10^{\log(FE)} = 10^{1.74 - 0.67 \log(ES)}$$

Use Property A to rearrange the right-hand side:

$$10^{\log(FE)} = 10^{1.74} 10^{-0.67 \log(ES)}$$

Then use Property I to move the power (-0.67) "back on top":

$$10^{\log(FE)} = 10^{1.74} 10^{\log(ES^{-0.67})}$$

Finally, use Property L to undo the logarithms:

$$FE = 10^{1.74}(ES^{-0.67})$$

Simplifying yields the power function relating fuel economy FE to engine size ES:

$$FE \approx 55.0 ES^{-0.67}$$

We plot this power function along with the original data (Figure 6-11). The power function model seems to be a good fit to the data.

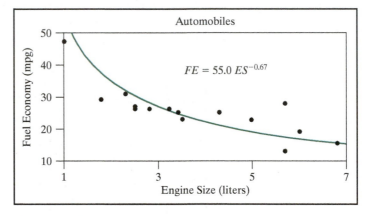

Figure 6-11   Power function from the straightedge method, along with the original engine size and fuel economy data

The process to convert a linear function through the transformed data into a power function through the original data may seem tedious, but there are two major shortcuts. The general form of a linear equation through the transformed data $(\log(x), \log(y))$ is

$$\log(y) = b + m \log(x)$$

The power function through the original data (x, y) has the form

$$y = kx^c$$

In our example, we found that $c = m$. Similarly, we found that $k = 10^{1.74}$ or $k = 10^b$. The general relationship between the constants in the two equations is therefore

$$\boxed{c = m \text{ and } k = 10^b}$$

The steps of the straightedge method for finding a power function to model nearly power law data can be summarized as follows:

1. Transform the data by taking the logarithms of both x and y.
2. Plot $\log(y)$ versus $\log(x)$. Using a straightedge, approximate the transformed data with a straight line.
3. The equation of the line has the form $\log(y) = b + m \log(x)$. Use two points on the line to calculate the slope m and the vertical intercept b.
4. Knowing that $c = m$ and $k = 10^b$, determine the power function of the form $y = kx^c$.

Least Squares Regression

As with linear and exponential data, we can use technology to find the power function that best fits a power-like data set. Programs on computers or calculators first transform the (x, y) data by taking the logarithms of both x and y (just as we did manually), then use **least squares regression** to find the best-fitting line through the transformed data. Recall from Chapter 4 that least squares regression minimizes the sum of the vertical distances from the data points to a line. This minimization procedure yields the best-fitting line through the data. The computer or calculator program then takes the slope and vertical intercept of the best-fitting line and determines the corresponding power function through the original, untransformed data using the mathematical steps described previously.

EXAMPLE 6-5 Power Regression of Automobile Data

We can use the same engine size and fuel economy data as in the previous example and compare the power regression results with those of the straightedge method. Output from the TI-83/84 calculator is shown below (detailed technology instructions to follow in the Using Technology: Power Regression section).

```
PwrReg
 y=a*x^b
 a=44.66839777
 b=-.4875453674
 r²=.7339896815
 r=-.8567319777
```

The calculator uses the variable a instead of k, and b instead of c in the power law formula. The best-fitting regression equation is approximately

$$y = 44.7x^{-0.4875}$$

Compare the regression equation with the equation from the straightedge method:

$$FE = 55.0\,ES^{-0.67}$$

The constants k and c from the two different methods are similar. We would choose the regression equation as being more accurate.

The calculator also provides the correlation coefficient $r = -0.857$. The negative sign indicates a general decrease in both the transformed and original data, which is correct. A value close to -1 indicates a fairly strong correlation between engine size and fuel economy. In other words, automobile engine size is a reasonable predictor of fuel economy for this model. See Chapter 4 for further discussion on the correlation coefficient.

USING TECHNOLOGY: POWER REGRESSION

Instructions for obtaining the power law regression equation on the TI-83/84 are given below. Visit the text Web site for instructions on how to use online computer applications that perform this task: **enviromath.com**.

Power Regression on the TI-83/84

Select **STAT>EDIT**, enter the 18 values of engine size and fuel economy (Table 6-3) in L1 and L2, respectively. If the number of values in L1 and L2 don't match, you will eventually get an error message: **DIM MISMATCH.**

L1	L2	L3	3
3.2	26		
2.8	26		
2.5	27		
2.8	26		
5.7	20		
5.7	13		
1.8	29		

L3(1)=

Select **STAT>CALC**, scroll down to **PwrReg**, press **ENTER** and type in **L1, L2** (don't forget the comma). Press **ENTER** to display the regression parameters.

```
PwrReg
 y=a*x^b
 a=44.66839777
 b=-.4875453674
 r²=.7339896815
 r=-.8567319777
```

POWER LAW FREQUENCY DISTRIBUTIONS

Histograms that illustrate the frequency of measured quantities are discussed in Chapter 3. Recall that the height of a histogram bar indicates the **frequency** or how many measurements fall within that **bin**. The bin boundaries are shown along the horizontal axis. When the measurements cluster near a central value, the corresponding frequency histogram is hump shaped. For example, temperatures of open-canopy streams in Washington State cluster near 14–20°C, with a few values that are substantially higher or lower (Figure 6-12).

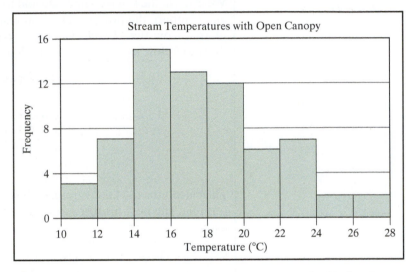

Figure 6-12 Foothills stream temperatures in June 1999. Notice the hump-shaped distribution of frequencies.

Other data sets show different distributions of frequencies with size or value. For example, there is a fairly steady decrease in the number of Bangladeshi females from the youngest cohort to the eldest. The frequency histogram of Bangladeshi females illustrates this uniform decrease from young to old (Figure 6-13).

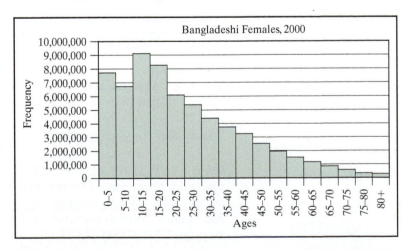

Figure 6-13 Frequency distribution of Bangladeshi females in the year 2000. Notice the fairly uniform decrease in frequency from bin to bin.

The preceding examples exhibit just two of the many possible histogram patterns. Another histogram shape (a hypothetical one) is shown in Figure 6-14. This histogram shows frequency decreasing very rapidly with increasing size. The data set contains *many small values, some medium-sized values, and few large values.*

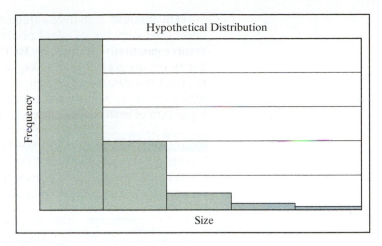

Figure 6-14   Frequency histogram with rapid drop-off

This special pattern of frequencies ("many small, few big") is quite common among both natural and artificial phenomena. Consider, for example, automobile accidents on the freeway. The most common accidents are small fender-benders, whereas multivehicle crashes are infrequent, and 10-car pileups are rather rare. This pattern of event sizes is typical of many environmental phenomena such as storms, earthquakes, wildfires, landslides, oil spills, nuclear accidents, and meteorite impacts. You can probably think of a different phenomenon that shows this same kind of frequency distribution of sizes.

Let's redraw Figure 6-14 by plotting a point at the top of each bar, directly at the midpoint of the bin. Each bar is thus represented by an (x, y) pair of the form (*bin midpoint, frequency*). The resulting set of points is a scatterplot of frequency versus size (Figure 6-15). One can imagine fitting an equation through these data points. The graph of this equation will have a steep negative slope at low values of size, and a shallow negative slope at high values of size. In short, the shape of the graph resembles that of a decreasing power function with $c < 0$.

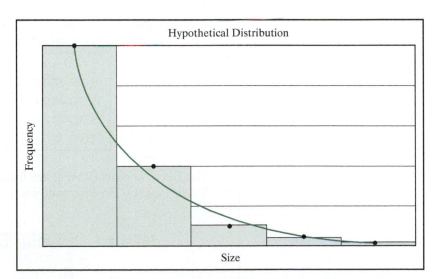

Figure 6-15   Histogram represented as a scatterplot of frequency versus bin midpoint. The solid line is the best-fitting power function.

Suppose we determine both the best-fitting power regression and correlation coefficient r using technology. The correlation coefficient tells us the strength of the power law trend in the data. If the data fit a decreasing power function well, with $r \approx -1$, then the distribution is referred to as a **power law frequency distribution**. Many environmental measurements, such as the acres burned by forest fires in a season or the volumes of oil spills by tankers, show a power law distribution of sizes.

The method to analyze a frequency distribution for power law behavior has one additional step. It's standard procedure to reformulate each frequency value as a **reverse cumulative frequency (RCF)**. The reverse cumulative frequency is the number of events greater than or equal to the size of the event (see Example 6-6). Expressing the data as reverse cumulative frequencies eliminates frequencies of zero, which cause problems when transforming the data with logarithms. Recall that the logarithm of zero is not defined.

EXAMPLE 6-6   Populations of Ohio Municipalities

Table 6-5 displays the year 2000 population values for 25 cities, towns, and villages in Wood County, Ohio.[6] Towns with less than 100 people have been omitted, and the data have been sorted by increasing population.

TABLE 6-5

Town	Pop.	Town	Pop.	Town	Pop.
Bairdstown	130	Haskins	638	Pemberville	1,365
Milton Center	195	Bloomdale	724	Weston	1,659
Custar	208	Fostoria	842	Walbridge	2,546
Hoytville	296	Wayne	842	North Baltimore	3,361
Tontogany	364	Luckey	998	Northwood	5,471
Portage	428	Grand Rapids	1,002	Rossford	6,406
Jerry City	453	Millbury	1,161	Perrysburg	16,945
Cygnet	564	Bradner	1,171	Bowling Green	29,636
Risingsun	620				

Source: Ohio State Univ. Extension Data Center

We collate the 25 values into 5 bins, each with a bin width of 6,000 people (Table 6-6). Notice that the fourth bin has a frequency of zero. To determine the reverse cumulative frequency for the first bin, add the frequencies starting with the first bin ($22 + 1 + 1 + 0 + 1 = 25$). To determine the RCF for the second bin, add the frequencies starting with the second bin ($1 + 1 + 0 + 1 = 3$). And so on.

TABLE 6-6

Bin (population)	Frequency	Reverse Cumulative Frequency (RCF)
0–6,000	22	$22 + 1 + 1 + 0 + 1 = 25$
6,000–12,000	1	$1 + 1 + 0 + 1 = 3$
12,000–18,000	1	$1 + 0 + 1 = 2$
18,000–24,000	0	$0 + 1 = 1$
24,000–30,000	1	$1 = 1$

The bin midpoints (BMPs) are listed with the reverse cumulative frequencies (RCFs) in Table 6-7. To determine a bin midpoint, add the bin boundaries and divide by 2.

TABLE 6-7

Bin Midpoint (BMP)	Reverse Cumulative Frequency (RCF)
3,000	25
9,000	3
15,000	2
21,000	1
27,000	1

We are now ready to construct a scatterplot of reverse cumulative frequency RCF versus bin midpoint BMP of the population data (Figure 6-16). Using technology, the best-fitting power regression can be calculated through these points:

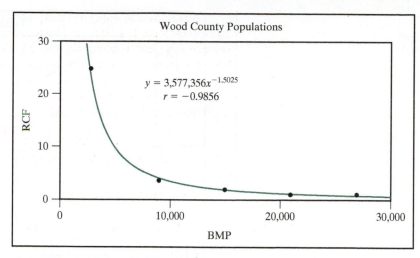

Figure 6-16 Scatterplot for the Wood County population data. The data show a power law frequency distribution of population sizes.

$$RCF = 3{,}577{,}356 BMP^{-1.5025}$$

A correlation coefficient of -0.9856 for the Wood County data is very close to -1, suggesting a very good fit to a power law model. Municipalities in Wood County show a power law frequency distribution of populations, with many small villages, a few towns, and the occasional large city. Such a frequency distribution may have arisen from settlement patterns in U.S. history, with larger cities forming near navigation routes or areas with plentiful natural resources. Demographic shifts from country to city may also influence the distribution of population sizes.

Suppose that the power function based on the smaller events holds true regardless of event magnitude. We could analyze the frequent small events, for which we often have information, and then make predictions about the very large, rare events. The following example illustrates this extrapolation.

EXAMPLE 6-7 Worldwide Tanker Spills

Environment Canada has compiled statistics on major petroleum spills by ocean-going tankers from 1993 to 1997. The data include worldwide spills with volumes greater than 1,000 barrels of petroleum; the maximum spill size is just under 600,000 barrels.[7] The original data (not shown here) were collated into 6 bins, each with a bin width of 100,000 barrels. The first bin is 1,000 to 101,000 barrels, the second bin is 101,000 to 201,000 barrels, and so on. The bin midpoints (BMPs), the reverse cumulative frequencies (RCFs), and the logarithms of those two quantities are given in Table 6-8.

TABLE 6-8

BMP	RCF	log (BMP)	log (RCF)
51,000	61	4.71	1.79
151,000	10	5.18	1.00
251,000	5	5.40	0.70
351,000	2	5.55	0.30
451,000	2	5.65	0.30
551,000	1	5.74	0.00

Source: Environment Canada

A scatterplot of the logarithms of the reverse cumulative frequencies versus the logarithms of the bin midpoints (Table 6-8) reveals that the transformed data show a very strong linear relationship between the two variables (Figure 6-17). Hence the data on petroleum spills by oil tankers can be characterized by a power law frequency distribution.

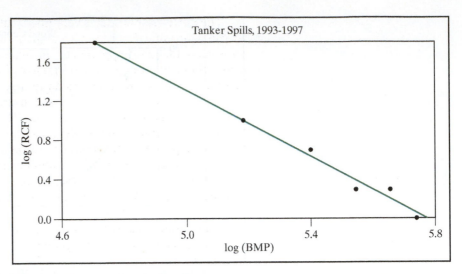

Figure 6-17   Scatterplot of the transformed data for oil tanker spills. The strong linear correlation indicates a power law frequency distribution.

Using the straightedge method, a line drawn through the data points in Figure 6-17 has the equation

$$\log(RCF) = 9.7 - 1.7\log(BMP)$$

The accuracy of this equation can be checked algebraically by examining the bin of 401,000 to 501,000 barrels. We substitute a bin midpoint of 451,000 barrels of petroleum into the equation:

$$\log(RCF) = 9.7 - 1.7\log(451,000)$$
$$\approx 0.09$$

To undo the logarithm, raise 10 to each side of the last equation using Property M in Chapter 5 (Table 5-4):

$$10^{\log(RCF)} = 10^{0.09}$$
$$RCF = 10^{0.09}$$

The reverse cumulative frequency of this size of spill is $10^{0.09} \approx 1$ spill. That is, the equation predicts that there was one spill worldwide in the interval 401,000–501,000 barrels or above during the period 1993 to 1997. According to Table 6-8, there were actually two spills greater than or equal to this size during the five-year period; the model checks out.

Let's examine a bin midpoint of 2,251,000 barrels, corresponding to a catastrophic spill between 2,201,000 to 2,301,000 barrels of petroleum. This spill is about four times larger than the biggest spill in Table 6-8. If the same power law also applies to larger spills, we can predict a reverse cumulative frequency for this catastrophe:

$$\log(RCF) = 9.7 - 1.7\log(2,251,000)$$
$$\approx -1.10$$

The reverse cumulative frequency is $10^{-1.10} \approx 0.08$ spills of this size or greater in a five-year period. To rephrase this prediction in a more friendly form, solve the following proportion for t:

$$\frac{0.08\,\text{spills}}{5\,\text{years}} = \frac{1\,\text{spill}}{t}$$

Cross multiplying and isolating t yields $t \approx 63$ years.

In other words, the power law model predicts a tanker spill of about 2.25 million barrels or greater once every 63 years. The largest tanker spill in history was the 1979 *Atlantic Empress* release of approximately 2 million barrels of crude oil.

The power law model is very powerful! By measuring a bunch of little oil spills, we can determine the frequency of giant ones, if the power law model is valid. Giant spills need not have happened to calculate their frequency of occurrence, if the spills truly adhere to a power law frequency distribution. Furthermore, the lack of giant oil spills in the past does not mean they won't happen in the future.

The frequency of petroleum accidents from tankers around the world is strongly dependent upon the size of the spill. This relationship is both unusual and a little creepy, if you think about it. We know that few of the petroleum spills from tankers had anything to do with the others (although tanker collisions do happen). We also know that the tankers were not communicating in some way. And the tankers had no memory of previous leaks. Yet there's something about the *system* of human error, lack of inspections, bad weather, and rusting single-hulled tankers that produces petroleum spills in a very distinctive fashion.

Many industries would argue that the abundance of small oil spills, minor radioactive discharges, or "manageable" sewage overflows indicates that small events are *characteristic* of their industry. Industries might argue that there's no need to worry about big events because the overwhelming majority of events are small. On the contrary! If these industrial incidents obey a power law frequency distribution, then there is no characteristic event size. It's just a matter of time before the "big one" happens.

POWER LAW DISTRIBUTIONS AND FRACTALS

Many phenomena, like earthquakes, nuclear accidents, and forest fires, show a power law distribution of event sizes. Now consider a frond from Barnsley's fern, shown on the left side of Figure 6-18. Notice the frequency distribution of sizes of frondlets; there are many small frondlets, some medium-sized ones, and one big frond. There seems to be a power law distribution of frondlet sizes. Now let's enlarge the frondlet enclosed by the rectangle and look at it in detail (see middle object).

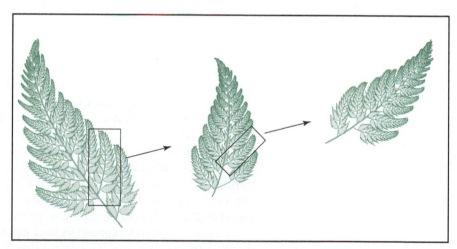

Figure 6-18   A frond from Barnsley's fern at three different magnifications. The fern's appearance is scale independent.

Figure 6-19   Lady fern showing a fractal distribution of frondlet sizes. *Photo: Langkamp /Hull.*

Magnification has not changed the appearance; the frondlet is just a small version of the big frond. We could repeat this exercise again; select a frondlet from the middle object, enlarge it, and examine the patterns of frondlets (right object). One of the characteristics of Barnsley's fern is that the appearance of fronds is independent of the scale of observation. No matter how closely you look at the fern, there are ever more frondlets, all with a similar pattern as the whole frond. Barnsley's fern is **fractal**, because its appearance is scale independent. By the way, Barnsley's fern is computer generated, but some real ferns are fractal too—up to a point! See Figure 6-19.

Oil spills, car wrecks, earthquakes, and wildfires are also fractal as they, too, are characterized by a power law distribution of sizes. Wildfires and Barnsley's fern are both fractal but not for the same reason. Many real ferns have a fractal appearance because the fronds require maximum surface area with a minimum of mass. Surface area is needed for light collection, and lots of holes prevents the fern from breaking under its own weight. **Biomechanics** may be driving the fractal structure of the fern. Many wildfires, on the other hand, are the product of a system at a critical state. The power law frequency distribution of wildfires results in part because of weather elements (lightning strikes and drought) and because of the linkages and connections between combustible patches of vegetation. A small event (someone dropping a cigarette butt) usually results in just a few acres of scorched grass, but on rare occasions, all the factors join to produce a major disaster.

CHAPTER SUMMARY

A **power function** or **power law** has the general form

$$y = kx^c$$

where x and y are the independent and dependent variables, and k and c are constants. The constant c is the power of the variable x. Some familiar power functions include $y = x^2$ and $y = x^{-1}$. Plots of power functions can take on a variety of shapes, depending on the values of k and c.

To solve for x when y is known requires just a small amount of algebra. For example, the relationship between the volume $v.c.$ and radius r of a sphere is given by the power function $v.c. = \frac{4}{3}\pi r^3$. If the volume is 42 cm^3, then we can solve for radius r as follows:

$$42 \text{ cm}^3 = \frac{4}{3}\pi r^3$$

$$42 \text{ cm}^3 \frac{3}{4\pi} = r^3$$

$$\left(42 \text{ cm}^3 \frac{3}{4\pi}\right)^{1/3} = r$$

$$2.2 \text{ cm} \approx r$$

Data that show a power-like behavior can be approximated with a power function using either the **straightedge method** or **least squares regression**. In the straightedge method, the logarithms of both the x- and y-values are determined and plotted, and a straight line is drawn through the $(\log(x), \log(y))$ values. The equation for this straight line has the general form $\log(y) = b + m\log(x)$. The power function through the original (x, y) data has the form $y = kx^c$. The relationships between the constants in the two equations are $c = m$ and $k = 10^b$. For example, if $\log(y) = 2.2 + 1.4\log(x)$, then $y = 10^{2.2}x^{1.4}$ or $y = 158.5x^{1.4}$.

The least squares regression uses calculator or computer technology to find the best fitting function. The calculator first transforms the original data using logarithms, then finds the best-fitting line through the transformed data by minimizing the sum of the vertical distances from the line to the $\log(y)$-values.

Some data sets show a distinctive relationship between size and frequency, with many small values of size, some medium-sized values, and a few large values. This pattern is known as a **power law frequency distribution** and is characteristic of phenomena such as wildfires, earthquakes, oil spills, and automobile accidents. This kind of distribution can be modeled with a power law using the **bin midpoints** and **reverse cumulative frequencies** for each bin. A power function can be determined using the straightedge method or least squares regression, and can be used to determine the frequency of large, rare events. Power law frequency distributions are also characteristic of **fractals** that are scale independent.

END *of* CHAPTER EXERCISES

Power Law Basics

1. Using technology, graph the power law functions $y = 2x^{1.3}$, $y = 10x^{0.65}$, and $y = 50x^{-0.4}$ in the same window for values of x between 0 and 20, and y from 0 to 100.
   a. Draw a quick sketch to illustrate their shapes; include labels.
   b. Which functions pass through the origin?
   c. What power will result in a straight line?

2. Using technology, graph the power law functions $y = 50x^{-0.5}$, $y = 50x^{0.25}$, and $y = 11x^{1.5}$ in the same window for values of x between 0 and 20, and y from 0 to 100.
   a. Draw a quick sketch to illustrate their shapes; include labels.
   b. Do any of these functions cross or intersect the x- or y-axes?
   c. Which functions are increasing? Which are decreasing?

3. For each of the following three functions, solve for x, given that $y = 33$.
   a. $y = 194x^{2.73}$
   b. $y = 14.01x^{0.34}$
   c. $y = 23x^{-0.11}$

4. For each of the following three functions, solve for x, given that $y = 176$.
   a. $y = 0.85x^{1.05}$
   b. $y = 4.98x^{0.63}$
   c. $y = 357.7x^{-1.84}$

Power Law Applications

5. Phosphorus is a key nutrient for plant growth; however, too much phosphorus from agricultural or sewage discharges can cause excessive growth of algae in lakes and streams. The amount of phosphorus discharged by rivers and streams from a drainage area or **watershed** is related to the size of that watershed. In a study of 35 watersheds in Wisconsin,[8] it was determined that the relation between the drainage area A (in square miles) and the amount of discharged phosphorus P (in pounds per year) is given by the power function

$$P = 490A^{0.79}$$

   a. What is the phosphorus output for a watershed with an area of 25 mi²? What is the phosphorus output for a watershed four times as large in area?
   b. Graph the function using technology. Trace along the graph to find the watershed area for a phosphorus discharge of approximately 10,000 pounds per year.

6. **Scaling laws** are mathematical relationships between parts of an object that maintain an object's shape or form. Scaling laws for geometric solids such as cubes and cones are well known. For example, the relationship between the radius and the volume of a sphere is $V = \frac{4}{3}\pi r^3$. Organisms from bacteria to whales often follow well-defined scaling laws.[9] For example, mammals show a power law relationship between heartbeat rate (HR, in beats per minute) and body mass (BM, in kilograms):

$$HR = 241BM^{-0.25}$$

   a. What is *your* body mass, in kilograms? Recall that 1 kilogram is approximately 2.2 pounds. Using your body mass, calculate your expected heartbeat rate. How does the expected rate compare to your actual rate?
   b. Using technology, graph the function for body masses less than 5 kilograms. Trace along the graph to find the heartbeat rate of a mammal whose body mass is 2 kilograms. Which mammal or mammals might this be?

7. **Paleoclimatology** (the study of past climate) depends upon clever techniques to acquire and analyze prehistoric climate data. **Proxies** or substitutes for climate variables such as precipitation and temperature are used quite often, as direct measurements of these climate variables are not typically available. For example, water vapor exits a plant through its leaves; therefore, plants in dry climates have small leaves to cut down on water loss. Leaf area is a proxy for precipitation. There is a strong correlation between the area of a leaf A (in square centimeters) and the mean annual precipitation MAP (in centimeters),[10] governed by the formula

$$MAP = 26.753\ A^{0.547}$$

Figure 6-20 A wind farm at Tehachapi, California. *U.S. Dept. of Energy.*

a. Suppose a fossil leaf, 5 million years old, has an area of 95 cm². What was the mean annual precipitation 5 million years ago? Convert this value to inches. How does this value compare with the average annual precipitation in your area?

b. Suppose you examine the vegetation in your area and note that there is a variety of trees with a variety of leaf sizes. The smallest leaf you collect is 14 cm² and the largest leaf is 27 cm². First find the average area of these two leaves; then calculate the mean annual precipitation.

8. The power generated by a windmill (Figure 6-20) is dependent on both the wind speed and the size of the blades.[11] The power output is related to the square of the blade length and the cube of the wind speed:

$$P = 1.9242L^2W^3$$

In this equation, P = power in watts, L = length of the blades in meters, and W = wind speed in meters per second.

a. Using technology, make two graphs of power versus wind speed on the same axes, one for a blade length of 10 meters and another for a blade length of 13 meters. Record on your homework paper.

b. For the 10-meter-long blade, how much power is generated at wind speeds of 10 m/sec and 5 m/sec? Express these results in kW (kilowatts).

Power Law Using the Straightedge Method

9. Glaciers are indicators of climate change, reflecting variations in temperature and precipitation. To contribute to our understanding of global warming and its effects, the sizes of 50 glaciers in the Southern Patagonia Icefield in Chile were measured in 1986.[12] Data for 7 of those glaciers are given in Table 6-9.

TABLE 6-9

Name	Length (km)	Area (km²)
HPS9	19	55
Ofhidro	26	116
Bernardo	51	536
Témpano	47	332
HPS31	23	161
Greve	51	438
HPS8	11	38

Source: Univ. of Chile Glaciology Laboratory

a. Transform the data using logarithms and construct a scatterplot (by hand) of the transformed data, with x = log(length) and y = log(area).

b. Using the straightedge method, determine a linear equation through the transformed data.

c. Calculate the power function between length and area using the equation found in part (b).

d. Glaciologists have reasoned that this power function should have a power of 2, because area is related to length². How well does your power c match this value?

10. As discussed in the introduction to this chapter, the number of bird species on an island is related to the area of that island. Biologists examined a number of islands in the Solomon Archipelago of the southwestern Pacific. Data on lowland bird species living on 8 of those islands with areas greater than 10 square miles are given in Table 6-10.[13]

TABLE 6-10

Island Name	Area (mi²)	Lowland Species
Bougainville	3,317	82
Guadalcanal	2,039	79
Ysabel	1,581	71
Choiseul	1,145	70
Buka	236	63
Nggela	142	61
Shortland	90	58
Fauro	27	51

Source: National Academy of Sciences

a. Transform the data using logarithms and place the values in a table.

b. Plot by hand the transformed data, with log(area) as the independent variable.

c. Using the straightedge method, determine both a linear equation through the transformed data and the corresponding power law equation through the original data.

d. Using the power law equation, find the number of lowland bird species for an island of 4,000 square miles.

e. Suppose an island in the Solomons of 1,000 square miles has 55 bird species. Does this number of bird species agree with the power law model? Give one explanation for the difference between the expected and the observed values.

11. Many marsupials are threatened or endangered due to loss of habitat and competition from introduced, nonnative species. **Dasyurids** are a group of marsupials in Australia that resemble mice and other small rodents. To develop a better understanding of the life history of dasyurids, biologists have measured some reproductive characteristics of these marsupials. Table 6-11 gives the body mass of females (in grams) and the time between conception and weaning of their offspring (tcw, in days) for 14 species of dasyurids.[14] By comparison, the average time between conception and weaning of a human is about 270 days (9 months to birth) plus an additional 180 days (for nursing), for a total of at least 550 days.

TABLE 6-11

Mass	tcw	Mass	tcw
9	81	150	141
11	90	180	141
22	100	500	137
30	82	900	139
27	97	1,047	156
60	93	4,000	157
120	156	5,400	276

Source: *Oecologia*

a. Calculate the logarithms of the body masses and the conception to weaning times. Create a scatterplot of the transformed data, either by hand or with technology that allows printing of the scatterplot.

b. Using the straightedge method, determine both a linear function through the transformed data and the corresponding power function through the unlogged data.

c. Make a quick sketch that shows the general shape of the power function. Briefly describe the relationship between the size of female dasyurids and the time from conception to weaning.

d. A female red kangaroo (*Macrapus rufous*) has a mass of about 35 kilograms using your function from part (b), determine the expected time from conception to weaning for a red kangaroo.

12. Lead, a toxic heavy metal, can be found on and adjacent to many streets and highways and can have a number of different automobile-related sources: lead wheel weights used to balance tires (Figure 6-21); lead from wear of automobile parts; and "legacy lead" left over from the era of leaded gasoline.

Figure 6-21 Lead weights used to balance tires, found along a busy road. *Photo: Langkamp/Hull.*

TABLE 6-12

Distance (meters)	Lead (ppb)
12.5	172
17.5	132
22.5	142
27.5	81
32.5	78
37.5	61
42.5	70

Source: *GSA Today*

As expected, higher lead levels are found closer to roads. Table 6-12 gives the lead concentration (in parts per billion) in surface soils as a function of distance (in meters) from a suburban street in Indianapolis, Indiana.[15]

a. Calculate the logarithms of the distances and concentrations. Create a scatterplot.

b. Using the straightedge method, determine both a linear function through the transformed data and the corresponding power function through the unlogged data.

c. Suppose the government sets a suburban lead soil standard of 400 ppb, which is half the standard for industrial and commercial areas.[16] Using the power function model, calculate the corresponding distance to the street.

d. How far from the curb is the edge of a typical suburban front yard? Will soil in the front yard meet or exceed a 400 ppb standard according to the model?

Power Law Regression

13. A lake of molten rock or **lava** 275 feet deep filled Makaopuhi Crater on the Big Island of Hawaii during an eruption in mid-March, 1965. About a month after the eruption, the crust that formed on the lava lake was thick enough to support a drilling rig. Geologists drilled through the thickening crust many times over the next year or so, measuring the thickness of the crusty layer.[17] The data are given in Table 6-13. Time is given in days since March 19, 1965, and thickness of the crusty layer is given in feet.

TABLE 6-13

Days	Thickness	Days	Thickness	Days	Thickness	Days	Thickness
31	6.50	61	10.30	115	12.90	196	17.70
33	7.20	66	9.90	119	13.90	235	19.70
34	7.00	68	10.20	119	14.20	269	20.90
40	8.00	68	10.40	136	14.65	306	22.25
48	8.80	80	10.85	151	16.00	321	23.00
59	10.00	84	11.30	164	15.70	430	26.00
59	9.30	89	11.25	180	16.90	496	29.15

Source: U.S. Geological Survey

a. Which variable is the independent variable? Determine the best-fitting power model for time and thickness using technology. What is the equation and what is the correlation coefficient? Is a power function a good model for these data? Explain.

b. What thickness does the model predict for a time of one day? Is this thickness reasonable?

c. Geologists came back about four years later (1,333 days since March 19, 1965) and found the crust to be 52 feet thick. Does the power law model still predict reasonable values for thickness after such a long time?

d. The geologists state that "cooling as a function of the square root of time is expected for lava cooling by conduction." Does your analysis support cooling of the lava lake by conduction? Explain.

14. On May 1, 1995, a moderate earthquake of magnitude 5.0 shook Duvall, Washington. A couple of days after the main shock, the U.S. Geological Survey put out seismic recording machines around Duvall, to record the **aftershocks**. Aftershocks are small earthquakes on the same fault after the "big one." Table 6-14 gives the number of aftershocks per day greater than magnitude 1.5 for up to 21 days following the main shock.[18]

TABLE 6-14

Days after May 1	Aftershocks per Day	Days after May 1	Aftershocks per Day	Days after May 1	Aftershocks per Day
3	75	10	5	16	3
4	26	11	5	17	3
5	22	12	5	18	3
6	12	13	4	19	2
7	7	14	4	20	2
8	1	15	4	21	1
9	6				

Source: Pacific Northwest Seismograph Network

a. Using technology, calculate the best-fitting power law regression for this time series of aftershocks. What is the regression equation and what is the correlation coefficient?

b. Is the power law regression a good model for the drop-off in aftershocks with time? Explain.

c. Scientists were not able to get recording machines to Duvall until three days after the main shock. Estimate the number of aftershocks per day for two days after the main shock (May 3).

d. Only one aftershock was recorded eight days after the main shock. How many aftershocks were expected on this day according to the power law model?

15. Smallmouth bass (*Micropterus dolomieui*) in Lake Whatcom near Bellingham, Washington, were captured by the Washington State Department of Ecology in a study of mercury in fish. The ages (in years) and weights (in grams) of the smallmouth bass are given in Table 6-15.[19]

TABLE 6-15

Age	Weight	Age	Weight	Age	Weight	Age	Weight
3	187	5	670	6	590	5	666
3	222	6	750	5	590	6	755
4	218	5	722	6	714	8	957
3	268	6	780	5	704	7	915
3	281	6	831	7	726	6	924
4	343	5	827	6	733	8	1,046
5	338	7	1,110	9	1,163	8	1,076
4	295	7	1,007	8	1,110	5	737
5	377	6	1,118	9	1,706	7	765
6	492	8	1,265	5	592	5	691
5	624	10	1,392	5	752	5	766
5	593	10	1,380	8	1,160	7	1,098
5	610	10	1,488	8	1,593	8	1,058
6	630						

Source: Washington State Dept of Ecology

a. Using technology, determine the best-fitting power law regression between age and weight. What is the best-fitting regression and what is the correlation coefficient?

b. Is the power law a good model for the growth of smallmouth bass from Lake Whatcom? Explain.

c. What is the weight of a two-year-old smallmouth bass, according to the model?

16. The volume of water in a stream flowing by a station is called **discharge**. Discharge has units of volume per time and is typically measured in cubic feet per second (cfs). The maximum discharge of Seneca Creek at Dawsonville, Maryland, was measured during each year from 1928 to 1958.[20] The peak discharges were ranked from 1 to 31 and then a "rank recurrence interval" (RRI) was calculated:

$$RRI = {}^{32\,\text{years}}\!/_{\text{rank}}$$

The highest discharge in this 31-year period (15,000 cfs) ranks first, with a rank recurrence interval of 32 years. The second highest discharge (9,300 cfs) has a recurrence interval of 16 years; two discharges of rank 2 are expected in this 31-year period. The data for all years are given in Table 6-16.

a. Using technology, determine the best-fitting power law regression equation for these data.

b. What is the value of the correlation coefficient? Is the power law model a good fit to the discharge data? Explain briefly.

c. What is the expected discharge from the regression equation for a rank 1 streamflow? How does the predicted value compare to the actual value? Give one explanation for this difference.

d. What is the predicted discharge for a rank recurrence interval of 100 years?

TABLE 6-16

RRI (years)	Discharge (cfs)	RRI (years)	Discharge (cfs)
1.03	959	2.13	2,280
1.07	1,240	2.29	2,280
1.10	1,300	2.46	2,410
1.14	1,380	2.67	2,420
1.19	1,420	2.91	2,610
1.23	1,450	3.20	2,620
1.28	1,460	3.56	2,660
1.33	1,600	4.00	2,810
1.39	1,730	4.57	2,940
1.45	1,740	5.33	3,620
1.52	1,990	6.40	3,640
1.60	1,990	8.00	3,800
1.68	2,020	10.67	7,330
1.78	2,110	16.00	9,300
1.88	2,150	32.00	15,000
2.00	2,240		

Source: *Water: A Primer*

Power Law Frequency Distributions

Exercises 17 & 18

Stream discharge is the volume of water passing by a measuring point in a certain amount of time. Discharge is usually measured in cubic feet per second (cfs). The U.S. Geological Survey maintains a network of stations where discharge is continuously monitored and recorded.

17. Average discharge for each day in May 2003, is given in Table 6-17 for Tonawanda Creek near Batavia, New York.[21]

TABLE 6-17

Day	cfs	Day	cfs	Day	cfs	Day	cfs
May 1	100	May 9	122	May 17	459	May 25	224
May 2	158	May 10	110	May 18	307	May 26	175
May 3	217	May 11	189	May 19	222	May 27	146
May 4	151	May 12	681	May 20	181	May 28	430
May 5	124	May 13	691	May 21	288	May 29	258
May 6	155	May 14	939	May 22	253	May 30	175
May 7	177	May 15	582	May 23	184	May 31	150
May 8	137	May 16	351	May 24	180		

Source: U.S. Geological Survey

a. What is the minimum discharge for this time period? Maximum discharge?

b. Group the discharge values; 5 to 10 bins are recommended. Record the bins and their frequencies in a table. Make a second table with the bin midpoints and reverse cumulative frequencies.

c. Create a scatterplot (either by hand or with technology) of reverse cumulative frequency versus bin midpoint. Briefly describe the trend or shape of the data.

d. Using technology, calculate the power law regression equation that best fits these data. Some streams have a power c between -1 and -5. How does the power c for Tonawanda Creek compare to these values?

e. What is the correlation coefficient r? Is the sign of the correlation coefficient correct? Explain briefly.

f. Do Tonawanda Creek discharges show a power law frequency distribution? Explain briefly.

18. Average discharge for each day in June 2003 is given in Table 6-18 for the Raging River in western Washington State.[22]

TABLE 6-18

Day	cfs	Day	cfs	Day	cfs	Day	cfs
June 1	30	June 9	21	June 17	18	June 25	22
June 2	29	June 10	22	June 18	17	June 26	20
June 3	27	June 11	22	June 19	19	June 27	18
June 4	26	June 12	21	June 20	29	June 28	17
June 5	25	June 13	21	June 21	43	June 29	16
June 6	24	June 14	22	June 22	34	June 30	16
June 7	22	June 15	20	June 23	33		
June 8	21	June 16	19	June 24	27		

Source: U.S. Geological Survey

a. What is the minimum discharge? Maximum discharge? Group the discharge values; between 5 to 10 bins are recommended. Record the bins and their associated frequencies in a table. Are there any zero values for frequency? In a second table, record the bin midpoints and reverse cumulative frequencies.

b. Create a scatterplot (either by hand or with technology) of reverse cumulative frequency versus bin midpoint. Briefly describe the trend or shape of the data.

c. Using technology, calculate the power law regression equation that best fits these data. What is the sign of the power c in the equation? Is this the correct sign? Explain briefly.

d. What is the correlation coefficient r? How close is r to a value of -1?

e. Do Raging River discharges show a power law frequency distribution? Explain briefly.

19. **Landslides** are downhill movements of soil, rock, vegetation, or other earth materials. Many factors contribute to landslides, such as steep slopes, weak materials, heavy rains, and undercutting by road construction. As part of a study of the role of logging roads in facilitating landslides, scientists measured the areas of over 3,000 landslides in Washington and Oregon.[23] Data for landslides up to 20,000 m^2 in area are given in Table 6-19.

TABLE 6-19

Bin (m^2)	Frequency	Bin (m^2)	Frequency
0–2,000	1,254	10,000–12,000	63
2,000–4,000	609	12,000–14,000	55
4,000–6,000	272	14,000–16,000	61
6,000–8,000	376	16,000–18,000	45
8,000–10,000	393	18,000–20,000	35

Source: *Hydrological Processes*

a. Make a table of bin midpoints and reverse cumulative frequencies for the landslide data.

b. Using technology, calculate the power law regression equation that best fits these data. What is the correlation coefficient r?

c. Do these landslides show a power law frequency distribution? Explain briefly.

d. Using technology, calculate the *exponential* regression equation that best fits the landslide data. What is the correlation coefficient?

e. Which of these two models best fits the data? Explain briefly.

20. The state of California operates a series of precipitation gauges throughout flood-prone areas of the state. One such area is the Eel River in northern California, which has experienced major floods in 1955, 1964, and 1995. Analyzing rainfall patterns is an important step in understanding the resulting flood patterns. Data on rainfall (in inches) in February 2000 from the station at Fort Seward near the Eel River are shown in Table 6-20.[24]

a. What is the minimum daily rainfall? Maximum daily rainfall? Collate the rainfall values into bins. Record the bins and their associated frequencies in a table. Add the bin midpoints and reverse cumulative frequencies to the table as well.

b. Using technology, calculate the power law regression equation that best fits these data. What is the correlation coefficient?

TABLE 6-20

Day	Rainfall (in.)	Day	Rainfall (in.)	Day	Rainfall (in.)
Feb. 1	0.08	Feb. 11	0.76	Feb. 21	0.12
Feb. 2	0.00	Feb. 12	0.64	Feb. 22	2.20
Feb. 3	0.04	Feb. 13	1.76	Feb. 23	0.00
Feb. 4	0.04	Feb. 14	1.92	Feb. 24	0.04
Feb. 5	0.64	Feb. 15	0.00	Feb. 25	0.20
Feb. 6	0.00	Feb. 16	0.04	Feb. 26	1.44
Feb. 7	0.00	Feb. 17	0.00	Feb. 27	1.00
Feb. 8	0.00	Feb. 18	0.00	Feb. 28	0.44
Feb. 9	0.16	Feb. 19	0.00	Feb. 29	0.52
Feb. 10	0.24	Feb. 20	0.12		

Source: California Dept of Water Resources

c. Do you think that every month in northern California will show a power law frequency distribution of rainfall? Explain briefly.

d. Find the bin midpoint that is about four times as large as the maximum bin midpoint in your table. Calculate the corresponding reverse cumulative frequency and interpret your result.

21. On the outside of river bends, the bank is typically cut away by fast-moving water, forming a steep bluff or cliff. These steep banks are prone to failure and landsliding, which can have a big impact on sediment supply, channel migration, and property loss. River bank failures on the North Fork of the Flathead River in northwest Montana were measured from 1945 to 1981. Data on the number of failure sites (frequencies) and failure volumes (in cubic meters) are given in Table 6-21.[25]

TABLE 6-21

Bin (m^3)	Frequency
0–5,000	112
5,000–10,000	32
10,000–15,000	22
15,000–20,000	5
20,000–25,000	7
25,000–30,000	4
30,000–35,000	5

Source: *Annals of the Association of American Geographers*

a. Record the bin midpoints and reverse cumulative frequencies in a table. Add the logarithms of both to the table as well.

b. Plot the logged data on graph paper. Using the straightedge method, find a linear model through the transformed data. Compute the power law function that best fits the original data.

c. Given a bin interval of 40,000 to 45,000 m^3, calculate the corresponding reverse cumulative frequency. Interpret this result.

22. Hundreds of small glaciers are found in the North Cascades National Park of Washington. To better understand the "health" of small glaciers, and changes in their areas with time, a series of air photos from 1958 were analyzed. Data on the frequencies of small glaciers and their areas (in km^2) are tabulated in Table 6-22.[26]

TABLE 6-22

Bin (km^2)	Frequency	Bin (km^2)	Frequency
0.0–0.1	100	0.7–0.8	7
0.1–0.2	93	0.8–0.9	8
0.2–0.3	47	0.9–1.0	3
0.3–0.4	16	1.0–1.1	2
0.4–0.5	10	1.1–1.2	4
0.5–0.6	11	1.2–1.3	1
0.6–0.7	4		

Source: Frank Granshaw

a. In a separate table, record the bin midpoints, the reverse cumulative frequencies, and their corresponding logarithms.

b. Using technology, graph the transformed values on a scatterplot. Do the transformed data show a strong linear trend?

c. Using the straightedge method, determine the linear function that approximates the transformed data. Calculate the corresponding power function.

d. The frequencies for the largest bins probably seem low, compared to your eyeballed line. Give one reason why larger glaciers might be underrepresented in the data set.

SCIENCE IN DEPTH

Earthquakes and Fractals

Faults are plane-like breaks in the Earth's crust. A fault is a break between adjoining **fault blocks**, which have moved relative to each other (Figure 6-22). The motion of one block of rock past another

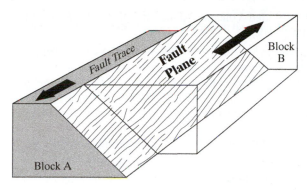

Figure 6-22 Two fault blocks sliding sideways

along active faults can be slow and constant, in which case no ground shaking is produced. But if the motion between blocks is fast, on the order of a few meters in a few minutes, then ground shaking results. As the two fault blocks slide jerkily past each other, vibrations or waves are produced, which radiate out in all directions from the fault and fault blocks. If these waves are big enough, and you are close enough to the source of the waves, you might feel them.[27]

What causes the fault blocks to move suddenly and without warning? The traditional explanation is that as fault blocks press steadily against each other, stress builds up. When the stress is high enough, the friction between the blocks is overcome, the fault blocks move relative to each other, and ground shaking is produced. So far, so good. According to this traditional explanation, the steady build-up of stress should result in a build-up and release pattern that occurs at regular intervals, in an episodic or cyclic pattern. But **seismologists** (earthquake scientists) have shown that earthquakes rarely show a repeated, cyclic pattern.[28] In fact, earthquakes are usually unpredictable in location, time, and size.

One nontraditional theory that involves a more random behavior for earthquakes is known as **self-organized criticality**.[29] To understand this theory, consider the simple example of a driver on the freeway during rush hour. Typically the driver swerves a bit, nothing happens, and the cars keep speeding ahead. But at other times, a driver swerves slightly, another driver jerks the wheel and his car fishtails, several cars bump each other, and a massive 17-car pileup results. During rush hour, the consequences of a small disturbance may die out to nothing or may propagate into a catastrophe—that's criticality. Traffic that arranges itself into this unpredictable state is said to be self-organized.

This concept of self-organized criticality applies to faults as well. A tiny amount of movement or slip takes place along a fault. If the initial movement dies out, then a tiny earthquake is produced. But if the initial movement grows in a cascading fashion, involving an ever larger area of the fault, a giant earthquake is produced. Either outcome is possible, and therefore the size of the resulting earthquake, just like the size of the auto accident, is unpredictable!

How does this tie in with fractals? Recall the earlier example of Barnsley's fern. There are many small frondlets that make up the fern, some medium-sized fronds, and a couple of big fronds. This distribution of sizes fits a power law. Earthquakes show a similar distribution of sizes. Along a big active fault, there are many small earthquakes, some medium-sized earthquakes, and a few "great earthquakes." Earthquake sizes also exhibit a power law frequency distribution, as seismologists discovered many years ago.

Earthquakes are fractal but not in the same way as objects such as fern fronds, roads, or blood vessels. Blood vessels form a network that reaches every cell in the body, to supply nutrients and take away wastes. Blood vessels have a space-filling hierarchy of conduits that maximizes the flow of blood. Earthquakes are fractal because of the connectedness of different areas along a fault. The earthquake propagates along the fault much like a wildfire spreading across the ground. If different areas or patches along a fault are well connected, the earthquake can mushroom into a giant event such as the Indonesian earthquake of December 2004.

CHAPTER PROJECT:
A NEW MODEL FOR EARTHQUAKES

Earthquake prediction has been a goal of scientists for many decades. While there have been a few claims of success, none of these alleged predictions has really stood up to close scrutiny. Billions of dollars have been spent investigating earthquakes. Is it possible that no amount of study and investigation will ever create a reliable earthquake prediction scheme?

In the companion exercise for this chapter found at this text's Web site (**enviromath.com**), you will work in teams to generate artificial earthquakes using a simple but powerful model. You'll explore the complexity of the artificial earthquakes and learn how the behavior of the earthquake model relates to real faults and real earthquakes. And you'll apply the results of your analysis to the hot topic of earthquake prediction. Maybe earthquakes are predictable after all! Or maybe not!

NOTES

[1] J. M. Diamond, "Colonization of Exploded Volcanic Islands by Birds: the Supertramp Hypothesis," *Science* 184 (1974): 803–806.

[2] U.K. National Grid, Electric and Magnetic Fields, http://www.emfs.info/Source_transmission.asp.

[3] D. A. Maguire and D. W. Hahn, "The Relationship between Gross Crown Dimensions and Sapwood Area at Crown Base in Douglas-Fir," *Canadian Journal of Forest Research* 19 (1989): 557–565.

[4] D. M. Gray, "Interrelationships of Watershed Characteristics," *Journal of Geophysical Research* 66 (1961): 1215–1233.

[5] U.S. Dept of Energy, Model Year 1999 Fuel Economy Guide, http://www.fueleconomy.gov

[6] Ohio State University, Extension Data Center, http://www.osuedc.org/profiles/

[7] Environment Canada, Environmental Technology Centre, http://www.etc-cte.ec.gc.ca

[8] J. C. Panuska and R. A. Lillie, "Phosphorus Loadings from Wisconsin Watersheds: Recommended Phosphorus Export Coefficients for Agricultural and Forested Watersheds," Wisconsin Dept of Natural Resources, *Research Management Findings*, April 1995.

[9] K. Schmidt-Nielsen, *Scaling: Why Is Animal Size So Important?* (Cambridge: Cambridge University Press, 1984), 241.

[10] P. Wilf, S. Wing, D. Greenwood, and C. Greenwood, "Using Fossil Leaves as Paleoprecipitation Indicators: An Eocene Example," *Geology* 26 (1998): 203–206.

[11] Danish Wind Industry Association, "Guided Tour on Wind Energy," http://www.windpower.org/en/tour.htm

[12] University of Chile, Glaciology Laboratory, http://www.glaciologia.cl/spi.html

[13] J. M. Diamond and E. Mayr, "Species-Area Relation for Birds of the Solomon Archipelago," *Proceedings of the National Academy of Sciences USA* 73 (1976): 262–266.

[14] S. D. Thompson, "Body Size, Duration of Prenatal Care, and the Intrinsic Rate of Natural Increase in Eutherian and Metatherian Mammals," *Oecologia* 71 (1987): 201–209.

[15] G. Filippelli, M. Laidlaw, J. Latimer, and R. Raftis, "Urban Lead Poisoning and Medical Geology: An Unfinished Story," *GSA Today* 15 (2005): 4–11.

[16] U.S. Environmental Protection Agency, Metals Workgroup, http://www.epa.gov/superfund/programs/lead/almfaq.htm

[17] T. L. Wright and R.T. Okamura, "Cooling and Crystallization of Tholeiitic Basalt, 1965 Makaopuhi Lava Lake, Hawaii," *U.S. Geological Survey Professional Paper* 1004 (1977), 78.

[18] The Pacific Northwest Seismograph Network, http://www.geophys.washington.edu/SEIS/PNSN

[19] Washington State Dept of Ecology, *Mercury Concentrations in Edible Muscle of Lake Whatcom Fish*, by D. Serdar et al., publication no. 01-03-012 (2001).

[20] L. Leopold, *Water: A Primer* (San Francisco: W. H. Freeman, 1974).

[21] U.S. Geological Survey, National Water Information System, http://nwis.waterdata.usgs.gov

[22] Ibid.

[23]D. R. Montgomery, K. Sullivan, and H. M. Greenberg, "Regional Test of a Model for Shallow Landsliding," *Hydrological Processes* 12 (1998): 943–955.

[24]California Dept of Water Resources, Division of Flood Management, http://cdec.water.ca.gov/queryDaily.html

[25]M. Fonstad and W. A. Marcus, "Self-Organized Criticality in Riverbank Systems," *Annals of the Association of American Geographers* 90, no. 2 (2003): 281–296.

[26]F. D. Granshaw, "Glacier Change in the North Cascades National Park Complex, Washington State, USA, 1958-1998" (Master's thesis, Portland State University, 2002).

[27]B. Bolt, *Earthquakes* (New York: W. H. Freeman, 1993), 331.

[28]R. Yeats, *Living with Earthquakes in the Pacific Northwest* (Corvallis: Oregon State University Press, 1998), 309.

[29]P. Bak, *How Nature Works: The Science of Self-Organized Criticality* (New York: Springer-Verlag, 1996), 212.

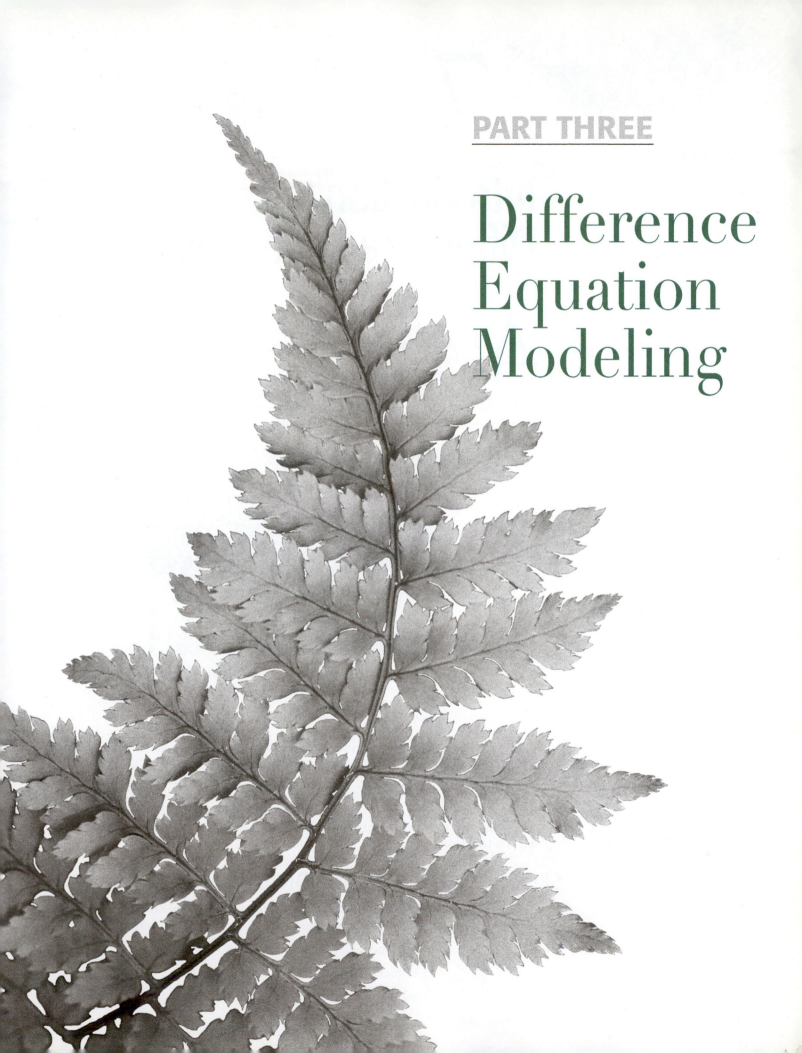

PART THREE

Difference Equation Modeling

7

Introduction to Difference Equations

Arriving in Mexico City, the view from an airplane is truly stunning: A system of buildings and roads extends in every direction, filling the enormous basin that long ago held several freshwater lakes (Figure 7-1). Crammed into this megalopolis live 20 million inhabitants, a cosmopolitan mixture of people that rivals any of the great world cities. One would think that there is no more room, and yet this population continues to expand because of a high birth rate and migration into the city for employment. Can we predict how Mexico City's population will grow in the future?

Figure 7-1 The majority of the global population now lives in large cities, a result of high birth rates and migration from rural areas. *Photo by Ken Karp/Prentice-Hall, Inc.*

In this chapter we look at an entirely new type of mathematical model called a **difference equation**. Using difference equations, we can model relatively simple phenomena that change linearly or exponentially, but also more complex phenomena such as the growth of the population of Mexico City. We start by introducing sequences, which form the mathematical foundation of difference equations. Then we discuss how to write and interpret difference equations and learn about their connection to linear and exponential functions.

SEQUENCES AND NOTATION

Underlying the mathematics of difference equations are **sequences**, which are no more than ordered lists of numbers. Consider the following three examples:

$$\text{sequence } A \quad \{100, 95, 90, 85, 80, 75, \ldots\}$$

$$\text{sequence } B \quad \{5, -3, 3, 21, 65, -14.2, 32, \tfrac{2}{3}\}$$

$$\text{sequence } C \quad \{3, 6, 12, 24, 48, 96, \ldots\}$$

Each number in a sequence is called a **term**. Sequence B has eight terms. Sequences A and C have an infinite number of terms, as the ellipsis (...) indicates that the sequence pattern continues on without end.

Terms have a position in the sequence identified by the **term number**. The actual number that you see in the sequence is called the **term value**. We'll always use the convention that the **initial term** has a term number of zero. The first five terms of sequences A, B, and C are displayed in Table 7-1.

TABLE 7-1

Sequence A		Sequence B		Sequence C	
Term Number	**Term Value**	**Term Number**	**Term Value**	**Term Number**	**Term Value**
0	100	0	5	0	3
1	95	1	−3	1	6
2	90	2	3	2	12
3	85	3	21	3	24
4	80	4	65	4	48

The initial term value in sequence A is denoted $a(0)$, the next term value is denoted $a(1)$, the next is $a(2)$, and so on. We pronounce $a(0)$ as "a of zero," $a(1)$ as "a of one," and so on. This type of notation is referred to as **function notation**. Beware that the parentheses do not indicate multiplication; that is, $a(2)$ does *not* mean that a is multiplied by 2.

From sequence A we obtain $a(0) = 100$, $a(1) = 95$, and $a(2) = 90$. Looking at the other sequences, we get $b(2) = 3$, $b(6) = 32$, $c(1) = 6$, and $c(4) = 48$. The first five terms of sequences A, B, and C are displayed using function notation in Table 7-2. The variable n represents the term number and can take on any integer value greater than or equal to 0.

TABLE 7-2

Sequence A		Sequence B		Sequence C	
$n =$ **Term Number**	$a(n) =$ **Term Value**	$n =$ **Term Number**	$b(n) =$ **Term Value**	$n =$ **Term Number**	$c(n) =$ **Term Value**
0	$a(0) = 100$	0	$b(0) = 5$	0	$c(0) = 3$
1	$a(1) = 95$	1	$b(1) = -3$	1	$c(1) = 6$
2	$a(2) = 90$	2	$b(2) = 3$	2	$c(2) = 12$
3	$a(3) = 85$	3	$b(3) = 21$	3	$c(3) = 24$
4	$a(4) = 80$	4	$b(4) = 65$	4	$c(4) = 48$

While sequences A and C have identifiable patterns, sequence B does not (although there could be an obscure pattern that we cannot detect). We would probably model sequence A linearly, because the term values are decreasing by the constant amount of 5. Similarly, we would probably choose an exponential model with a growth multiplier of 2 to describe sequence C.

Sequence A is displayed in Table 7-3 with greater embellishment. Notice that we have added extra rows to indicate that the sequence continues on forever and have included arrows indicating that we subtract 5 to jump from one term value to the next. Most importantly, there are additional rows for two successive "generic" terms with term numbers denoted by $n-1$ and n. The values of these terms are denoted $a(n-1)$ and $a(n)$.

TABLE 7-3

n = Term Number	a(n) = Term Value	
0	$a(0) - 100$	-5
1	$a(1) - 95$	-5
2	$a(2) - 90$	-5
3	$a(3) - 85$	-5
4	$a(4) - 80$	-5
5	$a(5) - 75$	-5
$\vdots$	$\vdots$	
$n-1$	$a(n-1)$	-5
n	$a(n)$	

Generic designation of two consecutive term numbers

To describe the pattern in sequence A with mathematics, we write the value of one term using the value of the previous term. For example, $a(1) = a(0) - 5$ and $a(2) = a(1) - 5$ (Table 7-4).

TABLE 7-4

n = Term Number	a(n) = Term Value
0	$a(0) = 100$
1	$a(1) = a(0) - 5$
2	$a(2) = a(1) - 5$
3	$a(3) = a(2) - 5$
4	$a(4) = a(3) - 5$
5	$a(5) = a(4) - 5$
$\vdots$	$\vdots$
$n-1$	$a(n-1)$
n	$a(n) = a(n-1) - 5$

Sequence A has a pattern, namely that any term value is the previous term value minus 5. The last entry in Table 7-4 summarizes this pattern mathematically: $a(n) = a(n-1) - 5$. We finish this example by giving a verbal and mathematical description of sequence A, indicating the initial value and the numerical pattern in the sequence.

Verbal description of sequence A   Let the initial value equal 100. To get the value of any term in the sequence, take the previous term value and subtract 5.

Mathematical description of sequence A $a(0) = 100$ and $a(n) = a(n-1) - 5$.

Let's now revisit sequence C. Table 7-5 lists term values using function notation.

TABLE 7-5

n = Term Number	c(n) = Term Value
0	$c(0) = 3$
1	$c(1) = 6$
2	$c(2) = 12$
3	$c(3) = 24$
$\vdots$	$\vdots$
$n-1$	$c(n-1)$
n	$c(n)$

In Table 7-6 sequence C is listed again, except that term values have been written in function notation as "two times the previous term." For example, $c(1) = 2c(0)$ and $c(2) = 2c(1)$.

TABLE 7-6

n = Term Number	c(n) = Term Value
0	$c(0) = 3$
1	$c(1) = 2\,c(0)$
2	$c(2) = 2\,c(1)$
3	$c(3) = 2\,c(2)$
⋮	⋮
$n-1$	$c(n-1)$
n	$c(n) = 2\,c(n-1)$

Now we can write verbal and mathematical descriptions for sequence C.

Verbal description of sequence C The value of the first term is 3. To get the value of any term in the sequence, take the previous term value and multiply by 2.
Mathematical description of sequence C $c(0) = 3$ and $c(n) = 2\,c(n-1)$.

The preceding mathematical descriptions include an **initial condition** and a **difference equation**. The initial condition is the value of the initial term, while the difference equation is the mathematical rule that tells how to jump from one term value to the next. For sequence C, the initial condition is $c(0) = 3$ and the difference equation is $c(n) = 2\,c(n-1)$.

Difference equations are sometimes written to highlight the *difference* between $a(n)$ and $a(n-1)$, thus the origin of their name. For example, the difference equation $a(n) = a(n-1) + 5$ can be written as $a(n) - a(n-1) = 5$. We will *not* be writing difference equations in this manner, but you may see this convention used in other texts.

TABLE 7-7

n	d(n)
0	25
1	31
2	37
3	43
4	49

EXAMPLE 7-1

Suppose that sequence D is given by the difference equation $d(n) = d(n-1) + 6$ and initial condition $d(0) = 25$. The verbal description of sequence D is as follows: "The starting term is 25. To find any term, take the previous term and add 6." Table 7-7 lists the first five terms of sequence D.

MODELING WITH DIFFERENCE EQUATIONS

We now give some practical examples on how to use difference equations to model environmental problems. We start with two examples examining different populations of moose—one that grows linearly, and one that decays exponentially.

EXAMPLE 7-2 A Moose Population Growing Linearly

Suppose biologists estimate that there are 150 moose in a wildlife park and predict that the number of moose will increase by 20 each year (Figure 7-2). How do we use this information to forecast the moose population in, for example, 10 years? With just a little effort you may solve this one in your head, but we'll first review a more formal approach that you've seen in Chapter 4 on linear functions.

Start by defining variables. Let $n =$ time in years, with $n = 0$ the current year, and let $p =$ the population of moose. In writing formulas for sequences, it is customary to use n for the independent variable.

Because the number of moose is increasing by a constant amount each year, the growth is linear and we seek a linear equation relating p to n. As you learned in Chapter 4, with an initial population of 150 and a rate of increase equal to 20 moose per year, the equation that models the linear growth is $p = 150 + 20n$. This equation can be written using function notation as $p(n) = 150 + 20n$. We can use this equation to find the population of moose in 10 years by substituting 10 directly for n. We get $p(10) = 150 + 20(10) = 350$ moose.

We can also find the number of moose in 10 years by using difference equations. Again we start by defining variables, letting n represent the number of years from now, and $p(n)$ the population of moose in year n. Using these variables, $p(0)$ represents the number of moose now, $p(1)$ is the number of moose one year from

Figure 7-2 A bull moose can weigh nearly 1,500 pounds, although its diet consists principally of dry woody materials and aquatic plants. The English word *moose* comes from the Algonquin *moz*, meaning "twig eater." *U.S. Fish and Wildlife Service/photo by Ronald L. Bell.*

now, $p(2)$ is the number of moose two years from now, and so on. The moose population sequence is displayed in Table 7-8. A graph of the sequence is obtained by plotting the (years, population) ordered pairs (Figure 7-3).

TABLE 7-8

n = Years since 1990	$p(n)$ = Moose Population
0	$p(0) = 150$
1	$p(1) = 170$
2	$p(2) = 190$
3	$p(3) = 210$
⋮	⋮
8	$p(8) = 310$
9	$p(9) = 330$
10	$p(10) = 350$

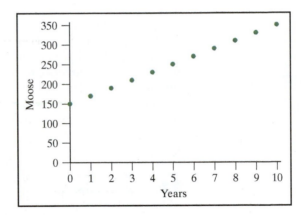

Figure 7-3   The linear growth of the moose population

We describe the moose population sequence verbally: "The initial value or initial condition is 150, and each term is found by adding 20 to the previous term." The mathematical description of the moose population is as follows:

$$p(0) = 150 \quad \text{and} \quad p(n) = p(n-1) + 20$$

EXAMPLE 7-3   A Moose Population Decaying Exponentially

Again suppose that the population of moose in the park today is 150. What if the biologists predict that the population will decrease each year by the fixed percentage 7.3%? As we discussed in Chapter 5, when a population grows or decays by a fixed percentage, it changes exponentially. We now derive the difference equation to describe the projected exponential decay in the moose population.

Using the same variables as in Example 7-2, we write that the initial population is $p(0) = 150$. To calculate a 7.3% decrease, let $r = -7.3\% = -0.073$ and multiply the initial population by $M = 1 + r = 0.927$. (Recall that the number $M = 0.927$ is called the **decay multiplier** and the number r the **decay rate**.) The population of moose after one year is

$$p(1) = 0.927p(0)$$
$$= 0.927(150 \text{ moose})$$
$$= 139.05 \text{ moose}$$
$$\approx 139 \text{ moose}$$

In two years, the population will equal

$$p(2) = 0.927p(1)$$
$$= 0.927(139.05 \text{ moose})$$
$$\approx 129 \text{ moose}$$

Note that in the last computation we multiplied by 139.05 moose, instead of the integer value of 139 moose, to avoid accumulating rounding errors.

To find the population three years from now, we would use the population from the previous year. To find the population n years from now, we would use the population $n-1$ years from now. This stepwise process is exactly how we generate new terms in a sequence. The mathematical formula for this pattern is the difference equation $p(n) = 0.927 p(n-1)$. The graph of the difference equation is given in Figure 7-4.

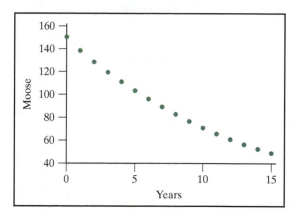

Figure 7-4 A moose population decaying exponentially.

Recall from Chapter 5 the basic form of an exponential function: $y = y_0 M^x$. Using the variables n and p to represent years and population, the exponential function modeling the moose population is $p(n) = 150(0.927)^n$. This function, which relates the population of moose directly to n, is called the **solution equation** to the difference equation $p(n) = 0.927 p(n-1)$ and initial condition $p(0) = 150$.

In Example 7-2 we modeled the linear growth of a moose population with an initial condition of $p(0) = 150$ and a difference equation of $p(n) = p(n-1) + 20$. The solution equation for this model is $p(n) = 150 + 20n$. Table 7-9 summarizes the two ways we can describe the linear growth and exponential decay of the moose populations.

TABLE 7-9

	Linear	Exponential
Initial Condition and Difference Equation	$p(0) = 150$ $p(n) = p(n-1) + 20$	$p(0) = 150$ $p(n) = 0.927 p(n-1)$
Solution Equation	$p(n) = 150 + 20n$	$p(n) = 150(0.927)^n$

It may seem strange that we call an equation such as $p(n) = 150 + 20n$ or $p(n) = 150(0.927)^n$ the "solution equation." When we investigate complex natural phenomena, the difference equation is often the first type of equation that is found. It is the solution equation that is often derived from the difference equation—hence the "solution."

An important distinction between difference equations and solution equations is that they are evaluated differently. Difference equations are evaluated by starting with an initial condition and finding the subsequent terms one at a time in a stepwise manner. Solution equations, on the other hand, allow us to find the population (or other quantity) directly by simply substituting a value for n.

Another important distinction between difference equations and solution equations is the following. A difference equation can only be evaluated for positive *integer* values of n ($n = 1, 2, 3, \ldots$). This type of equation is called a **discrete equation**, and its graph consists of distinct or isolated points. In contrast, a solution equation can be evaluated for *any* value of n, as long as we don't divide by zero or take square roots of negative numbers. The solution equations found in this text are **continuous equations**, and their graphs consist of connected sets of points. You can think of discrete equations as having graphs that are plotted "point-by-point" and continuous equations as having graphs plotted "without your pencil leaving the paper."

In Figure 7-5 the left graph shows the linearly increasing moose population, and the right graph the exponentially decaying moose population. In each graph both the difference equation (discrete points) and solution equation (continuous curve) are plotted. In each graph the difference equation and solution equation "overlap" for integer values of *n*.

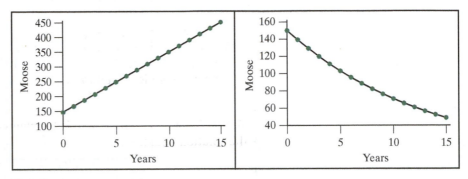

Figure 7-5 The two moose population models, linear (left) and exponential (right). The isolated points in each graph are from the difference equation models. The continuous curves are from the solution equations.

Let's now look at one more example to help you better understand the connection between difference equations and solution equations.

Figure 7-6 The Seeds of Change research farm in New Mexico tests over 1,000 varieties of organically grown fruits and vegetables each year. *Courtesy of Scott Vlaun/Seeds of Change.*

EXAMPLE 7-4 Exponential Growth of Organic Farmland

Organic crops are raised without the use of synthetic fertilizers and pesticides (Figure 7-6). In 2002, a United States Department of Agriculture (USDA) publication reported the following:[1]

Acreage of certified organic farmland is increasing to meet growing consumer demand. According to the most recent USDA estimates, U.S. certified organic cropland doubled between 1992 and 1997, to 1.3 million acres.

Suppose that the growth in organic farmland continues at the same pace. How do we describe future growth with a difference equation? Using methods discussed in Chapter 5, we first determine that a quantity which doubles in five years has an annual multiplier of $M = 2^{1/5} \approx 1.149$. This multiplier, incidentally, tells us that organic farmland acreage was increasing about 14.9% each year.

Next we define variables by letting *f* represent the amount of farmland in millions of acres, and *n* the number of years since 1997. The amount of farmland at time *n* is represented by $f(n)$, and the amount of farmland in the previous year by $f(n-1)$. Using the growth multiplier approximated above, the difference equation that relates $f(n)$ to $f(n-1)$ is

$$f(n) = 1.149\, f(n-1)$$

Notice that this difference equation is an equation about *growth*. In fact, the amount of organic farmland mentioned in the publication, namely 1.3 million acres, is *not* part of the difference equation! That quantity, of course, is the initial condition $f(0) = 1.3$ million acres.

Starting with that initial condition, and using the difference equation to describe the growth, we can forecast future acreage assuming that the growth rate remains the same (Table 7-10). Notice that the predicted acreage continues to double every five years indicating that our multiplier was computed correctly.

TABLE 7-10 Projected acreage of organic farmland in the United States

Year	n	$f(n)$ ($\times 10^6$ acres)
1997	0	1.3
2002	5	≈ 2.6
2007	10	≈ 5.2
2012	15	≈ 10.4

Instead of finding acreage projections through a stepwise process involving the difference equation, we could have more easily used the exponential solution equation. With a growth multiplier of $M = 1.149$, and an initial acreage of 1.3 million, the exponential solution equation is

$$f(n) = 1.3(1.149)^n$$

To predict the amount of organic farmland in year 2012, we simply plug $n = 15$ into the solution equation:

$$f(15) = 1.3(1.149)^{15}$$
$$= 10.44104055$$
$$\approx 10.4 \text{ million acres}$$

We mentioned earlier that difference equations can only be evaluated for integer values of n. If the year corresponds to a noninteger value of n, then we *must* use the solution equation. For example, to predict the amount of farmland in the middle of 2012 ($n = 15.5$), we find

$$f(15.5) = 1.3(1.149)^{15.5} \approx 11.2 \text{ million acres}$$

WHY USE DIFFERENCE EQUATIONS?

At this stage you might be asking, "Why use difference equations if we can use solution equations?" One reason is that difference equations are often easier to write than solution equations. This might not be the case for phenomena that change linearly or exponentially, but it is true for many other examples. We will investigate two such examples in the next section.

In addition to being easier to work with, difference equations are used because they better model or represent phenomena that have inherently discrete characteristics. Let's return to the moose population to understand this point more fully. Moose, especially those living in habitats with extreme winters, give birth primarily during late spring or early summer. At that time of year, there is a discrete "pulse" or "jump" in the moose population (Figure 7-7). To describe this type of population change, a difference equation is the model of choice. Other populations, such as that of humans,

Figure 7-7 Many animal populations are best approximated with discrete difference equation models because of the population "pulse" that occurs in the birthing season. *National Park Service photos.*

reproduce continuously throughout the year—for these populations, continuous versions of difference equations called **differential equations** are the preferred models. Differential equations involve the mathematics of calculus, so, rest assured, we won't be working with those equations in this text.

Knowing when to use difference equations (discrete) as opposed to differential equations (continuous) requires an intimate knowledge of the population, or other phenomena, under study and is one aspect of the "art" of mathematical modeling. These subtleties, of course, will be left for more advanced texts. Without knowing calculus and differential equations, you might think that you are somewhat limited in the types of environmental examples that can be studied. As it turns out, most continuous phenomena modeled with differential equations can be *closely approximated* using discrete models—in other words, with difference equations! Consequently, we are able to describe and investigate all types of phenomena—both discrete and continuous—with difference equations. With this in mind, let's move on to those more complicated examples.

AFFINE DIFFERENCE EQUATIONS

So far in this chapter, we have looked at difference equation models for environmental phenomena that exhibit linear or exponential change. We now investigate situations in which exponential change and linear change are *both* present. Our first example examines the growth of Mexico City's population, accounting for both the exponential "natural" growth and the linear growth due to migration.

EXAMPLE 7-5   Mexico City Population

The population of Mexico City's metropolitan area exploded in the latter part of the twentieth century and numbered about 20 million people in the year 2002. Demographers predict that the population will grow with an annual rate of natural increase (birth rate minus death rate) hovering near 1.7%, and an annual net migration (immigration into minus emigration out of) close to 750,000 people per year.[2]

Human populations change continually, but we can approximate the changes using discrete equations—in other words, difference equations. Before modeling the future population of Mexico City, we need to make a few assumptions. First, we assume that the annual rate of natural increase and net annual migration do not change. Second, we assume (for computational purposes) that during any given year, the 1.7% natural increase takes place first, followed by the influx of 750,000 or 0.75 million people. The order in this second assumption is arbitrary; we could assume that migration occurs followed by natural growth—that topic will be examined in a chapter exercise.

Let n represent the number of years since 2002, and $p(n)$ the population in millions in year n. The initial condition is $p(0) = 20$ million people. To compute the population for 2003, year $n = 1$, we first multiply the initial population by $1 + r = 1.017$ and then add 0.75 million people:

$$p(1) = 1.017p(0) + 0.75$$
$$= 1.017(20) + 0.75$$
$$= 21.09 \text{ million people}$$

To obtain the population for the year 2004, we multiply the 2003 population by 1.017 and then add 0.75 million people:

$$p(2) = 1.017p(1) + 0.75$$
$$= 1.017(21.09) + 0.75$$
$$\approx 22.20 \text{ million people}$$

Using a similar process, we predict future population values for Mexico City (Table 7-11).

TABLE 7-11	*n* = Years since 2002	*p(n)* = Population in Millions
	0	20
	1	21.09
	2	22.20
	3	23.33
	4	24.47
	5	25.64
	6	26.82

The difference equation that summarizes the year-to-year population growth is

$$p(n) = 1.017p(n-1) + 0.75$$

This equation accounts for the natural, exponential growth of 1.7%, followed by the migratory, linear growth of 0.75 million people each year. A difference equation that combines both exponential and linear components of growth or decay is called an **affine difference equation**.

To predict the population in year 2020 ($n = 18$) requires us to create the entire sequence of population values from $p(1)$ through $p(17)$. We would then find the population in year $n = 18$ by computing $p(18) = 1.017p(17) + 0.75$. We won't go through this work here, but we mention this as a reminder that the difference equation doesn't allow us to project far into the future very easily, unless we can do many computations quickly. Performing numerous, patterned computations is one thing that calculators and computers do exceptionally well—we'll explore that in the last section of this chapter.

An alternative model that is more efficient for projecting into the future is the solution equation. If we had a solution equation for Mexico City's population, we could predict the population in year 2020 by simply plugging in $n = 18$. Unfortunately, finding the solution equation to an affine difference equation takes some work, as we'll see in the next chapter.

We end this example by graphing the discrete output from the Mexico City affine difference equation over a 100-year period, plotting the data for every fifth year (Figure 7-8). Notice the nonlinear, exponential-like growth in Mexico City's modeled population.

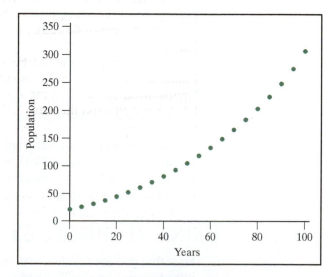

Figure 7-8 The predicted Mexico City population, 2002–2102. There appears to be an exponential-like increase to the affine difference equation.

Figure 7-9 A water treatment facility in Parwan, Afghanistan, is powered by a Bergey microturbine. Microturbines are also used for off-grid homes, for rural electrification, and for boosting the performance of solar electric systems. *Courtesy of Bergey Windpower Co.*

EXAMPLE 7-6   Fighting Poverty with Microloans

A paper from the International Institute of Environment and Development, which discusses the link between poverty and the environment, states:[3]

> *The environment is the source of what every one of us needs to survive—air, water and food. . . . The absence or denial of these basic necessities constitutes absolute poverty. Unequal access to basic necessities and other environmental resources is the foundation of relative poverty. In addition to being excluded from access to basic resources, the poor are also most likely to be subjected to the degrading or polluting impacts of the consumption patterns of others. In industrial and post-industrial societies this may take the form of exposure to higher levels of toxicity in the air, water and earth.*

One way in which poverty can be reduced is through **microloans**, which are typically small loans with below-market borrowing rates. These loans can assist poor communities in starting small businesses, health services, reforestation initiatives, and water supply projects. Microloans help communities develop sustainable energy sources based on solar power, wind power, and compost (Figure 7-9). All loans, whether microloans or your personal car loan, are based on simple mathematical principles that can be modeled by difference equations. For information about an organization that specializes in microloans, search for "Grameen Bank" on the Internet.

Suppose that a community borrows $1,500 from a bank to purchase a small windpowered electric generator. The bank charges 3% annually, or $r = 0.25\%$ monthly. Each month the community is charged $50. Some of this amount pays for interest that accrues during the month, while the remainder is applied toward the loan principal. Let n represent the number of months since the loan was issued, and $u(n)$ the loan balance at the end of month n. The loan balance is simply the amount still owed on the loan. When the loan was first issued the loan balance was $u(0) = \$1,500$.

After the first month, and without making any payments, the community would owe $1.0025(\$1,500) = \$1,503.75$. With a payment of $50 at the end of the month, the loan balance after the first month will equal.

$$u(1) = 1.0025(1,500) - 50$$
$$= \$1,453.75$$

At the end of the second month, the balance on the loan will equal

$$u(2) = 1.0025(1,453.75) - 50$$
$$\approx \$1,407.38$$

In general, the change in the loan balance from one month to the next is given by the difference equation

$$u(n) = 1.0025\,u(n-1) - 50$$

Because there is both exponential growth and linear decay of the loan balance each month, the difference equation is affine.

We now explore several graphing calculator features that make investigating difference equations easier. We'll use these features to forecast Mexico City's population in the year 2020, and determine the total number of payments necessary to pay off our community's microloan.

USING TECHNOLOGY: DIFFERENCE EQUATIONS

Now that you are familiar with the fundamentals of difference equations, we introduce features on the TI-83/84 calculator that make working with these equations much easier. Visit the text Web site for instructions on the use of an online computer application that performs the same tasks: **enviromath.com**

Setting the Correct Modes on the TI-83/84

To put the calculator into *sequence* mode: Press **MODE**. Use the arrow keys to move the cursor over **Seq**, found on the fourth line. Press **ENTER** to highlight **Seq**. Press **CLEAR** to exit the mode menu.

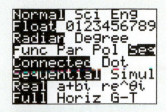

Put the calculator window into *time* format: Press **FORMAT** (**2nd ZOOM**). If **Time** is not highlighted on the first line, use the arrow keys to move the cursor over **Time**, and press **ENTER** to highlight it.

Entering Difference Equations on the TI-83/84

Press **Y=**. The first line asks for the value of *n***Min**. In this text, the initial term of a sequence always corresponds to $n = 0$, so keep *n***Min** equal to 0 .

Enter the difference equation $u(n) = u(n-1) + 3$ using **u** (**2nd 7**) as the dependent variable. The independent variable *n* is on the **X,T,θ, n** key. Parentheses **( )** keys are located above **8** and **9**. To the right of **u(*n*Min)=** enter the initial condition, for example, 12. Note that braces **{ }** will be automatically placed around 12 after you press **ENTER**.

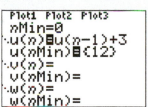

Difference equations can also be entered under **v(*n*)** and **w(*n*)**. The variables **v** and **w** are **2nd 8** and **2nd 9**, respectively.

Creating Tables on the TI-83/84

Enter the difference equation $u(n) = u(n-1) + 3$ and initial condition $u(0) = 12$ (see directions in the preceding box). Then press **TBLSET** (**2nd WINDOW**) and set the parameters as displayed to the right.

Press **TABLE** (**2nd GRAPH**). Use the up and down arrow keys to scroll through the table.

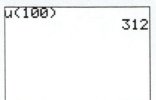

Instead of scrolling down to find, for example, $u(100)$, go to **TBLSET** and set **TblStart = 100**. Then press **TABLE**.

A faster way to find $u(100)$ is to go to the homescreen (**QUIT**) and input its expression directly. Then press **ENTER.**

Graphing a Difference Equation on the TI-83/84

Enter the difference equation $u(n) = u(n-1) + 3$ and initial condition $u(0) = 12$ (see directions in the second box in this section). Press **WINDOW** and set the parameters as displayed to the right. Note the following:

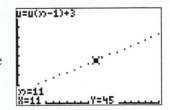

- *n***Min** is the initial n value to be plotted. In this text we always set *n***Min = 0**.
- *n***Max** is the largest n value to be plotted.
- **PlotStart** is the first term to be plotted, but the calculator refers to *n***Min** as term #1. In this text we always set **PlotStart = 1**.
- **PlotStep** is the "jump" in plotted n values.
- **Xmin, Xmax, Xscl, Ymin, Ymax, Yscl** define the window size. Normally we set **Xmin = 0** and **Xmax =** *n***Max**.

Press **GRAPH** to view the difference equation. To find coordinates of points, press **TRACE** and use the left and right arrows to move along the line.

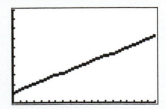

To graph in **connected mode**, go to **Y=**. Move the cursor to the left of u(n), over the three slanted dots `∴`.. Press **ENTER** once for connected mode; press **ENTER** again for bold, connected mode. Press **GRAPH** to view the results.

EXAMPLE 7-7 Mexico City Revisited

Recall that for the Mexico City population model in Example 7-5, the initial condition is $p(0) = 20$ million people and the affine difference equation is $p(n) = 1.017p(n-1) + 0.75$. Calculator screens illustrating window parameters and the corresponding graph of population values are shown next. Plotted values range from $n = 0$ to $n = 20$.

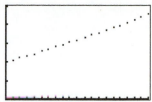

To find the population for year 2020, trace along the graph until $n = 18$ (see next screen). We estimate that in the year 2020, Mexico City will have a population of 42.7 million.

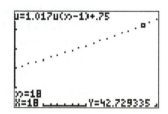

EXAMPLE 7-8   The Microloan Revisited

Let's return to Example 7-6, in which we investigated the financial mathematics of a community microloan. We started with an initial loan amount of $u(0) = \$1,500$ and developed the difference equation $u(n) = 1.0025\, u(n-1) - 50$ to reflect the 0.25% monthly interest rate and \$50 monthly payment. Using technology, we can generate a table to observe the loan balance at the end of each month. We notice that after 31 months of \$50 payments, the amount owed on the loan is about \$11.16.

n	$u(n)$
25	308.39
26	259.16
27	209.81
28	160.33
29	110.73
30	61.008
31	11.162

$u(n) = 11.16199036$

The community will have to make one additional payment at the end of the 32nd month equal to \$11.16 plus interest. The amount of the 32nd payment will equal $(1.0025)(\$11.16) = \$11.19$. Thus the total amount paid on the loan is $(31)(\$50) + \$11.19 = \$1,561.19$. This means that the bank will assess a total of \$61.19 in finance charges over the life of the loan.

CHAPTER SUMMARY

A **sequence** is an ordered list of numbers called **terms**. Each term has a term number and a term value. Consider sequence A, given by $\{12, 17, 22, 27, \ldots\}$. The initial term has a term number of 0 and a term value of 12. Term number 2 in sequence A has a term value of 22. **Function notation** is a convenient way to associate term numbers and term values. In sequence A, $a(0) = 12$, $a(1) = 17$, $a(2) = 22$, and so on. Sequence A is listed in Table 7-12. The variable n is used to represent the term number.

TABLE 7-12

n	$a(n)$
0	12
1	17
2	22
3	27
$\vdots$	$\vdots$
$n-1$	$a(n-1)$
n	$a(n)$

Sequences can be described verbally and mathematically. For example, a verbal description of sequence A could be "The initial term is 12, and to get any term value, take the previous term value and add 5." The mathematical description of sequence A consists of $a(0) = 12$, which describes the **initial condition**, and $a(n) = a(n-1) + 5$, which describes the "jump" from one term value to the next. This last equation is called a **difference equation**.

The difference equation $a(n) = a(n-1) + 5$ is **linear** because of the constant increase of 5 from one term value to the next. A difference equation such as $b(n) = 1.15\, b(n-1)$ is **exponential** because of the fixed 15% percent increase from one term value to the next. A difference equation that combines linear change and exponential change is called **affine**. An example of an affine difference equation is $c(n) = 1.09\, c(n-1) - 6$, in which there is an increase of 9% followed by a decrease of 6, going from each term value to the next.

A **solution equation** can be written to describe a linear or exponential sequence. Solution equations are functions and can be evaluated directly to find any term value of the sequence. Sequence A, described by the initial condition $a(0) = 12$ and linear difference equation $a(n) = a(n-1) + 5$, can also be described by the solution equation $a(n) = 12 + 5n$. The initial condition $b(0) = 20$ and exponential difference equation $b(n) = 1.15\, b(n-1)$ have a corresponding solution equation of $b(n) = 20(1.15)^n$. The graph of an initial condition and difference equation will be a **discrete** (disconnected) collection of points. Solution equations plot as **continuous** lines or curves. See Figure 7-10.

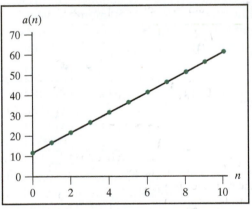 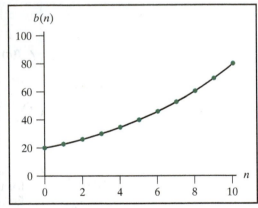

Figure 7-10　Two sequences, linear (left) and exponential (right). The isolated points in each graph are from the difference equation models. The continuous curves are from the solution equations.

END *of* CHAPTER EXERCISES

Basic Practice with Sequences

For Exercises 1–4, use sequences A, B, C, D, and E defined as follows. *Note: It is not necessary to write a difference equation for any of these sequences.*

$$A = \{\, 2, 6, 10, 14, 18, 22, 26, \ldots \}$$
$$B = \{\, 1, -1, 2, -2, 3, -3, 4, -4, \ldots \}$$
$$C = \{\, 6.2, -4.8, 1.9, 2.4 \,\}$$
$$D = \{\, 1, 2, 4, 8, 16, \ldots \}$$
$$E = \{\, -4, -1, 2, 5, 8, 11, \ldots \}$$

1. Find the values of the following expressions.
   a. $a(1)$
   b. $3\, a(4)$
   c. $b(5)$
   d. $a(1)^2$
   Note: In this text, the meaning of $a(1)^2$ is $a(1) \times a(1)$.

2. Find the values of the following expressions.
   a. $d(4)$
   b. $b(4) - b(5)$
   c. $4c(2)$
   d. $2b(3)^2$
   Note: The meaning of $b(3)^2$ is $b(3) \times b(3)$.

3. Evaluate the expressions based on the patterns in sequences A and B. Use any shortcuts you can come up with!
   a. $a(10)$
   b. $b(11)$
   c. $b(210) + b(211)$
   d. $a(40)$

4. Evaluate the expressions based on the patterns in sequences D and E. Again, use any shortcuts.
   a. $d(10)$
   b. $d(20)$
   c. $e(8)$
   d. $e(18)$

Basics of Difference Equations

5. Following is a table for sequence F. Complete the remainder of the table and then give a verbal and mathematical description of the sequence.

n	0	1	2	3	4	5
$f(n)$	10	18	26	34		

6. Sequence G is described by the difference equation $g(n) = 1.5\,g(n-1)$ and initial condition $g(0) = -6$. Describe the sequence verbally and make a table showing the first five terms.

7. Sequence H is given by the difference equation $h(n) = 0.75\,h(n-1) + 50$ and initial condition $h(0) = 2{,}000$. Describe the sequence verbally and make a table showing the first five terms.

8. The initial term value of sequence J is 50. To get any term value, take the previous term value, multiply by -1, and then add 75. Write the initial condition and difference equation for the sequence using correct mathematical notation. Then make a table showing the first five terms.

9. Consider the affine difference equation $a(n) = 0.6\,a(n-1) + 20$.
   a. Suppose that $a(0) = 60$. Find $a(1)$ through $a(5)$ and record in a table.
   b. Repeat part (a) using $a(0) = 40$ as the initial condition.
   c. Graph the two sequences that you found on the same set of axes.
   d. What do you notice about the sequence values as n gets bigger and bigger?

10. Let $a(n) = a(n-1)^2 + 2$ with $a(0) = 5$.
    Note: In this text $a(n-1)^2$ equals $a(n-1) \cdot a(n-1)$.
    a. Give a verbal description of the difference equation and initial condition.
    b. Find $a(1)$ through $a(3)$.

11. For each exercise, (i) identify the difference equation as either linear or exponential, (ii) find the solution equation, and (iii) sketch the graph of the solution equation, showing the vertical intercept and general shape.
    a. $u(n) = u(n-1) + 10; u(0) = 50$
    b. $u(n) = 1.25\,u(n-1); u(0) = 2{,}000$
    c. $u(n) = 0.8\,u(n-1); u(0) = 100$
    d. $u(n) = u(n-1) - 3; u(0) = 25$

12. For each exercise, (i) identify the difference equation as either linear or exponential, (ii) find the solution equation, and (iii) sketch the graph of the solution equation, showing the vertical intercept and general shape.
    a. $v(n) = 0.95\,v(n-1); v(0) = 1{,}000{,}000$
    b. $v(n) = 1.06\,v(n-1); v(0) = 10$
    c. $v(n) = v(n-1) - 20; v(0) = 2{,}000$
    d. $v(n) = v(n-1) + 5.6; v(0) = 0.0$

13. Consider the difference equation given by $a(n) = a(n-1) + 5$.

    a. If $a(0) = 10$, write the solution equation.

    b. If $a(0) = 15$, write the solution equation.

    c. Using technology, graph both solution equations (simultaneously) from $x = 0$ to $x = 10$. Adjust the vertical scale so that you can see the complete graphs. Sketch on your homework paper; label each graph with its solution equation.

    d. For the initial condition given in part (a), find the value of $a(4)$ using two different equations. Explain your work.

14. Suppose the difference equation $a(n) = 1.5\, a(n-1)$ has an initial condition of $a(0) = 2$.

    a. Is this a linear, exponential, or affine difference equation?

    b. Make a table listing a(0) through a(9).

    c. Graph the $(n, a(n))$ ordered pairs from your table.

    d. What is the solution equation to this difference equation? Explain how you can check that the solution equation is correct.

15. Suppose that a population has a natural monthly growth rate of $r = 5\%$ and a net migration loss of 20 individuals each month.

    a. Model the population with an affine difference equation.

    b. If the initial population is 2,000, determine the population after six months.

16. Suppose that a population has a natural yearly decay rate of $r = -1.5\%$ and a net migration gain of 60 individuals each year.

    a. Model the population with an affine difference equation.

    b. If the initial population is 5,000, determine the population after five years.

Applications

17. From 1991 to 2001, meat consumption in China rose by an average of 2 kg per capita per year.[4] (*Note*: *Per capita* means "per person.") Using this statistic only, which of the following can you write: the initial condition, difference equation, or solution equation? Explain.

18. In the example concerning the growing acreage of organic farming (Example 7-4), suppose we let $n = 0$ refer to the year 1992 rather than 1997. Which of the following would change: the initial condition, the difference equation, or the solution equation? Explain. Would any of these changes affect the predicted values for organic farmland acreage?

19. Worldwide thinning of glaciers is a symptom of global warming. Surveys taken by the North Cascades Glacier Climate Project indicate that North Cascade glaciers in Washington State have lost an average of 0.30 m of ice thickness each year between 1984 and 1991. This thinning is equivalent to losing about 10% of the volume of all glaciers in the North Cascades over the 17-year period.[5]

    a. Write a linear difference equation relating glacier thickness in year n to that in year $n-1$.

    b. Suppose that a glacier averages 60 meters in thickness in 1984. Find the thickness of the glacier for each of the next five years.

    c. For the glacier in part (b), determine the solution equation. Use the solution equation to find the thickness in year 2001. Would this glacier lose "about 10% of its volume" over the 17-year period from 1984 to 2001? Explain.

20. In the so-called corn belt of the United States, topsoil loss has negatively impacted corn production. A linear model that quantifies this impact is $P(t) = 100 - 5.65t$, where P represents production of corn (as a percentage of the maximum production) and t is the topsoil loss in inches.[6]

    a. Find the value of P when $t = 0$, and interpret the result.

    b. Interpret the meaning of the slope of the equation.

    c. Rewrite the linear equation in the form of an initial condition and difference equation.

    d. By what percentage will corn production decrease if soil loss equals 10 inches? Explain how you can determine the answer using two different methods.

21. Renewable energy sources include hydroelectric, wind, solar, wave and tidal, geothermal, and combustible renewables. Table 7-13 shows renewable energy production data.[7]

    a. What are the *yearly* growth multipliers and *yearly* growth rates in renewable energy production for industrialized and less industrialized countries? (*Note: Consult Chapter 5 for assistance if necessary.*)

TABLE 7-13

	1997 Renewable Energy Production (× 1000 metric tons of oil energy equivalent)	Percent Increase since 1987
Industrialized countries	335,929	27
Less industrialized countries	924,052	19

Source: World Resources Institute

b. Use the yearly growth multipliers to write two exponential difference equations, one for industrialized and one for less industrialized countries. Also state the initial conditions, using 1997 as the base year.

c. Use the initial conditions and difference equations that you wrote to estimate renewable energy production in year 2004 for both categories of countries.

d. Assuming current growth rates remain fixed, estimate when renewable energy production for industrialized and less industrialized countries will be equal. List two factors that might make this assumption unlikely.

22. Environmental exposure to lead poses many health risks, including nausea, vomiting, and palsy. In children, mental retardation and other learning disabilities are also symptoms of lead poisoning. Children's risk of lead poisoning is higher than that of adults because their gastrointestinal (GI) tract absorbs lead more readily. Lead from paint chips, pottery, and lead dust often enter the body through the GI tract, although lead can also enter through the skin and lungs. Lead is stored in the body in blood, soft tissue, and bones; most doctors monitor lead in the blood (Figure 7-11). *Note: In Chapter 8 there is an interesting Science in Depth article about the lead industry, and in Chapter 10 there is an extended exercise modeling lead in the body using multiple difference equations.*

a. Estimates vary for the elimination half-life of lead stored in the blood. One source indicates that the half-life is about 25 days, while another claims 36 days.[8,9] Which of these two numbers produces the slowest elimination rate? What is that rate?

b. Suppose that a person's bloodstream has a lead concentration of 80 μg/dL (micrograms per deciliter) and that there is no additional exposure to lead. Write a difference equation and initial condition that describe the lead concentration in the bloodstream over time. Base your difference equation on the slowest elimination rate from part (a).

c. A safe blood lead concentration in adults is thought to be at or below 25 μg/dL. In how many days will the person's blood lead concentration reach this level?

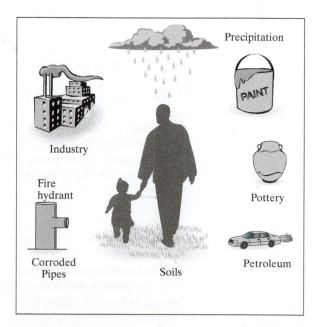

Figure 7-11 Lead is a heavy metal dangerous to human health. Some common sources of lead are depicted in this diagram.

23. In Example 7-5 concerning the population of Mexico City, the second assumption in the modeling process was to let the natural increase take place first, followed by the migration increase (in each given year). Assume that, for computational purposes, the order of these two growth processes is switched.

   a. Rewrite the difference equation. Then recalculate Table 7-11 using the new difference equation.

   b. What are the similarities and differences between the two modeling approaches? Is one approach better than the other? Explain.

24. Suppose that the microloan in Example 7-6 was described by $u(0) = 1,200$ and $u(n) = 1.003\,u(n-1) - 75$.

   a. Describe in your own words the details of the loan, assuming that n represents months.

   b. How much is owed after five months?

   c. After how many months will the loan be repaid? What is the amount of the last payment?

   d. What is the total finance charge for the microloan? Note: The total finance charge is the amount paid in addition to the original loan amount.

25. Deer populations in many regions have exploded as predators, such as wolves and coyotes, have been eliminated. Suppose that a deer population is estimated to number 2,000. Each season it grows by 4%. At the end of the season hunters are allowed to kill 200 deer.

   a. Write the difference equation describing the deer population in season n to that in season $n-1$. What type of difference equation did you write?

   b. Make a table listing the deer population over 14 years.

   c. Make a graph of the deer population over 14 years.

   d. What number of deer should wildlife managers allow to be killed each season if they plan to keep the population steady at 2,000?

   e. How many deer should wildlife managers allow to be killed each season if they plan to have the population number 2,400 after 10 seasons? This is a "trial and error" problem —just estimate an answer, observe the results, then revise your estimate accordingly.

26. In a laboratory study in 1934 by the famous experimental biologist G. F. Gauss, two microorganisms of the species *Paramecium caudatum* were placed in a test tube and allowed to multiply.[10] The food supply and environmental conditions were kept constant. The bacteria's growth is modeled by the difference equation $p(n) = 1.1\,p(n-1) - 0.000267\,p(n-1)^2$, where $p(n)$ is the population in hour n.

   a. Find $p(10)$, $p(20)$, $p(30)$, . . . , $p(120)$ and record in a table, using one decimal place.

   b. Make a graph which includes the initial condition and the 12 data values from part (a).

   c. Why do you suppose that the graph levels off? What appears to be the maximum population level permissible in the test tube's environment? This level is called the **carrying capacity** for the environment.

SCIENCE IN DEPTH

The Politics of Immigration

In the spring of 2004, the Sierra Club, one of the world's largest and most influential environmental organizations, held an election to select new members to the club's board of directors.[11] In addition to candidates recommended by the board, a group of outsiders ran for board seats in an unsuccessful attempt to change the Sierra Club's position on a single issue. That issue was immigration.

The challengers included prominent environmentalists such as David Pimental and the former governor of Colorado, Richard Lamm. This group called for stricter limits on immigration to help limit population growth. The challengers were met by stiff opposition from a variety of interest groups, and none were elected to the board of directors of the Sierra Club. What were some of the immigration issues raised during this election?

The anti-immigration challengers asserted that immigration is the major source of population growth in the United States today.[12] The rate of legal immigration has risen steadily in the United States since World War II, and currently amounts to over 1 million people per year.[13] In 2003, 35 million immigrants resided in the United States, representing over 10% of the U.S. population.

The challengers also noted that California's population grew from 30 million people in 1990 to 34 million people in 2000. In California, the number of children born to women during their lifetimes (the **fertility rate**) is approximately 2.1 on average. This value is the same as the **replacement rate**, the number of children necessary to replace their parents. A fertility rate equal to the replacement rate will eventually result in a stable population. Migration out of California to other states exceeds migration into California from other states by about 150,000 people annually. Therefore, all of the recent growth in California's population is due to foreign-born immigrants (60%) and birth of new citizens by foreign-born women (40%).[14]

Anti-immigration groups also believe that the country with the largest per capita consumption of natural resources on planet Earth should curtail, not expand, its population. The United States represents about 5% of the world population but uses about 25% of the world's energy.

Pro-immigration groups counter these arguments by claiming that immigration to the United States does not represent a gain in population on a planetary scale, but just a shift in population from Central America and eastern Asia to the United States. They argue that migration from less-industrialized countries such as Guatemala and China to a highly industrialized nation such as the United States will lower the death rate in their families, raise their standard of living, and increase their level of education (especially among women), all of which will lower the global fertility rate.

Pro-immigration forces also contend that over-consumption of natural resources by the United States should not be blamed on immigrants, and that banning immigration will not solve the problem of excess consumption in the United States. The United States needs more fuel-efficient transportation, agricultural practices that use less water, cleaner manufacturing processes, and a shift toward a more frugal lifestyle. Forbidding immigration will have no impact on solutions to over-consumption.

Except for indigenous Native Americans, the United States is a nation of immigrants and their descendants. Early immigrants to the United States mostly came from Europe, but over 80% of current immigrants come from Central and South America and eastern Asia. Pro-immigration forces argue that xenophobia or racism anchors the anti-immigration viewpoint.

CHAPTER PROJECT:
HUMAN POPULATION AND MIGRATION

Which countries have birth rates exceeding death rates? Which countries have more people migrating in than out? How can population pyramids be used to study the dynamics of a population over time? In the companion project to Chapter 7 found at the text's Web site (**enviromath.com**), you will use difference equations and Web-based population data to predict population growth or decline through the year 2050 for both Belgium and a country of your choice. You'll compare your model predictions with those of the U.S. Census Bureau. Also, you will explore how both immigration and emigration affect population predictions, and how mathematics and science can shine light on the politics of population growth.

NOTES

[1]U.S. Dept of Agriculture, Economic Research Service, *Recent Growth Patterns in the U.S. Organic Foods Market*, by Carolyn Dimitri and Catherine Greene, Agricultural Information Bulletin No. AIB777, September 2002.

[2]U.S. Bureau of the Census International Database, http://www.census.gov/ipc/www/idbnew.html; G. Tyler Miller, Jr., *Sustaining the Earth*, 5th ed. (Pacific Grove, CA: Brooks Cole, 2002), 91.

[3]International Institute for Environment and Development, *Poverty and Environment*, by Damian Killeen and A. Atiq Rahman, May 2001, http://www.iied.org/pubs/pdf/full/11010IIED.pdf

[4]World Resources Institute, Earth Trends, *Carnivorous Cravings: Charting the World's Protein Shift*, by Karen Holmes and Wendy Vanasselt (ed.), July 2001, http://earthtrends.wri.org/features/view_feature.cfm?theme=8&fid=24

[5]North Cascades Glacier Climate Project, http://www.nichols.edu/departments/Glacier/

[6]J. H. Stallings 1964. "Phosphorus and Water Pollution," *Journal of Soil Water and Conservation* 22 (1964): 228–231. As cited in "Soil Erosion and Crop Productivity: Topsoil Thickness," *Integrated Pest Management Newsletter*, January 29, 2001, Iowa State University, http://www.ipm.iastate.edu/ipm/icm/2001/1-29-2001/topsoilerosion.html

[7]World Resources Institute, *World Resources 2000–2001: People and Ecosystems: The Fraying Web of Life*, Table ERC. 4, http://pubs.wri.org/pubs_pdf.cfm?PubID=3027

[8]Massachussetts Division of Occupational Safety, "Health Hazards of Lead for Workers and Children," http://www.mass.gov/dos/leaddocs/lead_hlthhaz.htm

[9]Mount Sinai Medical Center, Bone Lead Measurement Laboratory, "Why Measure Lead in Bone?," http://www.mssm.edu/cpm/xrf/why.html

[10]G. F. Gause, *The Struggle For Existence* (1934; repr., New York: Hafner Publishing, 1964).

[11]Terence Chea, "Immigration a Hot Topic for Sierra Club," *The Seattle Post-Intelligencer,* April 10, 2005.

[12]Californians for Population Stabilization, *California's Population Growth 1990–2002: Virtually all from Immigration*, http://www.cap-s.org/main.html

[13]U.S. Dept of Homeland Security, Office of Immigration Statistics, http://uscis.gov/graphics/shared/statistics/index.htm

[14]U.S. Census Bureau, California QuickFacts, http://quickfacts.census.gov/qfd/states/06000.html

8

Affine Solution Equations and Equilibrium Values

I n the 1940s, whooping cranes in North America numbered about 20, and the population was teetering on the brink of extinction. Fortunately, efforts by conservation biologists to reintroduce the cranes into former ranges have been successful, and today the population numbers over 400. The efforts have been imaginative, to say the least! In 2001, a flock of recently hatched whooping cranes was led by an ultralight airplane between Wisconsin and Florida (Figure 8-1). What is the fate of the whooping crane population? Will it continue to grow, or will it stabilize at some equilibrium level?

In the previous chapter we explored the connection among linear and exponential difference equations, initial conditions, and solution equations. Recall that affine difference equations combine both linear change and exponential change. This chapter begins by exploring how to find the solution equation to an affine difference equation model. Then we examine equilibrium values and stability, which help us understand why some populations remain at stable levels while others go extinct.

For students new to the study of difference equations, Chapter 7 introduced quite a few fundamental ideas: describing sequences verbally and with difference equations, calculating solution equations, and more. As we investigate advanced difference equation problems in this and future chapters, it will be worth the time and effort to review those fundamentals in Chapter 7.

Figure 8-1 A flock of whoopers (*Grus americana*) led by an ultralight. The pilot is dressed in a full whooping crane costume so the birds maintain their natural fear of people. *Photo courtesy of U.S. Fish and Wildlife Agency.*

THE SOLUTION EQUATION TO THE AFFINE MODEL

Chapter 7 discussed the concept of a **solution equation**, which is the functional equation that corresponds to a difference equation and initial condition. Solution equations for linear and exponential difference equations are fairly straightforward to determine. Now we introduce a method for finding the solution equation to an affine difference equation. Our first example will show the method in its entirety. Because the method gets a bit long, we will later make an important assumption that will simplify the method substantially.

Suppose that a country has a population of 100 million people and has a **natural decay rate** (birth rate minus death rate) of $r = -5\%$ per year. In other words, deaths are exceeding births, and the population is decreasing at a rate of 5% annually. To offset this population decrease, the country allows a net influx of 2 million migrants each year (immigrants moving into the country minus emigrants moving out). These numbers have been selected to make this example mathematically easier. Nonetheless, this population scenario (negative natural growth rate and positive migration) is realistic for several countries around the world today, including Germany, Italy, Russia, and Sweden.[1]

Because the population of our fictitious country is decreasing exponentially, with an annual decay multiplier of $M = 0.95$, as well as increasing linearly by 2 million people each year, an affine difference equation will best model the population. Let p represent the population (in millions of people) and n represent years, with $n = 0$ representing the initial year. The affine difference equation and initial condition are

$$p(n) = 0.95p(n-1) + 2; \quad p(0) = 100 \text{ million people}$$

Our task is to find a solution equation for this difference equation and initial condition. We start by examining the graph of the population over an 80-year period (Figure 8-2). We notice that the vertical intercept on the graph has the value $p = 100$, which agrees with the initial condition. The shape of the graph is similar to that of an exponential decaying function, but the graph levels off at about 40 million people—well *above* the horizontal axis. We know that the solution equation cannot be a standard exponential function, because exponential graphs approach the horizontal axis (i.e., keep getting closer and closer to $y = 0$).

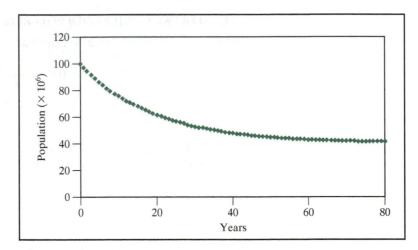

Figure 8-2 The population appears to decay exponentially but levels off around 40 million people.

We conjecture that the function is exponential but that it has been *shifted upward* from the horizontal axis. A standard exponential function has the form $p(n) = p_0 M^n$ while a vertically shifted exponential function looks like $p(n) = a M^n + b$. In the shifted formula, n and p are the independent and dependent variables, M is the multiplier, and a and b are constants. A positive value for b will make a standard exponential graph shift upward, while a negative value for b will make it shift downward. In this example, b must be positive, and its value should be about 40. Note that in the shifted version of the exponential equation, the initial condition is not the constant a (we'll hear more about this later).

Our task is to determine the *exact* values of a, M, and b. The method we use involves substituting three data points into the equation $p(n) = a M^n + b$ to create a system of three equations. By solving these equations simultaneously, we'll find values for a, M, and b. Remember, when finished we will be able to take some shortcuts. *So hang in there as we work through this!*

Population values over several years, as derived from the difference equation, are listed in Table 8-1. We'll use the first three points (ordered pairs) to find the solution equation.

TABLE 8-1

n	$p(n)$
0	100
1	97
2	94.15
3	91.4425
4	88.870375
5	86.4268563

First Ordered Pair:

Substituting the first point $(n, p) = (0, 100)$ into $p(n) = a M^n + b$ gives $100 = aM^0 + b$. Because $M^0 = 1$, this equation simplifies to $100 = a + b$. Solving for b yields $b = 100 - a$.

Second Ordered Pair:

Substituting the second point $(n, p) = (1, 97)$ into $p(n) = a M^n + b$ gives the new equation $97 = a M + b$. Using the formula $b = 100 - a$ from before, we substitute and simplify as follows:

$$97 = a M + b$$

$$97 = a M + 100 - a$$

$$-3 = a M - a$$

$$-3 = a(M - 1)$$

We would like to solve for a in the last equation above by dividing by $M - 1$. This is only allowed if $M - 1$ is not equal to 0 (remember, division by zero is undefined or meaningless). But $M - 1 = 0$ only if $M = 1$, and this is not the case with either exponential decay or growth. We're OK to divide by $M - 1$, so we get

$$a = \frac{-3}{M - 1} \qquad \textbf{Equation 1}$$

Previously we've expressed a in terms of M, and we do the same for b:

$$b = 100 - a$$

$$b = 100 - \frac{-3}{M - 1}$$

$$b = 100 + \frac{3}{M - 1} \qquad \textbf{Equation 2}$$

Using Equation 1 and Equation 2, substitute for a and b in the equation $p(n) = a M^n + b$, and simplify as follows:

$$p(n) = a M^n + b$$

$$p(n) = \frac{-3}{M - 1} M^n + 100 + \frac{3}{M - 1}$$

$$p(n) = \frac{-3 M^n}{M - 1} + 100 + \frac{3}{M - 1}$$

Third Ordered Pair:

To find the value of M, substitute the third point $(n, p) = (2, 94.15)$ into this last equation:

$$94.15 = \frac{-3 M^2}{M - 1} + 100 + \frac{3}{M - 1}$$

Subtract 100 from each side to get

$$-5.85 = \frac{-3 M^2}{M - 1} + \frac{3}{M - 1}$$

Clear denominators by multiplying each side by $M - 1$:

$$-5.85(M - 1) = -3 M^2 + 3$$

Simplify and set equal to zero:

$$3 M^2 - 5.85 M + 2.85 = 0$$

The last result is a quadratic equation. Recall from intermediate algebra that solutions to a quadratic equation $A x^2 + B x + C = 0$ can be found by using the quadratic formula:

$$x = \frac{-B \pm \sqrt{B^2 - 4AC}}{2A}$$

So the solutions to our equation are

$$M = \frac{5.85 \pm \sqrt{5.85^2 - 4(3)(2.85)}}{6}$$

$$M = \frac{5.85 \pm 0.15}{6}$$

$$M = 0.95 \text{ or } M = 1$$

As mentioned earlier, the multiplier M cannot equal 1, so the only feasible solution is $M = 0.95$. We now know that the solution equation will be of the form $p(n) = a(0.95)^n + b$. To determine values for a and b, substitute $M = 0.95$ into Equation 1 and Equation 2:

Solving for a	**Solving for b**
$a = \dfrac{-3}{M-1}$	$b = 100 + \dfrac{3}{M-1}$
$a = \dfrac{-3}{0.95-1}$	$b = 100 + \dfrac{3}{0.95-1}$
$a = 60$	$b = 40$

With $M = 0.95$, $a = 60$, and $b = 40$, the solution equation is

$$p(n) = 60(0.95)^n + 40$$

Note that the value of b, which is the upward shift in the exponential graph, is consistent with our earlier graphical estimate. Also, when $n = 0$, the population is equal to $p(0) = 60(0.95)^0 + 40 = 100$ million people—this is exactly the initial condition.

We started with a population growing *linearly* by 2 million immigrants each year and decaying *exponentially* by 5% each year. We found that the value of M in the solution equation was equal to the value of M in the difference equation. This means that the exponential decay in the solution equation was equal to the exponential decay in the difference equation *as if there were no linear growth!* This is rather surprising, but it happens to be true for all affine difference equations and their solution equations. In future problems, we will take it as fact that M in the solution equation is identical to M in the difference equation.

> **The value of M in the solution equation is the same as the value of M in the affine difference equation.**

Knowing the value of M for the solution equation saves us quite a bit of time, for we will only need to determine the values of a and b. Let's examine another problem using this new approach.

EXAMPLE 8-1   The Population of Bolivia

Bolivia's population in 2003 was equal to 8.6 million people, and its natural growth rate was approximately 1.8%. The number of migrants leaving the country exceeded the number entering by about 12,000 people.[2] Model the population with an affine difference equation and initial condition, and determine the solution equation.

Solution: First identify variables. Let n represent the number of years since 2003, and p the population in millions. The population in the year 2003 is $p(0) = 8.6$, which is the initial condition. With an annual natural growth rate of $r = 1.8\%$, the annual growth multiplier is $M = 1.018$. The number of migrants leaving each year is $12{,}000/10^6 = 0.012$ million people, so the affine difference equation modeling Bolivia's population is

$$p(n) = 1.018p(n-1) - 0.012$$

The solution equation to the affine model will be of the form $p(n) = aM^n + b$. The value of M in this equation will be the same as that in the difference equation, so the solution equation has the form $p(n) = a(1.018)^n + b$. Because we already know M, we need only two data points to determine the values of a and b. One data point is the initial condition: $(0, 8.6)$. A second data point can be readily chosen from a list of predicted population values (Table 8-2). We'll use the second point listed, $(1, 8.7428)$.

TABLE 8-2

n	$p(n)$
0	8.6
1	8.7428
2	8.8881704
3	9.03615747
4	9.18680830
5	9.34017085

Substituting the point $(0, 8.6)$ into $p(n) = a(1.018)^n + b$ gives

$$8.6 = a(1.018)^0 + b$$

which simplifies to

$$8.6 = a + b \qquad \textbf{Equation 1}$$

Substituting the point $(1, 8.7428)$ into $p(n) = a(1.018)^n + b$ yields

$$8.7428 = a(1.018)^1 + b$$

which simplifies to

$$8.7428 = 1.018a + b \qquad \textbf{Equation 2}$$

Equation 1 and Equation 2 can be solved simultaneously to determine values for a and b. One way to solve the system is through further substitution. We rewrite Equation 1 as $b = 8.6 - a$ and then substitute this into Equation 2:

$$8.7428 = 1.018a + 8.6 - a$$

Now collect the two terms involving a, and subtract 8.6 from each side:

$$0.1428 = 0.018a$$

After dividing each side by 0.018, we obtain $a \approx 7.93$. Finally, substitute this value of a into the formula $b = 8.6 - a$ to get $b \approx 0.67$. Using these values for a and b, we can write the solution equation as

$$p(n) = 7.93(1.018)^n + 0.67$$

Figure 8-3 displays projected population values for Bolivia based upon the difference equation and initial condition (discrete points) and the solution equation (continuous curve). We see that the graphs overlap at integer values of years, verifying that we have found the correct solution equation.

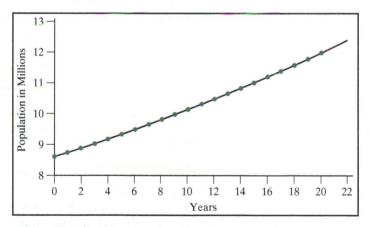

Figure 8-3 Bolivian population since 2003, as modeled by both the difference equation (discrete) and solution equation (continuous) graphs

EXAMPLE 8-2   Combined Sewage Overflows

In some U.S. municipalities, sewage (untreated human and commercial waste) and surface water (water and pollutants draining from streets) flow together through underground pipes to treatment plants. During rainy conditions, the combined flows can exceed the capacity of the treatment facilities in what is called a **combined sewage overflow**. When this occurs, the combined overflow is discharged into nearby water bodies, compromising water quality with the untreated human and industrial waste and surface water pollutants (Figure 8-4).

Today, many municipalities are rerouting sewage and surface water flows through separate pipes to prevent sewage overflows. These projects are extremely expensive, so many state governments offer discounted loans as a means of aiding

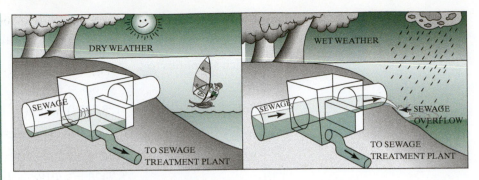

Figure 8-4 Combined sewer overflows in dry and wet conditions

communities. For example, the city of Detroit borrowed approximately $60 million from the Michigan Department of Water Quality in 2002 to upgrade part of its sewer system. The annual interest rate on this loan is 2.5%.[3] If Detroit repays this loan in 30 annual installments, determine the difference equation and solution equation that models the amount due at the end of each year.

Solution: Let's assume that Detroit's annual payments are made at the end of each year. During the year, the amount owed on the loan increases by 2.5% and then is reduced after the loan payment has been made. Let $u(n)$ represent the loan balance (the amount due on the loan) in millions of dollars at the end of the nth year. The initial condition is $u(0) = 60$ million dollars. The difference equation will be of the form

$$u(n) = 1.025\, u(n-1) - \text{annual payment}$$

To determine the annual payment, we use a trial-and-error approach. We choose some payment, then display the difference equation output in a table. If the loan does not "zero out" after year 30, we'll keep adjusting the payment until it does. For example, with an annual payment of $2 million, the difference equation becomes $u(n) = 1.025\, u(n-1) - 2$, and the loan balance is over $38 million after 30 years. See the following calculator screen.

n	$u(n)$
0	60
5	57.372
10	54.398
15	51.034
20	47.228
25	42.921
30	**38.049**

$u(n)=38.04864842$

The annual payment will have to be greater than $2 million. By trial and error, we find that a payment of $2.866 million (a difference equation of $u(n) = 1.025\, u(n-1) - 2.866$) makes the 30-year balance close to zero. See the following calculator screen.

n	$u(n)$
0	60
5	52.82
10	44.696
15	35.505
20	25.106
25	13.34
30	**.02891**

$u(n)=.028907479$

Now that we've estimated the correct difference equation, we can proceed to find the solution equation, which will be of the form $u(n) = a(1.025)^n + b$. Substituting the points $(0, 60)$ and $(1, 58.634)$ into this equation produces the two equations

$$60 = a + b \quad \text{and} \quad 58.634 = 1.025a + b$$

The solution to these equations is $a = -54.64$ and $b = 114.64$ (that work is left for the student as an exercise). The amount due at the end of each year n is given by the solution equation

$$u(n) = -54.64(1.025)^n + 114.64$$

We can check that the solution equation is correct by calculating the amount due after 30 years:

$$u(30) = -54.64(1.025)^{30} + 114.64$$

$$= 0.02890748 \text{ million dollars}$$

$$= \$28{,}907.48$$

This amount agrees with the value produced by the difference equation and displayed in the last calculator screen. We would need to adjust the annual loan payment to the nearest penny to have the 30-year balance zero out exactly. Often, the last payment on a loan is a bit higher or lower than the other payments; in this example, the city of Detroit would need to include an additional \$28,907.48 with its 30th payment.

In the two preceding examples, we found the solution equation to an affine difference equation and initial condition using a systems of equations approach. The following is a list of the mathematical steps that were used:

Systems Approach to Finding the Affine Solution Equation

1. Start with the affine difference equation $u(n) = M \cdot u(n-1) + k$ and initial condition $u(0)$. The solution equation will be of the form $u(n) = a M^n + b$, with the value of M in the solution equation equal to the value of M in the difference equation.

2. Find two points (two ordered pairs) generated by the difference equation and initial condition. It's easiest to use the points corresponding to $n = 0$ and $n = 1$.

3. Substitute each point into $u(n) = a M^n + b$ to create a system of two equations involving a and b.

4. Solve the system to find a and b. Use these values to write the solution equation.

EQUILIBRIUM VALUES

We now investigate long-term behavior of affine difference equations and note the patterns that emerge when various initial conditions are used. We start by introducing a simplified example from wildlife management.

Suppose that an otter population currently totals 100 otters but is decreasing by 25% each year because of habitat destruction and disease. Wildlife managers plan to boost the population by introducing 125 otters from a captive breeding program into the population each year. The affine model for the population change from year $n-1$ to year n is given by the difference equation

$$p(n) = 0.75p(n-1) + 125$$

This model assumes that the otters bred in captivity are introduced at the end of each year (after the 25% reduction) and that they, too, will die off at the annual rate of 25% in future years. So what will happen to the otter population in the long term if this management strategy is carried out for many years? A graph over a 12-year period shows the population increasing rapidly at first, and then more slowly, approaching a level of 500 otters (Figure 8-5).

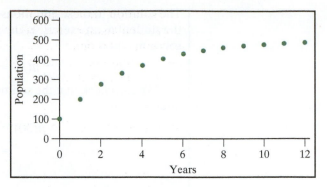

Figure 8-5 Projected otter population under the management plan. The population grows quickly at first, then levels off near 500 otters.

For the same difference equation, we can investigate what would happen to the population if the initial condition were equal to 500 otters, that is, $p(0) = 500$. Then the number after one year would equal

$$p(1) = 0.75p(0) + 125$$
$$= 0.75(500) + 125$$
$$= 500 \text{ otters}$$

Likewise, the number after the second year would equal

$$p(2) = 0.75p(1) + 125$$
$$= 0.75(500) + 125$$
$$= 500 \text{ otters}$$

We speculate that if the population ever equals 500 otters, it will remain at that level. In fact, we can *prove* that this is true using the difference equation. Assume that the population in year $n-1$ is 500 otters; then the population in year n will equal

$$p(n) = 0.75p(n-1) + 125$$
$$= 0.75(500) + 125$$
$$= 500 \text{ otters}$$

Biologically speaking, we say the otter population is in **equilibrium** when it numbers 500 individuals, meaning it will stay at that level (Figure 8-6). Mathematically speaking, we say the difference equation $p(n) = 0.75p(n-1) + 125$ has an **equilibrium value** or **equilibrium level** of 500 otters. An equilibrium value for a difference equation is a value for $p(n-1)$ that produces the same value for $p(n)$.

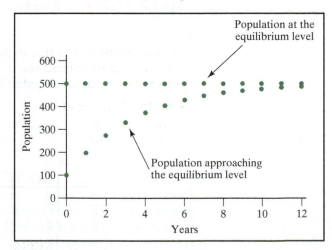

Figure 8-6 Otter populations starting with $p(0) = 100$ and $p(0) = 500$

Is there a way to find the equilibrium value for a difference equation using algebra? The answer is yes, and here's how. Suppose that the equilibrium value we're searching for is denoted by the variable E. When we plug E into $p(n-1)$ in the difference equation, we should get E back for $p(n)$. For the otter population difference equation $p(n) = 0.75p(n-1) + 125$ these substitutions result in

$$E = 0.75E + 125$$

Finding the value of E is now rather easy. First, subtract $0.75E$ from each side to obtain

$$0.25E = 125$$

Then divide each side by 0.25 to get the equilibrium value of $E = 500$. This is exactly the value that we expected. Let's now use this simple algebraic technique to find equilibrium values for a few additional difference equations.

EXAMPLE 8-3   Finding an Equilibrium Value

Consider the difference equation $p(n) = 1.2p(n-1) - 50$. Find the equilibrium value.

Solution: The equilibrium value is the number E such that if $p(n-1) = E$, then $p(n) = E$. These substitutions result in the following equation and solution:

$$E = 1.2E - 50$$
$$-0.2E = -50$$
$$E = 250$$

Check: If $p(n-1) = 250$, then $p(n) = 1.2(250) - 50 = 250$.

EXAMPLE 8-4   Finding an Equilibrium Value

Find the equilibrium value for the difference equation $u(n) = u(n-1) + 50$.

Solution: We let $u(n) = E$ and $u(n-1) = E$ and get the equation $E = E + 50$. Subtracting E from each side gives the new equation: $0 = 50$. This equation has no solution, which implies that there is no equilibrium value for this difference equation. Can you explain, using some other approach, why this difference equation has no equilibrium value? What kind of difference equation is this?

EXAMPLE 8-5   Finding an Equilibrium Value

Find the equilibrium value for the following difference equation:

$$u(n) = 2\,u(n-1) - 0.5\,u(n-1)^2$$

Solution: Let $u(n) = E$ and $u(n-1) = E$. The resulting equation is

$$E = 2E - 0.5E^2$$

Subtract E from each side to produce $0 = E - 0.5E^2$. Factoring out E from both terms on the right-hand side gives $0 = E(1 - 0.5E)$. Setting each factor equal to zero results in the two equilibrium values $E = 0$ and $E = 2$.

Check: If either of these values is used for an initial condition, then all subsequent outputs from the difference equation will be the same. Tables of values confirm this fact. See the following calculator screens.

checking $E = 0$	checking $E = 2$

n	$u(n)$
0	0
1	0
2	0
3	0
4	0
5	0
6	0

$u(n)=0$

n	$u(n)$
0	2
1	2
2	2
3	2
4	2
5	2
6	2

$u(n)=2$

CLASSIFICATION OF EQUILIBRIUM VALUES

In the otter population example, the difference equation describing the 25% annual decrease combined with the constant 125 otter increase is $p(n) = 0.75p(n-1) + 125$. We found that with an initial population numbering 100, the population increases until it levels off at 500 otters. We also found that if the initial population were equal to 500 otters, the population would stay at that level. What if the initial population is more than 500 otters? For example, suppose that $p(0) = 800$. To investigate what happens, we use the difference equation and this initial condition to create a table of values.

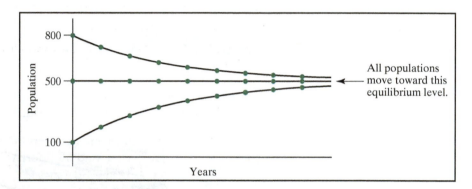

With an initial population of 800 otters, we find that the population decreases and levels off at 500 otters. As a matter of fact, *all* populations governed by this difference equation eventually move toward the equilibrium value $E = 500$. This type of equilibrium value is called **stable** or **attracting** (Figure 8-7).

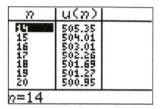

Figure 8-7 All populations approach $E = 500$ otters

Suppose a wolf population is modeled by the difference equation given in Example 8-3, which is $p(n) = 1.2p(n-1) - 50$. In that example, we showed algebraically that the equilibrium value is $E = 250$. So if a wolf population is modeled by this difference equation, and the current population equals 250 wolves, then future populations will remain at 250 wolves.

If the initial population was less than 250 wolves, what would happen in the future? With this difference equation and an initial population of 245 wolves, the population *decreases* and eventually reaches zero between the 21st and 22nd year. See the following screen.

With an initial population of 255 wolves, the population grows and grows. After 60 years it reaches about 282,000 wolves (see the following calculator screen). Most likely this would not happen and our model would need to be adjusted for such long-term projections.

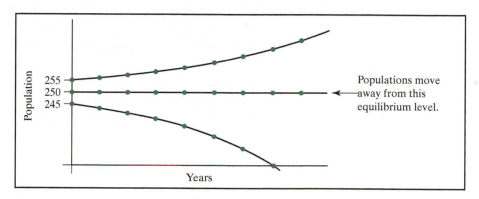

n	$u(n)$
0	255
10	280.96
20	441.69
30	1436.9
40	7598.9
50	45752
60	281988

$u(n)=281987.5718$

If the population is not originally at the equilibrium value of 250 wolves, then it *moves away from* the equilibrium value (Figure 8-8). We call this type of equilibrium value **unstable** or **repelling**. If a wolf population gets too high or too low, wildlife managers might redistribute wolves from one population to another (Figure 8-9).

Figure 8-8 Population values move away from the unstable equilibrium value of $E = 250$

Figure 8-9 To rebuild self-sustaining populations, wildlife biologists move animals from one population to another. In this photo, a gray wolf (*Canis lupus*) is tranquilized and prepared for transport. *Courtesy of U.S. Fish and Wildlife Service/photo by Ronald L. Bell.*

EXAMPLE 8-6   Classifying Equilibrium Values

Find the equilibrium value for the difference equation $u(n) = 2.5\,u(n-1) - 3{,}000$. Then classify the equilibrium value as stable or unstable.

Solution: By solving $E = 2.5E - 3{,}000$, we find that the equilibrium value is $E = 2{,}000$. If the initial condition is slightly larger than E, for example, $u(0) = 2{,}010$, then the difference equation outputs grow larger. If the initial condition is smaller than E, for example $u(0) = 1,990$, the outputs get smaller and smaller (see the following screens). Thus the equilibrium value of $E = 2,000$ is unstable.

Testing the stability of $E = 2,000$:

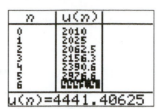 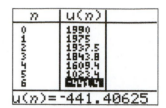

EXAMPLE 8-7 Classifying Equilibrium Values

The equilibrium values for the difference equation $u(n) = 2u(n-1) - 0.5u(n-1)^2$ are $E = 0$ and $E = 2$ (see Example 8-5). Classify the equilibrium values as stable or unstable.

Solution: To test the equilibrium value $E = 0$, we use nearby values as initial conditions, such as $u(0) = -0.1$ and $u(0) = 0.1$ (Of course, if u represents a biological population, then $u(0)$ could not be negative.) We find that the difference equation outputs in either case move away from the equilibrium value, so $E = 0$ is an unstable equilibrium value (see the following screens).

Testing the stability of $E = 0$:

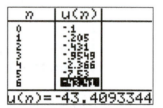 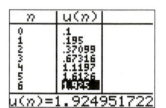

Now look at the equilibrium value $E = 2$. Pick initial conditions a bit smaller and a bit larger that the equilibrium value, such as $u(0) = 1.8$ and $u(0) = 2.2$. In both cases, the difference equation outputs are attracted to the equilibrium value, so $E = 2$ is a stable equilibrium value (see the following screens).

Testing the stability of $E = 2$:

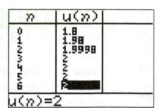 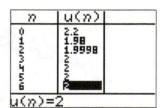

REVISITING THE AFFINE SOLUTION EQUATION

We again return to the otter population example in which each year there is a 25% decrease followed by a constant 125 otter increase. We found that the difference equation describing this population is $p(n) = 0.75p(n-1) + 125$, and the equilibrium value is $E = 500$.

Recall that the solution equation will be of the form $p(n) = a(0.75)^n + b$. For any initial condition, we can find the values of a and b using the system of equations method described earlier. Without showing the details (they are left for a student exercise) we get the following results:

- If $p(0) = 100$, the solution equation is $p(n) = -400(0.75)^n + 500$.
- If $p(0) = 350$, the solution equation is $p(n) = -150(0.75)^n + 500$.
- If $p(0) = 600$, the solution equation is $p(n) = 100(0.75)^n + 500$.
- If $p(0) = 875$, the solution equation is $p(n) = 375(0.75)^n + 500$.

So what patterns do we observe? The following box summarizes our findings.

> In the affine solution equation $p(n) = a(0.75)^n + b$,
> - b is equal to the equilibrium value E.
> - $a + b$ is equal to the initial condition $p(0)$.

These twin observations are true for all affine solution equations, although they are a bit difficult to prove. We'll just take them as fact, which makes the problem of finding the affine solution equation that much easier. Let's look at a few examples.

EXAMPLE 8-8 Back to Detroit's Loan

Earlier we investigated a loan of $60 million obtained by the city of Detroit (Example 8-2). The difference equation describing the amount (in millions of dollars) owed on the loan after year n was given by $u(n) = 1.025u(n-1) - 2.866$. Determine the solution equation by using the equilibrium value.

Solution: The solution equation will be of the form $u(n) = a(1.025)^n + b$. By solving the equation $E = 1.025E - 2.866$, we find that the equilibrium value is $E = 114.64$. But this is also the value for b, so $b = 114.64$.

We know that $a + b$ is equal to the initial condition, so $a + 114.64 = 60$. This simplifies to $a = -54.64$. The solution equation is given by

$$u(n) = -54.64(1.025)^n + 114.64$$

In the preceding example, what is the *significance* of the equilibrium value $E = 114.64$? If the initial loan amount were $u(0) = \$114.64$ million, then the amount due in future years would neither go up nor go down. This is because the annual amount of interest (2.5% × $114.64 million) would exactly equal the amount of principal paid off each year ($2.866 million). In this case, Detroit would never pay off its loan! Create a table or graph on your calculator or computer to confirm that this is true.

EXAMPLE 8-9 Salmon Hatcheries

Along many U.S. streams and rivers that empty into the Pacific, and a few that flow into the Atlantic, hatcheries have been constructed to breed and introduce salmon to aid declining natural populations (Figure 8-10). The population dynamics of salmon are quite complex, as is the interplay between natural and hatchery-bred fish. Models describing these mixed populations are well beyond the scope of this book, but we can examine a simple scenario with the mathematics developed in this chapter.

Suppose that a river's salmon population numbers 1,200 fish but is decreasing by 20% each year. Hatchery managers want to introduce the same number of hatchery salmon each year so that the river's population reaches 5,000 by the end of 10 years. How many salmon should they introduce each year?

Solution: This is a problem that could be solved by trial and error, as we demonstrated with sewage overflows in Example 8-2. Instead, we'll use an algebraic method by letting k represent the number of hatchery salmon introduced annually. The affine difference equation describing the river's salmon population is

$$p(n) = 0.8p(n-1) + k$$

and the initial condition is $p(0) = 1,200$.

The equilibrium value for the difference equation is found by solving the equation

$$E = 0.8E + k$$

Figure 8-10 A bucket of eggs from chinook salmon (*Oncorhynchus tshawytscha*) is artificially inseminated in a hatchery.
Photo: Langkamp/Hull.

Simplifying this equation results in $E = \dfrac{k}{0.2}$ or $E = 5k$.

The affine solution equation has the form $p(n) = a(0.8)^n + b$. But b is equal to the equilibrium value, so $b = E = 5k$. Also, $a + b = 1{,}200$, which results in $a = 1{,}200 - 5k$. Thus the solution equation is

$$p(n) = (1{,}200 - 5k)(0.8)^n + 5k$$

Now we use the fact that in 10 years the salmon population is required to equal 5,000. Substituting these numbers results in the equation

$$5{,}000 = (1{,}200 - 5k)(0.8)^{10} + 5k$$

Being careful in rearranging the preceding equation, we can isolate k. Note that $(0.8)^{10} = 0.1073741824$, but using this decimal version is cumbersome, so we just leave $(0.8)^{10}$ alone until later. Start by removing parentheses:

$$5{,}000 = 1{,}200(0.8)^{10} - 5(0.8)^{10}k + 5k$$

Now collect numbers on the left and factor out k on the right:

$$5{,}000 - 1{,}200(0.8)^{10} = (-5(0.8)^{10}+5)k$$

Divide each side by the quantity $-5(0.8)^{10} + 5$ to isolate k:

$$\frac{5{,}000 - 1{,}200(0.8)^{10}}{-5(0.8)^{10} + 5} = k$$

This last fraction can be entered into a calculator in one line to avoid round-off error. Just remember to put parentheses around both the numerator and denominator. The resulting solution is $k = 1{,}091.42059$ salmon. See the following calculator screen.

```
(5000-1200*.8^10
)/(-5*.8^10+5)
        1091.42059
```

Notice that if we use the full decimal version of k, then the difference equation produces 5,000 salmon in year 10 (Table 8-3). Rounding off to the nearest whole fish, we conclude that the hatchery must release 1,091 salmon each year.

TABLE 8-3	n	$p(n)$	n	$p(n)$
	0	1,200	6	4,341
	1	2,051	7	4,564
	2	2,733	8	4,743
	3	3,277	9	4,886
	4	3,713	10	5,000
	5	4,062		

CHAPTER SUMMARY

A difference equation that combines linear change and exponential change is called **affine**. Affine difference equations can model simple population change where both exponential change (births and deaths) and linear change (immigration and emigration) are present. The general form of an affine difference equation is $u(n) = M \cdot u(n-1) + k$. The affine solution equation has the form $u(n) = a M^n + b$, with the value of M in the solution equation equal to the value of M in the difference equation. The values of a and b in the solution equation can be found by solving a **system**

of equations. First substitute two data points into $u(n) = a M^n + b$ to create a system of two equations; then solve the system to find a and b.

An **equilibrium value** or **equilibrium level** for a difference equation is a value for $u(n-1)$ that produces the same value for $u(n)$. Once a biological population is at an equilibrium level, it will remain at that level. To find an equilibrium value E for a difference equation, substitute E for both $u(n-1)$ and $u(n)$; then solve the resulting equation. For example, in the affine difference equation $u(n) = 0.60 u(n-1) + 160$, the equilibrium value can be found by solving the equation $E = 0.60E + 160$. The solution is $E = 400$.

Equilibrium values can be classified as either **stable** (attracting) or **unstable** (repelling). If populations move toward an equilibrium value, the equilibrium value is stable. If populations move away from an equilibrium value, the equilibrium value is unstable (Figure 8-11).

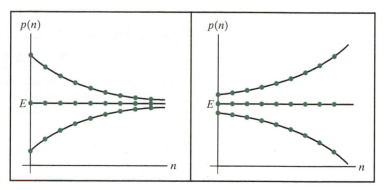

Figure 8-11 Equilibrium values that are stable (left) and unstable (right)

The equilibrium value of an affine difference equation can be used to find the solution equation. If the solution equation is written in the form $u(n) = a M^n + b$, then b is equal to the equilibrium value E, and $a + b$ is equal to the initial condition $u(0)$. Consider the affine difference equation $u(n) = 0.60 u(n-1) + 160$ and initial condition $u(0) = 700$. The solution equation will be of the form $u(n) = a(0.60)^n + b$. As shown previously, the equilibrium value for this difference equation is $E = 400$, which makes $b = 400$. Because $a + b$ equals the initial condition of 700, the value of a must equal 300. Therefore, the solution equation is $u(n) = 300(0.60)^n + 400$.

END *of* CHAPTER EXERCISES

Affine Solution Equations

1. Solve the system given in Example 8-2: $60 = a + b$ and $58.634 = 1.025a + b$.

2. Suppose that the points $(n, p) = (0, 315)$ and $(n, p) = (1, 303)$ lie on the graph of the solution equation $p(n) = a(1.24)^n + b$. Determine the values of a and b; then write the solution equation. Check that the two points solve the solution equation.

3. For each affine difference equation and initial condition, find the solution equation. Use the system of equations method outlined in the first few examples of this chapter.
   a. $p(n) = 2.5p(n-1) + 900$; $p(0) = 400$
   b. $p(n) = 0.75p(n-1) - 45$; $p(0) = 510$
   c. $p(n) = 1.03p(n-1) - 75$; $p(0) = 75$

4. Find the solution equation for each affine difference equation and initial condition using the system of equations method.
   a. $u(n) = 0.6u(n-1) + 80$; $u(0) = 300$
   b. $u(n) = 1.05u(n-1) + 10$; $u(0) = 50$
   c. $u(n) = 2.2u(n-1) - 600$; $u(0) = 100$

5. Label each difference equation as linear, exponential, or affine. For each difference equation and initial condition, find the solution equation. For affine difference equations, use the system of equations method to find the solution equation.

   a. $u(n) = 0.64u(n-1)$; $u(0) = 4$

   b. $v(n) = v(n-1) - 6$; $v(0) = 100$

   c. $w(n) = 0.2w(n-1) + 4$; $w(0) = 7$

6. Label each difference equation as linear, exponential, or affine; then find the solution equation. For affine difference equations, use the system of equations method.

   a. $u(n) = u(n-1) + 11$; $u(0) = -20$

   b. $v(n) = 1.6v(n-1) - 12$; $v(0) = 4$

   c. $w(n) = 1.07w(n-1)$; $w(0) = 325$

Equilibrium Values and Stability

7. A population is modeled by the difference equation $p(n) = 0.80p(n-1) + 200$.

   a. Determine the solution equation for each of the following initial conditions: $p(0) = 600$, $p(0) = 800$, and $p(0) = 1,300$. Use systems of equations.

   b. Using technology, graph the three solution equations simultaneously. (You will need a function grapher rather than a difference equation grapher.) Make a quick sketch of the graphs on your homework paper. Indicate the most likely location of the equilibrium level with a horizontal dashed line.

   c. Is the equilibrium value stable or unstable?

8. A population is modeled by the difference equation $p(n) = 1.1p(n-1) - 40$.

   a. What type of difference equation is this? Describe a biological scenario that might warrant using values 1.1 and –40.

   b. Using the systems of equations method, find the solution equation for each of the following initial conditions: $p(0) = 300$, $p(0) = 500$, and $p(0) = 650$.

   c. Using technology, graph the three solution equations simultaneously. (See technology directions in the previous exercise.) Sketch the three solution equations on a single graph. Indicate the most likely location of the equilibrium level with a horizontal dashed line.

   d. Is the equilibrium value stable or unstable?

9. Find the equilibrium value(s) for each of the following difference equations. If there is no equilibrium value, say so. When you find equilibrium values, check them directly in the difference equation.

   a. $u(n) = 2.2u(n-1) - 48$

   b. $u(n) = 3u(n-1)$

   c. $v(n) = 0.75v(n-1) + 20$

   d. $v(n) = v(n-1) + v(n-1)^2 - 25$

10. Find the equilibrium value(s) for each difference equation. If there is no equilibrium value, say so. When you find equilibrium values, check them.

    a. $p(n) = p(n-1) + 20$

    b. $p(n) = 0.98p(n-1) + 40$

    c. $p(n) = 1.025p(n-1) - 6,000$

    d. $p(n) = p(n-1)^2 - 7p(n-1) + 7$

11. Find the equilibrium value for each affine difference equation. Then classify the equilibrium value as stable or unstable.

    a. $u(n) = 1.25u(n-1) - 2,000$

    b. $v(n) = 0.5v(n-1) + 60$

12. Find the equilibrium value for each affine difference equation. Then classify the equilibrium value as stable or unstable.

    a. $p(n) = 0.64p(n-1) + 1,800$

    b. $p(n) = 3.6p(n-1) - 3,120$

Equilibrium Values and Solution Equations

13. For each affine difference equation, find the equilibrium value. Then using the initial condition, find the solution equation (see Example 8-8 and Example 8-9).

    a. $u(n) = 1.5u(n-1) - 45$; $u(0) = 300$

b. $v(n) = 0.98v(n-1) + 40$; $v(0) = 102$

c. $w(n) = 1.025w(n-1) + 200$; $w(0) = 2,000$

14. Use the equilibrium value of each affine difference equation to find the solution equation.

a. $p(n) = 0.85p(n-1) + 60$; $p(0) = 175$

b. $p(n) = 1.75p(n-1) - 150$; $p(0) = 400$

c. $p(n) = 0.96p(n-1) - 1$; $p(0) = 50$

Applications

15. Suppose that a tree farm has 6,000 live trees. Farm managers implement a plan to annually harvest 25% of the trees and then plant 500 new ones (Figure 8-12).

Figure 8-12 A tree farm in Iowa *Photo: USDA National Resource Conservation Service.*

a. Define variables and write a difference equation modeling the number of live trees on the farm.

b. Find the solution equation to the difference equation, given that there were originally 6,000 trees. Use the system of equations method.

c. Make a graph of this function over a 20-year period.

d. At what value does the number of trees eventually level off? How many trees will be harvested annually once the tree population is at that level?

16. Russia's 2005 population was estimated to equal 143 million people, and its natural decay rate was approximately $r = -0.5\%$. The number of migrants entering the country exceeded the number exiting by about 150,000 people.[4]

a. Model the population with an affine difference equation and initial condition.

b. Use your difference equation and initial condition to estimate the population in the year 2015.

c. Determine the solution equation using the systems of equations approach.

d. Use your solution equation to estimate the population in the year 2015. Does the answer agree with that for question (b)?

17. In Chapter 7, the mathematics of loans were introduced in a scenario in which a community borrows $1,500 to purchase a small wind-powered generator. The community is charged an interest rate of 3% annually, or 0.25% monthly. Each month the community paid $50, until the loan was paid off. Using n to represent months, and u to represent the loan balance, the difference equation and initial condition are given by $u(n) = 1.0025u(n-1) - 50$ and $u(0) = 1,500$ dollars.

a. Find the equilibrium value for this difference equation. Explain what the equilibrium value means in terms of this loan.

b. Use the equilibrium value to find the solution equation.

c. Verify that the difference equation and solution equation produce the same loan balance after 10 months.

18. Follow-up to previous exercise.

    a. Suppose the community borrows the same amount of money at the same interest rate, but the monthly payments are half as large. Write the new difference equation and initial condition.

    b. How many months will it take to pay off the loan? What is the payment on the last month? *Note:* Assume that the last payment is a bit more than $25.

    c. What is the finance charge (total interest paid) for the loan in this scenario?

    d. Find the equilibrium value for the difference equation. Explain, using a common sense argument, why the equilibrium value is less than the one in the previous exercise (when paying $50 monthly).

    e. Use the equilibrium value to find the solution equation.

19. There is concern that dental amalgams (fillings) pose a significant health threat because they contain mercury. Mercury is released from the amalgams in the form of mercury vapor and passes through the cell walls of the lungs and mouth into the bloodstream. Once in the bloodstream, mercury moves to, and is stored in, the kidneys, liver, and central nervous system. One scientific report claims that an individual with eight amalgam fillings can absorb up to 120 μg (micrograms) of mercury per day. This same report states that mercury's half-life is 15–30 years in the central nervous system and that mercury causes "psychological, neurological and immunological problems in humans."[5] The American Dental Association disputes the danger of mercury from dental amalgams, claiming "amalgam restorations remain safe and effective."[6]

    a. Show that with a half-life of 15 years, mercury is eliminated from the body with an annual decay multiplier of $M = 0.9548$.

    b. If 120 μg of mercury per day are absorbed into the body, how many micrograms are absorbed each year? By the way, how many grams is that?

    c. Using results from parts (a) and (b), write a difference equation that models the amount of mercury in the body (in micrograms) in year n compared to the amount in year $n-1$.

    d. Find the equilibrium value to the difference equation. Round your answer to the nearest 1,000 micrograms. Show that the equilibrium value is stable. What does a stable equilibrium value mean for the human body? Explain.

    e. Suppose that a person's body has 0.5 grams of mercury when $n = 0$. Find the solution equation to the affine difference equation.

20. In order to grow, plants need nutrients such as nitrogen, phosphorus, potassium, magnesium, calcium, and sulfur. If nutrients are scarce, plant growth will be diminished. In freshwater lakes, phosphorus is often the nutrient that is most lacking and is therefore called a **limiting nutrient**. During rainy weather, combined sewer overflows (see Example 8-2) can pour raw sewage into lakes, adding a tremendous amount of phosphorus to the water. Additional phosphorus commonly causes an **algal bloom**, which is an explosion in the population of algae.

    Suppose that a small lake has an area of 6 ha (hectares), an average depth of 15 m (meters), and an initial phosphorus concentration of 100 μg/L (micrograms per liter).

    a. Determine the number of kilograms of phosphorus in the lake initially. *Note:* 1 ha = 10,000 m^2, 1 m^3 = 1,000 liters, and 1 kg = $10^9 \mu$g. Let p represent the total amount of phosphorus in the lake in kilograms and n the number of months. Write the initial phosphorus amount as an initial condition.

    b. Phosphorus can be removed from a lake through physical processes such as flushing and settling, through biological uptake, and through chemical reactions. Experiments have shown that in a typical freshwater lake, phosphorus amounts decrease at a rate of approximately 5.6% per day.[7] If this is the case, what is the daily decay multiplier? What is the monthly decay multiplier, assuming 30 days in a month? *Write using five decimal places.*

    c. A town neighboring the lake cuts it budget for sewage treatment, resulting in frequent sewage overflows. The town's combined sewer overflow system dumps sewage water into the lake, on average, once a month. The typical volume of these sewage overflows is 80,000 m^3. Studies have shown that the mean phosphorus concentration of sewage water is 10 mg/L (milligrams per liter).[8] Determine the number of kilograms of phosphorus that the town dumps into the lake each month. *Note:* 1 mg = 10^{-6} kg.

    d. Use your answer from parts (b) and (c) to write an affine difference equation modeling the phosphorus amount in the lake water. *Assume that all of the sewage phosphorus is added at the end of the month, with no change in the total volume of the lake.*

    e. Using algebra, find the equilibrium phosphorus amount, rounded to the nearest kilogram. Use your answer to write a solution equation.

    f. Determine the amount of phosphorus in the lake after 2.5 months.

21. Rework Example 8-9 on salmon hatcheries using the following changes: The river's salmon population numbers 2,000 fish and is decreasing by 25% each year. Fish managers want the salmon population to reach 6,000 by the end of 15 years.

22. Suppose that a deer population is growing by 2% annually. Biologists want to manage the herd by allowing hunting so that the population in 20 years is around 2,500 deer, or half the current population.

    a. Let the yearly number of culled deer be represented by the letter k. Write an affine difference equation modeling the deer population in year n compared to that in year $n-1$. What is the value of the initial condition?

    b. Find the equilibrium value of the difference equation in terms of k.

    c. Write the solution equation for the difference equation and initial condition, using the variable k where appropriate.

    d. Determine the value of k, the annual deer culling. Use as much precision as possible during your computations. Round your final answer appropriately.

23. Sustainable microenergy systems built around photovoltaic (solar) cells are being installed and financed by Grameen Shakti in rural Bangladesh, where more conventional energy sources are too costly to install. Photovoltaic energy systems can power low-wattage appliances, such as soldering irons, lights, phones, and computers. Access to these products has been extremely successful in improving the lives of the rural poor. A typical photovoltaic system costs about $450. Grameen Shakti offers one option in which the customer pays 15% of the total price as a down payment. The remaining 85% of the cost is repaid through 36 monthly payments with a 12% annual (1% monthly) service charge.[9]

    a. Assume that a customer has obtained a loan for a photovoltaic system under the given payment plan. Let $u(n)$ represent the loan balance after month n. What is the value of $u(0)$?

    b. Let the monthly payment be represented by the letter k (see Example 8-9). Write a difference equation modeling the amount due in month n compared to that in month $n-1$.

    c. What is the equilibrium value of the difference equation, in terms of k?

    d. Write the solution equation for the difference equation and initial condition, using the variable k where appropriate.

    e. Determine the value of k, the monthly payment. Use as much precision as possible during your computations; round your final answer to the nearest penny.

    f. If 35 monthly payments are made (equal to the rounded value of k), what is the amount of the last payment?

24. A second financing option provided by Grameen Shakti is the following: The customer pays 25% of the total price as a down payment. The remaining 75% of the cost is to be repaid in 24 months with an 8% annual (0.67% monthly) service charge.

    a. Rework parts (a) through (e) in the previous exercise to determine the monthly payment for purchasing a $450 photovoltaic system under this financing option.

    b. How much more, or less, will the photovoltaic system cost over the lifetime of the loan under this plan as compared to the plan in the previous exercise?

SCIENCE IN DEPTH

Get the Lead Out

Lead is a naturally occurring element that is used in batteries, metal alloys, paints, and gasoline additives. Breakdown of these manufactured products releases lead into the air, water, and soil. Lead is also released during combustion of coal, garbage, and tobacco. Lead is quite toxic; long-term exposure can damage the brain, kidneys, and intestines. The toxicity of lead has been known for hundreds of years through the death and injury of workers in lead-handling industries. Children are very susceptible to lead poisoning.

Lead was commonly used as an additive in paints and glazes to add flexibility and durability. Lead paints were promoted by the lead industry with hyperbolic advertising; one print advertisement was titled "Lead helps to guard your health." Lead-based paints were widely used indoors and were promoted by the paint industry for use on toys, cribs, and other applications that would bring the toxic lead paint in direct contact with children (Figure 8-13). Small children who sucked or chewed on objects greatly increased their blood lead levels. The paint industry simultaneously blamed

Lead takes part in many games

Figure 8-13   Advertisement by the National Lead Company, 1923.

children for indiscriminant behavior and claimed that lead paints were safe. In 1971 the U.S. government took a small step forward in regulating lead paints, by banning their use on interior surfaces in public housing projects. Lead paints are still found on both interior and exterior surfaces of many older homes in the United States.

While the medical community and consumer groups in the United States were working hard to remove lead from paint, another source of lead in the environment was being introduced. Starting in the 1920s, tetraethyl lead was added to gasoline in order to increase fuel economy and horsepower, and reduce engine knock. With the explosion in the number of automobiles in the twentieth century, automobile emissions became the number one source of lead pollution in the United States.

Tetraethyl lead is just as toxic as lead in paints; when manufacturing processes were lax, workers died from just a few hours of exposure to high concentrations of tetraethyl lead. Despite the well-known harmful effects of ingesting lead, the U.S. lead industry responded to criticism of leaded gasoline by launching a propaganda campaign to convince the public that tetraethyl lead was safe. In the late 1980s, the U.S. government finally banned lead in gasoline. However, lead from past automobile emissions ("legacy lead") can be found in soils and sediments, especially along roadsides. Leaded gasoline is still being used in Central America, Asia, and Africa and was only recently banned in Europe.

While high concentrations of lead along roadsides have been traditionally blamed on former automobile emissions, another source may contribute to roadway lead: wheel weights (see Figure 6-21 in Chapter 6). Lead weights are attached to wheel rims to balance the tires and give a smoother ride. These weights fall off vehicles onto roadways and are ground down by passing traffic, producing lead particles and dust. A study in Albuquerque, New Mexico, found that 1 kilometer of highway accumulated at least 50 kilograms of wheel weight lead in a single year.[10] With over 60 million kilograms of lead consumed worldwide each year just to balance wheels, some countries are considering tin as a substitute for lead weights.

To learn more about the U.S. lead industry, read *Deceit and Denial: The Deadly Politics of Industrial Pollution* by Gerald Markowitz and David Rosner (Berkeley: University of California Press, 2002).

 ## CHAPTER PROJECT:
LEAD IN THE BODY

Lead pipes have been phased out and replaced in the United States and no longer constitute a significant source of lead in drinking water. However, drinking fountain water can be contaminated with lead from the solder used to connect copper pipes. Solder is a combination of lead and zinc that melts at low temperatures, and the lead can leach from the solder in acidic water. In schools, high quantities of lead can accumulate when the water in fountains sits without being flushed.

In the companion project to Chapter 8 found at this text's Web site (**enviromath.com**), you will use difference equations to model a child's lead intake from a school's water system. You'll look at the long-term trend in blood lead concentration and compare the model values with acceptable levels suggested by health professionals. Just how dangerous is a little lead in the water supply? Has the water in your school been tested for lead and other heavy metals?

NOTES

[1]Population statistics for all countries can be found at the U.S. Census International Database, http://www.census.gov/ipc/www/idbprint.html

[2]U.S. Census International Database.

[3]Michigan Dept of Environmental Quality, *The Loan Arranger*, Winter 2002, http://www.michigan.gov/deq/0,1607,7-135-3308_3579-18646--,00.html

[4]U.S. Census International Database.

[5]Joseph Mercula and Dietrich Klinghardt, "Mercury Toxicity and Systemic Elimination Agents," *Journal of Nutritional and Environmental Medicine* 11, no. 1 (2001), 53–62.

[6]John E. Dodes, "The Amalgam Controversy: An Evidence-Based Analysis," *Journal of the American Dental Association* 132 (2001): 348–356.

[7]E. B. Welch, *Ecological Effects of Wastewater: Applied Limnology and Pollution Effects*, 2nd ed. (London: Cambridge University Press, 1980), 66.

[8]Ibid., 84.

[9]Grameen Shakti, "Financing Policy of Grameen Shakti," http://grameen-info.org/grameen/gshakti/programs.html

[10]R.A. Root, "Lead Loading of Urban Streets by Motor Vehicle Wheel Weights," *Environmental Health Perspectives* 108 (2000): 937–940.

9

Logistic Growth, Harvesting, and Chaos

In 1859, Charles Darwin published his influential work *The Origin of Species*, laying the foundation for the biological theories of competition among species, natural selection, and evolution. Surprisingly, it was not until the early 1900s that biologists tested some of these theories in their laboratories. One such researcher, a young Russian named G. F. Gause, placed several bacteria (*Paramecium caudatum*) in a test tube with ample food and allowed them to multiply. Over six days of observation, Gause found that the population grew rapidly at first, and then leveled off at 375 bacteria as the test tube became more and more crowded (Figure 9-1).[1] The S-shaped graph of the bacteria population is called a **logistic curve**.

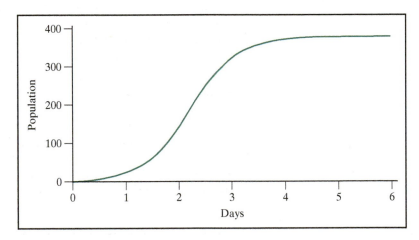

Figure 9-1 Bacteria growing according to the logistic curve.

In many examples in past chapters, we assumed that a population grew by the same percentage each time period. This fundamental assumption is the basis of exponential growth. But as Gause verified, real populations in closed environments grow quite differently. Typically, as populations increase in size, growth rates decrease due to constraints on habitat, nutrients, or other vital needs. In this chapter, we modify the exponential difference equation with a decreasing growth rate; the result is the **logistic difference equation**. We also examine equilibrium values and harvesting strategies and finish with an introduction to periodic behavior and chaos.

MODELING LOGISTIC GROWTH WITH DIFFERENCE EQUATIONS

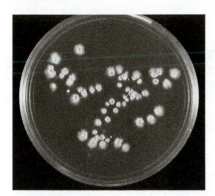

Figure 9-2 The bacteria *Nocardia asteroides* growing in a plate of agar. The population is constrained by the amount of space and nutrients. *Photo: CDC/Dr. William Kaplan.*

Suppose that a biologist puts 1 million bacteria into a petri dish that contains sufficient nutrients for the population to grow by 5% each hour. If we let n represent time in hours and p the population of bacteria, then the exponential difference equation and initial condition which model the growth of the bacteria are

$$p(n) = 1.05p(n-1); \qquad p(0) = 1,000,000 \text{ bacteria}$$

Exponential models often produce unrealistic results when extrapolated too far into the future. According to the preceding model, after 480 hours (20 days) there would be $p(480) \approx 1.5 \times 10^{16}$ bacteria in the petri dish. For typical-sized bacteria (0.005 mm in diameter), the total volume would be close to 7 cubic meters—much bigger than the petri dish itself!

Exponential models predict that populations will grow without limit. But in environments with habitat and nutrient constraints, populations can only get so large (Figure 9-2). Scientists call the maximum level which a population can attain the **carrying capacity** of the environment. For example, a small pond might have a carrying capacity of 10 frogs, while an Antarctic island might have a carrying capacity of 50,000 penguins. Another example concerns people on planet Earth; estimates for Earth's carrying capacity of *Homo sapiens* range from 10 to 30 billion people.

The basic exponential model assumes that the growth rate stays the same no matter how large the population. The **logistic model** assumes that the growth rate gets smaller as the population gets larger. More specifically, the logistic model assumes the following:

- With zero individuals present, the population has the greatest capacity for growth. That is, the percent growth rate r will take on its largest value. We call this value the **unrestricted growth rate** or r_{max}.
- As the population gets larger, the percent growth rate will get smaller in a *linear* manner.
- At the carrying capacity, the population's percent growth rate r will equal 0%.

Let's assume that the petri dish mentioned earlier has a carrying capacity of 6 million bacteria. When the population numbered 1 million, the rate of growth was $r = 5\%$. Under the logistic assumptions, the growth rate decreases linearly as the population gets larger, eventually reaching $r = 0\%$ at the carrying capacity. This information is displayed graphically in Figure 9-3.

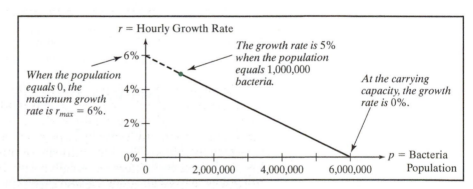

Figure 9-3 The logistic model assumes that the percent growth rate decreases linearly as the population increases in size.

By extending the line backward on the graph (dashed line), we determine that when the population is 0, the growth rate is $r = 6\%$ per hour. This is the maximum or unrestricted growth rate, so $r_{max} = 6\%$ per hour. We could have found r_{max} using an algebraic process, using points $(p, r) = (1,000,000, 5\%)$ and $(p, r) = (6,000,000, 0\%)$. That work is left as a student exercise.

To build the logistic difference equation model, we need the equation of the line in Figure 9-3. With a vertical intercept of $b = 6\%$ and a slope $m = \frac{\text{rise}}{\text{run}} = \frac{-6\%}{6,000,000}$, the linear equation in the form $y = b + mx$ is

$$r = 6\% - \frac{6\%}{6,000,00} p$$

With 6% written in decimal form, the linear equation becomes

$$r = 0.06 - \frac{0.06}{6,000,000} p$$

Note that the fraction in the preceding equation can be simplified, but we leave it alone so the values for the unrestricted growth rate and carrying capacity are "visible." This formula tells us how the hourly growth rate decreases as the population increases. But in difference equations, calculations are based on values from the previous time step. So to convert to difference equation form, we replace p with $p(n-1)$, resulting in

$$r = 0.06 - \frac{0.06}{6,000,000} p(n-1) \quad \textbf{Equation 1}$$

A logistic difference equation is simply a modified exponential difference equation. Recall that a basic exponential difference equation is written as

$$p(n) = M \cdot p(n-1)$$

or

$$p(n) = (1 + r) \cdot p(n-1) \quad \textbf{Equation 2}$$

In exponential growth models, the value of r is fixed at the same number (e.g., $r = 0.05$). But in the logistic model, the value of r will decrease according to a linear formula. For the petri dish, that linear formula is exactly the one found in Equation 1. To get the logistic difference equation, we replace r in Equation 2 with the formula for r given by Equation 1:

$$p(n) = \left(1 + 0.06 - \frac{0.06}{6,000,000} p(n-1) \right) p(n-1)$$

or, equivalently,

$$p(n) = \left(1.06 - \frac{0.06}{6,000,000} p(n-1) \right) p(n-1)$$

Recall from Chapter 7 that in this text we write $p(n-1) \cdot p(n-1)$ as $p(n-1)^2$. Using this notation, we expand our last equation and write

$$p(n) = 1.06p(n-1) - \frac{0.06}{6,000,000} p(n-1)^2 \quad \textbf{Equation 3}$$

Equation 3 is a **logistic difference equation**—it describes the growth of the bacteria population constrained by the petri dish environment. The algebraic techniques that we used to create Equation 3 are probably new to you. Don't worry; you will not have to commit those techniques to memory. What's more important is to understand that the logistic model is an exponential model modified so that the percent growth rate decreases as the population increases.

The connection between exponential and logistic difference equations is quite apparent. Notice how the first term of the logistic model looks like the familiar exponential difference equation (Figure 9-4). The second term, in which $p(n-1)$ is squared, is a quadratic **damping term**. This second term, which is *subtracted* from the first, holds the exponential growth "in check."

$$p(n) = \underbrace{1.06\, p(n-1)}_{\text{Exponential term}} - \underbrace{\frac{0.06}{6,000,000}\, p(n-1)^2}_{\text{Quadratic damping term}}$$

Figure 9-4 The exponential and damping terms of the logistic difference equation

It is easiest to observe the damping effect by displaying the population data in a table and graph. Recall that the biologist initially placed 1,000,000 bacteria in the petri dish, making the initial condition $p(0) = 1,000,000$. Using a graphing calculator or computer software, we can generate a table giving population values every 10 hours.

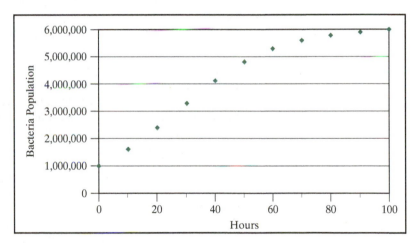

n	$u(n)$
0	1E6
10	1.59E6
20	2.37E6
30	3.26E6
40	4.11E6
50	4.8E6
60	5.28E6

$u(n)=5283977.832$

n	$u(n)$
70	5.59E6
80	5.77E6
90	5.87E6
100	5.93E6
110	5.96E6
120	5.98E6
130	5.99E6

$u(n)=5989241.467$

The population of bacteria, initially equal to 1,000,000, grows and grows, eventually approaching the carrying capacity of the petri dish (6,000,000 bacteria). Figure 9-5 displays a graph of the bacteria population values plotted every 10 hours. Notice how the population grows rapidly at first (high growth rate) but then levels off (low growth rate) as the damping takes effect.

Figure 9-5 Initially numbering 1,000,000 bacteria, the population grows logistically and levels off at 6,000,000 bacteria.

We found the logistic difference equation modeling the population of bacteria living in a petri dish. Because we'll be constructing logistic difference equations for many other populations, the general form of the equation is provided in the following box.

For a population with an unrestricted growth rate of r_{max} and a carrying capacity of K, the logistic difference equation is

$$p(n) = (1 + r_{max})p(n-1) - \frac{r_{max}}{K}p(n-1)^2 \quad \textbf{Equation 4}$$

Equation 4 has two constants—the unrestricted growth rate (r_{max}) and the carrying capacity (K). It is important to understand that these numbers depend only upon the type of organism and the type of environment. They do not depend upon the initial number present (the initial condition). In the previous example, the unrestricted growth rate ($r_{max} = 6\%$) and the carrying capacity ($K = 6,000,000$) depend only on the type of bacteria and the amount of food, space, and so on. Therefore, the logistic model (Equation 3) should describe the growth of any number of bacteria placed in the petri dish—any initial condition. We investigate this idea by plotting the same logistic difference equation using an initial condition of $p(0) = 5,000$ bacteria (Figure 9-6).

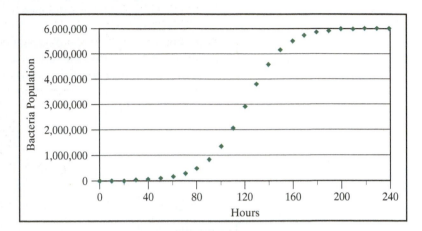

Figure 9-6 Logistic bacteria growth with initial population of 5,000.

The bacteria population again climbs toward the carrying capacity, although it takes longer to reach that level. The "**S-shape**" in Figure 9-6 is typical of logistic growth graphs.

The following are a few additional examples illustrating how to construct and use logistic difference equations.

EXAMPLE 9-1 A Rabbit Population

Suppose that a prairie grassland ecosystem has a carrying capacity of 240,000 rabbits and that the rabbit population has an unrestricted annual growth rate of 25%. If there are 16,000 rabbits initially, in how many years will the population reach the carrying capacity?

Solution Use $r_{max} = 0.25$ and $K = 240,000$ in Equation 4. The logistic difference equation becomes

$$p(n) = 1.25p(n-1) - \frac{0.25}{240,000} \, p(n-1)^2$$

We create a table of values starting with the initial condition $p(0) = 16,000$. After 58 years the population equals approximately 240,000 rabbits (the carrying capacity).

n	$u(n)$	
53	239998	
54	239999	
55	239999	
56	239999	
57	239999	
58	240000	
59	240000	

$u(n)=239999.5683$

EXAMPLE 9-2 Pacific Ocean Halibut

A study of Pacific halibut (*Hippoglossus stenolepis*) found that the mass of all halibut had an unrestricted yearly growth rate of approximately 71%, and that the total mass of the population had an upper limit in the ocean of 80,000,000 kg.[2] Suppose that at time $n = 0$ years there are 20,000,000 kg of halibut in the population.

a. Write the difference equation that models the mass of halibut (in millions of kilograms).

b. Find the mass of halibut in year $n = 4$.

c. After how many years will the halibut mass reach 95% of the carrying capacity?

Solution

a. Let $u(n)$ represent the mass of halibut in millions of kilograms in year n. The difference equation is $u(n) = 1.71u(n-1) - \frac{0.71}{80}u(n-1)^2$.

b. Using $u(0) = 20$ million kg, we find that $u(4) = 69.4$ million kg.

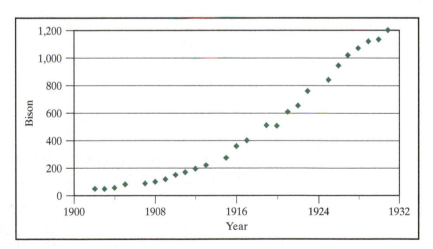

c. We calculate that 95% of 80,000,000 kg is 76,000,000 kg. According to the preceding table, this amount is reached after approximately $n = 5$ years.

EXAMPLE 9-3   Yellowstone National Park Bison

American bison (*Bison bison*, commonly called buffalo) once roamed North America in vast herds with populations numbering in the millions. From 1830 to 1890, these herds were brought to the brink of extinction as human migrants invaded Native American lands. Although many bison were killed for their furs or for sport, habitat destruction and livestock diseases were also to blame. Yellowstone National Park, established in 1872, was one area where a small population of bison survived. Although fewer than 50 bison existed there in 1902, bison numbers grew significantly over the next 30 years, as Figure 9-7 displays.[3]

Figure 9-7 Yellowstone bison population, 1902–1931.

Analysis of the 1902–1931 data (not provided here) indicates that the yearly growth rate (r) of the bison population is related to the population (p) according to the linear formula $r = 0.1879 - 0.0001298p$.

a. Use this formula to create a logistic difference equation model for the bison population.

b. According to this formula, what is the unrestricted growth rate and the theoretical carrying capacity?

Solution

a. We rewrite the formula: $r = 0.1879 - 0.0001298p$ as $r = 0.1879 - 0.0001298p(n-1)$ and then substitute for r in Equation 2:

$$p(n) = (1 + 0.1879 - 0.0001298p(n-1)) \cdot p(n-1)$$

Expanding this equation gives the logistic difference equation

$$p(n) = 1.1879p(n-1) - 0.0001298p(n-1)^2$$

b. The unrestricted growth rate is simply the growth rate when the population equals zero: $r_{max} = 0.1879$ or 18.79%. The carrying capacity is the population level at which the growth rate becomes zero. By substituting 0 for r in the formula $r = 0.1879 - 0.0001298p$, we can solve to find this population p:

$$0 = 0.1879 - 0.0001298p$$

$$0.0001298p = 0.1879$$

$$p \approx 1{,}448 \text{ bison}$$

According to our model, Yellowstone's carrying capacity is 1,448 bison. This looks reasonable considering the shape of Figure 9-7. We've made the point, though, that projecting too far into the future from a set of data is dangerous, and this is certainly the case with the Yellowstone bison. Since 1931, the population has grown well above 1,448 bison. In fact, today's herd in Yellowstone National Park numbers almost three times that value![4]

LOGISTIC EQUILIBRIUM VALUES

The prairie ecosystem described in Example 9-1 has a carrying capacity of 240,000 rabbits, and an unrestricted annual growth rate of 25%. We assume that these values describe the ecosystem in its typical state, when nutrients such as grasses and predators such as hawks are at normal levels. We found that the difference equation describing the rabbit population in this typical state is

$$p(n) = 1.25p(n-1) - \frac{0.25}{240{,}000} p(n-1)^2$$

Now suppose that one season proves to be exceptionally favorable for rabbits, and the population "spikes" to 360,000 individuals (well above the standard carrying capacity). When the ecosystem returns to its normal state, what will happen to the population? We answer this by graphing the difference equation using the initial condition $p(0) = 360{,}000$ rabbits (Figure 9-8). The graph shows that the spike in the rabbit population was just temporary, for the population decreases rapidly and returns to the carrying capacity of 240,000. This suggests that the carrying capacity is a stable equilibrium value. In fact, that is exactly the case.

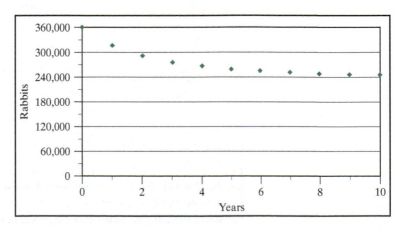

Figure 9-8 After the population spikes to 360,000 rabbits, it returns to the carrying capacity of 240,000 rabbits.

Recall that an equilibrium value for a difference equation is the number E that, when substituted for $p(n-1)$, produces the same number for $p(n)$. For the logistic difference equation modeling the rabbit population, the equilibrium value is the solution to the equation

$$E = 1.25E - \frac{0.25}{240,000}E^2$$

To solve this quadratic equation, first subtract E from each side to get

$$0 = 0.25E - \frac{0.25}{240,000}E^2$$

Now factor out E from each term on the right hand side of the equation:

$$0 = E\left(0.25 - \frac{0.25}{240,000}E\right)$$

This equation is true if either factor equals zero; that is, if $E = 0$ or if

$$0.25 - \frac{0.25}{240,000}E = 0$$

By clearing fractions and then isolating E, we find that the solution to this last equation is $E = 240,000$. Therefore, the difference equation has two equilibrium values, $E = 0$ rabbits and $E = 240,000$ rabbits. The first equilibrium value is rather meaningless—it tells us that if there are no rabbits initially, there will be no rabbits in the future. The more interesting equilibrium value is the second one—its value is the same as the ecosystem's carrying capacity.

The graph of the population difference equation using various initial conditions (Figure 9-9) indicates that the equilibrium value of $E = 0$ rabbits is unstable, for populations slightly larger than 0 move away from this level. The equilibrium value of $E = 240,000$ rabbits is stable, as larger or smaller populations converge toward this level.

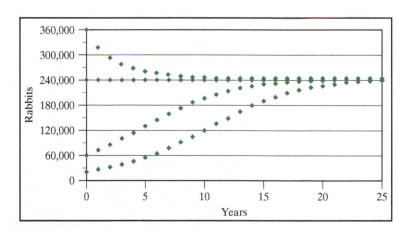

Figure 9-9 The logistic difference equation for rabbits plotted using four different initial conditions: 20,000; 60,000; 240,000; and 360,000 rabbits

HARVEST MODELS

In Chapters 7 and 8 we explored populations that not only grew exponentially, but also had some fixed amount added or subtracted each time period. These populations were modeled with affine difference equations. In a similar manner, populations growing logistically can be modified to account for additions or subtractions. Let's investigate an example of one of these modified logistic models, which are typically called **harvest models**.

EXAMPLE 9-4 Fixed Harvest

Suppose that a wildlife park holds 200 elk but has a carrying capacity of 1,500 elk. Also suppose that the elk's unrestricted annual growth rate is $r_{max} = 30\%$. With n representing years, the logistic difference equation and initial condition for the population are given by

$$p(n) = 1.30p(n-1) - \frac{0.30}{1,500}p(n-1)^2; \quad p(0) = 200 \text{ elk}$$

The logistic model predicts that the population will grow to 1,500 elk in about 30 years.

Now suppose that park managers allow hunters to harvest (i.e., kill) a fixed number of elk at the end of each year, to slow the impact of the elk on the park's ecosystem. If managers allow the harvest of 50 elk each year, how long will it take for the elk herd to reach the carrying capacity of the park?

Solution By harvesting 50 elk each year, the logistic difference equation is

$$p(n) = 1.30p(n-1) - \frac{0.30}{1,500}p(n-1)^2 - 50$$

Now we find that the elk population *never* reaches the carrying capacity of the park. In fact, the population appears to level off at about 1,309 elk, reaching this level after about 55 years (see the following calculator screens). This maximum level is an equilibrium value and can be found using a similar technique to that discussed earlier. We leave that task for a homework exercise.

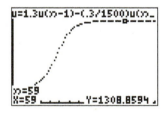

EXAMPLE 9-5 Proportional Harvest

Instead of harvesting a fixed number each year from a population, wildlife managers often prescribe harvesting an amount that is proportional to the population. In other words, harvest a smaller number when the population is small, and a larger number when the population is large. For example, suppose that for the park described in Example 9-4, wildlife managers allow 10% annual removal of the elk population. What would happen to the elk population in the long run?

Solution The modified logistic difference equation becomes

$$p(n) = 1.30p(n-1) - \frac{0.30}{1,500}p(n-1)^2 - 0.10p(n-1)$$

The right-hand side can be simplified by combining the first and third terms involving $p(n-1)$:

$$p(n) = 1.20p(n-1) - \frac{0.30}{1,500}p(n-1)^2$$

Using this difference equation and the initial condition $p(0) = 200$ elk, we find that this management strategy results in the population leveling off in the long run at about 1,000 elk.

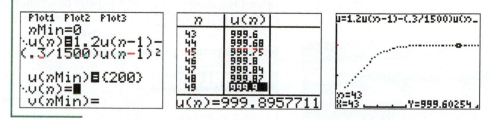

It is easy to modify logistic difference equations so that they model fixed or proportional harvests. Wildlife biologists rely on these types of quantitative models as a basis for developing harvest management plans. Yet the wise biologist knows that real-world populations don't behave exactly like mathematical models, and only field monitoring can tell if a management plan is working or not. Sometimes these plans are changed in midstream because of improved science, political whim, or public input. A good example involves the management of the Yellowstone bison. In Example 9-3, we modeled the logistic growth of the population from 1902 to 1931, during the initial period of herd restoration. But, as the U.S. National Park Service has documented,

> *Starting in 1925, concerns about brucellosis [an infectious animal disease] and how many bison Yellowstone could support led to periodic culling [killing]. During the next 40 years, park staff also reduced the elk herds in order to limit winter mortality and maintain a presumed "balance" between bison, elk, and their forage. However, by the 1960s, public opposition and evolving views of wildlife management brought herd reductions to an end. Instead of focusing on individual plants and animals, park managers now try to preserve the environmental processes that shape an ecosystem over time.*[5]

Figure 9-10 illustrates the fluctuating nature of the Yellowstone bison population from 1900 to 2002. Notice how management strategies have impacted the herd size, but also observe that natural factors, such as the severe winter of 1996–1997, can cause even more dramatic swings in the population. Are population fluctuations such as those recorded for Yellowstone bison predictable? Are there mathematical models that forecast wildly changing population values with no apparent pattern? We'll explore these questions and more in the next few sections.

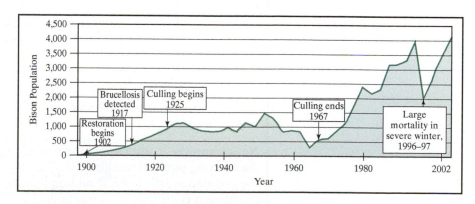

Figure 9-10 Yellowstone bison population, 1900–2002

PERIODIC BEHAVIOR

For small organisms such as insects, the capacity for population growth is often quite high. For example, the red locust (*Nomadacris septemfasciata*, similar to a grasshopper) can produce enough offspring so that population numbers increase 100-fold annually.[6] Such fecundity can lead to population explosions or plagues and often cause great environmental damage. A locust plague infesting the Sahel

Figure 9-11 A locust swarm threatens crops in Africa. *Photo: USDA.*

of West Africa in 2004 devoured an estimated 3 to 4 million hectares of cropland (Figure 9-11).[7]

High potentials for growth are, of course, related to large unrestricted growth rates. One might expect the logistic difference equation model would produce typical **S**-shaped graphs even with large unrestricted growth rates. Surprisingly, this is not the case. For unrestricted growth rate greater than 200% ($r_{max} > 2$), the logistic difference equation produces very different results. Let's consider an example.

EXAMPLE 9-6 Locust Population with $r_{max} = 2.2$

Suppose that a locust population has an unrestricted growth rate of $r_{max} = 2.2$ per year, a carrying capacity of $K = 100$ million locusts, and an initial population of 7 million locusts. Using p to denote population in millions and n the number of years, the logistic difference equation and initial condition are

$$p(n) = 3.2p(n-1) - \frac{2.2}{100} p(n-1)^2, \qquad p(0) = 7 \text{ million locusts}$$

Values for this locust population over a 20-year period are displayed in a series of calculator screens and in graphical form (Figure 9-12).

n	$u(n)$
0	7
1	21.322
2	58.229
3	111.74
4	82.882
5	114.1
6	78.718

$u(n)=78.71486002$

n	$u(n)$
7	115.57
8	75.973
9	116.13
10	74.917
11	116.26
12	74.625
13	116.28

$u(n)=116.2801803$

n	$u(n)$
14	74.633
15	116.28
16	74.626
17	116.28
18	74.625
19	116.28
20	74.625

$u(n)=74.62468941$

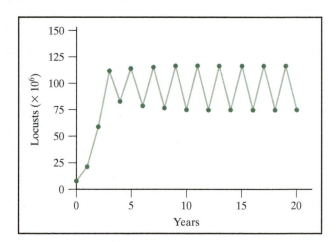

Figure 9-12 Locust population cycling above and below the carrying capacity

Notice how the locust population begins to grow logistically, but then overshoots the carrying capacity, reaching 111.74 million locusts by year $n = 3$. By year $n = 4$, the population falls to 82.88 million locusts, well below the carrying capacity. The population continues to alternate between values above and below the carrying capacity of 100 million locusts, eventually approaching the two values of 74.62 and 116.28 million locusts. We call this sequence of values a **2-cycle**. In general, alternating behavior such as this is called **periodic** or **cyclical**.

EXAMPLE 9-7 Locust Population with $r_{max} = 2.48$

Suppose that the locust population from the previous example has an unrestricted growth rate of $r_{max} = 2.48$, which is slightly higher than before. Describe the long-term behavior of the population assuming the same initial condition and carrying capacity.

Solution A calculator graph of the locust population over a 20-year time period is displayed next.

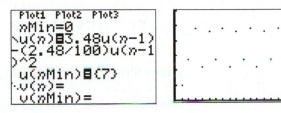

To make the graph more clear, set the plotting style to **connected mode**. (Detailed instructions on how to do this are given in the "Using Technology" section of Chapter 7.)

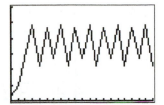

In the connected graph, we notice how the population again cycles above and below the carrying capacity, but with less regularity. To examine the long-term behavior, we create a table starting with a large n value, such as $n = 100$. See the following calculator screen. The population eventually cycles between the four values 122.0, 55.4, 116.7, and 68.4 million locusts. The sequence of population values forms a **4-cycle**.

n	$u(n)$
100	55.436
101	116.7
102	68.359
103	122
104	55.436
105	116.7
106	68.359

$$u(n)=68.35944736$$

Periodic behavior in animal populations has been well documented. Insects often have population cycles of just 2 years, but these cycles can reach 17 years, as with some species of cicadas.[8] Perhaps you saw or heard of the 17-year locusts that appeared in many U.S. locations in the spring of 2004. Actually, they were always in those locations but were living underground in the early stage of their lives!

Mammal populations can also exhibit cyclical patterns. The population of Canadian lynx (*Lynx canadensis*), approximated with trappings data, shows periodical behavior with a cycle of about 10 years (Figure 9-13).[9] Periodic behavior in organisms is not always due to high unrestricted growth rates; it is often the case that external environmental factors come into play. For example, the abundance or scarcity of rainfall

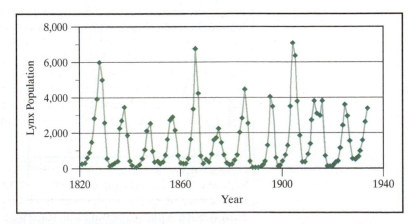

Figure 9-13 Annual number of lynx trapped, MacKenzie River, 1821–1934

is often the cause of the highs and lows of insect populations. For the Canadian lynx, it is often the populations of predators, such as wolves, that drive the fluctuations.

CHAOTIC BEHAVIOR

As r_{max} is *slowly* increased, starting just above 2, the logistic difference equation progresses through 2-cycles, then 4-cycles, then 8-cycles, and so on. In other words, the number of cycles keeps doubling and the population patterns get more and more complicated. But when r_{max} exceeds approximately 2.57, something quite strange happens—the cyclical patterns disappear. We can observe this strange behavior by letting the locust population model (from the previous examples) have an unrestricted growth rate of $r_{max} = 2.8$. The results are graphed in Figure 9-14 over a time period of 50 years.

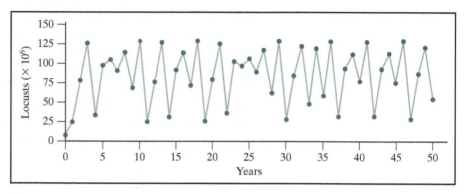

Figure 9-14 A locust model with $r_{max} = 2.8$, $K = 100$, and $p(0) = 7$

Notice how the periodic pattern that was evident with smaller values of r_{max} has now disappeared. But the population values are not fluctuating randomly—they all fall within a limited range (approximately from 25 to 125) and seem to "hover" near the carrying capacity of $K = 100$ million locusts.

Difference equations that exhibit **chaos** (or chaotic behavior) are characterized by more than just wild-looking graphs. They also exhibit **sensitivity to initial conditions**. To understand what this means, consider the following example.

EXAMPLE 9-8 A Population of Moths

Suppose we try to predict the population of moths in a small field 16 months from now using the logistic difference equation and the following information: The environmental carrying capacity is $K = 500$ moths and the unrestricted monthly growth rate is $r_{max} = 2.95$. We head into the field and estimate that there are currently 250 moths. But our field partner disagrees; she says, "Wait! I think there are 251 moths."

With a carrying capacity of 500 moths, does it really make a difference in our prediction if the starting population is 250 or 251 moths? Surprisingly, with chaotic behavior the answer is "yes!," as the two graphs in Figure 9-15 illustrate. The left graph was generated using the initial condition $p(0) = 250$, the right using $p(0) = 251$.

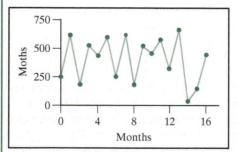 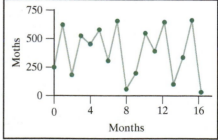

Figure 9-15 Moth populations with initial conditions $p(0) = 250$ (left) and $p(0) = 251$ (right). A slight difference in initial conditions produces wildly different results over time.

Over the first several months, the graphs look quite similar. But starting around month 8, the differences between the graphs becomes more and more noticeable. At 16 months, the left graph predicts a population of 442 moths; the right, 33 moths.

It is amazing that the two graphs in Figure 9-15 are generated using the same difference equation and nearly the same initial condition. With an equation that exhibits chaos, *small differences in the initial condition can lead to large differences in future predictions*. This is what is meant by sensitivity to initial conditions.

The topic of chaos is a very exciting and important field of study today. Chaotic behavior has been found in a multitude of examples in biology, geology, chemistry, economics, physics, and engineering. All of these fields play some role in the study of the environment. For example, scientists conjecture that biological populations that are chaotic (not patterned) are more healthy than those that are periodic (patterned). Chaotic populations promote diversity in the age structure of species and may make it difficult for predators and disease to "track" a population.

Another example involves weather models built upon equations similar to difference equations. In the 1950s, Edward Lorenz discovered sensitivity to initial conditions in these models and coined the term **the butterfly effect**. The idea behind this term is that a butterfly could flap its wings in Brazil, change the wind speed (an initial condition) ever so slightly, and instead of having sunny skies in Miami, a hurricane would roar in. To learn more about chaos theory, consider reading the popular, nontechnical book *Chaos: Making a New Science*, by James Gleick.[10] Another good source is the Public Broadcasting Service NOVA special *The Strange New Science of Chaos*.

CHAPTER SUMMARY

The **logistic model** assumes that with zero individuals present, a population has the greatest capacity for growth. The growth rate at that population level is called the **unrestricted growth rate** and is denoted r_{max}. As the population increases, the percent growth rate decreases in a *linear* manner. The **carrying capacity** is the maximum population that an environment will support. At the carrying capacity, denoted by K, the population's growth rate is $r = 0\%$. These conditions lead to the **S**-shaped logistic graph, and the logistic difference equation given by

$$p(n) = (1 + r_{max})p(n-1) - \frac{r_{max}}{K}p(n-1)^2$$

In the equation $p(n) = 1.3p(n-1) - \frac{0.3}{6,000}p(n-1)^2$, for example, the unrestricted growth rate is $r_{max} = 0.3 = 30\%$, and the carrying capacity is $K = 6,000$. Equilibrium values can be found using methods from Chapter 8. The two equilibrium values for the logistic difference equation above are the "zero population" $E = 0$ and the carrying capacity $E = 6,000$.

Logistic difference equations that are modified to account for additions or subtractions to a population are called **harvest models**. In a **fixed harvest** model, the same population amount is added or subtracted each time period. In a **proportional harvest** model, a fixed percentage of the population is added to or removed from the logistically growing population.

For an unrestricted growth rate greater than 200% ($r_{max} > 2$), logistic difference equations produce graphs very different from the standard **S** shape. For example, with an unrestricted growth rate of $r_{max} = 2.3 = 230\%$ and a carrying capacity of $K = 40$, a population eventually exhibits **periodic behavior** by alternating between population values of 27.5 and 47.3 (Figure 9-16). This is called a **2-cycle**.

As the value of r_{max} is slowly increased, a population will move toward 4-cycles, 8-cycles, 16-cycles, and so on. When r_{max} exceeds about 2.57, the cyclic patterns disappear and the difference equations exhibit **chaos**. Chaos is also characterized by **sensitivity to initial conditions**, in which small differences in the initial condition can lead to large differences in predicted values.

n	p(n)
0	5
1	15.0625
2	36.6607
⋮	⋮
15	47.2662
16	27.5181
17	47.2681
18	27.5140
19	47.2675
20	27.5152

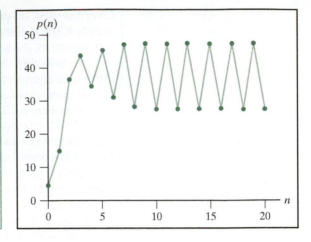

Figure 9-16 A population 2-cycle

END *of* CHAPTER EXERCISES

Warm-Up to Logistic Difference Equations

1. Consider the following logistic difference equation:

$$p(n) = 1.09p(n-1) - \frac{0.09}{1{,}000}p(n-1)^2$$

   a. What is the carrying capacity K? What is the value of r_{max}?

   b. Using technology, graph the difference equation using the initial condition $p(0) = 50$. Plot from $n = 0$ to $n = 100$. Use trace to confirm the value of K. Record on graph paper with axes labeled.

2. Consider the following logistic difference equation:

$$p(n) = 1.125p(n-1) - \frac{0.125}{800}p(n-1)^2$$

   a. What is the carrying capacity K? What is the value of r_{max}?

   b. Using technology, graph the difference equation using the initial condition $p(0) = 1{,}600$. Plot from $n = 0$ to $n = 40$. Trace along the graph to confirm the value of K. Record on graph paper with axes labeled.

3. Suppose that a population has a carrying capacity and unrestricted growth rate as given. Write the logistic difference equation.

   a. $K = 60$; $r_{max} = 0.02$

   b. $K = 4{,}000$; $r_{max} = 0.075$

   c. $K = 20{,}000$; $r_{max} = 5\%$

   d. $K = 2{,}000$; $r_{max} = 0.3\%$

4. Suppose that a population grows logistically with a carrying capacity of 3,000 individuals and an unrestricted growth rate of 3%.

   a. Write the logistic difference equation.

   b. Using technology, graph the difference equation using the initial condition $p(0) = 200$. Plot from $n = 0$ to $n = 250$. Record on graph paper; label axes.

   c. Graph the same difference equation using the initial conditions $p(0) = 2{,}000$ and $p(0) = 4{,}000$. Record on the same graph that you made in part (b).

Exploring Assumptions Behind the Logistic Model

5. Refer to the graph in Figure 9-3. Two points were given to create this graph:

$$(p, r) = (1{,}000{,}000, \, 5\%) \text{ and } (p, r) = (6{,}000{,}000, \, 0\%).$$

   a. Write the percentages in decimal form; then determine the slope between these points.

   b. Determine the equation of the line (see Chapter 4 for review, if necessary).

c. What is the vertical intercept of the equation that you found in part (b)? Interpret the meaning of this number. Does it agree with the graph?

d. In the text, the equation of the line given for Figure 9-3 is $r = 0.06 - \frac{0.06}{6,000,000} p$. Is this the same equation that you found?

6. Explore the formula $r = 0.06 - \frac{0.06}{6,000,000} p$.

a. Make a two-column table with the headings *population* and *growth rate*. In the first column, list population values from $p = 0$ to $p = 6,000,000$, jumping every 1,000,000 bacteria. Use the equation given above to determine growth rate values for the second column.

b. Explain how the patterns in the table match Figure 9-3 and agree with the basic assumptions of the logistic model.

7. Suppose that a population's growth rate r gets smaller as the population p gets larger according to the formula $r = 0.125 - 0.000125p$.

a. Make a two-column table with the headings *population* and *growth rate*. In the first column, list population values from $p = 0$ to $p = 1,000$, jumping by 100 individuals. Complete the second column using the formula.

b. What is the unrestricted growth rate of the population? How do you know?

c. What is the carrying capacity of the population? How do you know?

d. Rewrite the equation $r = 0.125 - 0.000125p$ so that it is in the format $r = r_{max} - \frac{r_{max}}{K} p$.

8. Review Example 9-1 on the logistic growth of rabbits in a prairie ecosystem.

a. Complete the following table, making use of the basic assumptions behind logistic growth models.

b. Graph the values in the table. What do you observe?

Number of Rabbits	Yearly Growth Rate of Rabbits	Number of Rabbits	Yearly Growth Rate of Rabbits
0		144,000	
24,000		168,000	
48,000		192,000	
72,000		216,000	
96,000		240,000	
120,000			

Logistic Applications

9. A graph of Earth's human population from 1850 to 2150 is given in Figure 9-17. Past data and future projections are from a United Nations' report titled *The World at Six Billion*, published in 1999.[11]

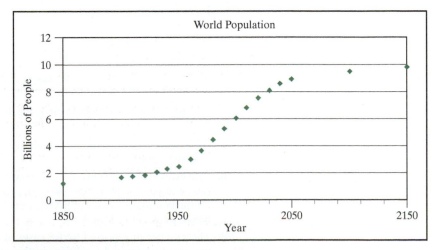

Figure 9-17 World Population, 1850–2150

a. What is the projected human carrying capacity of planet Earth, according to this graph?

b. One estimate for the unrestricted growth rate is 1.6% per year. Use this rate and your answer from part (a) to write a logistic difference equation. In your equation, let p represent Earth's population in billions of people and n the number of years after 1850.

c. The United Nations' estimate for the world population in 1850 is 1.26 billion people. With this number as the initial condition, make a table of world population values every 30 years from 1850 to 2150.

d. According to your difference equation and initial condition, when will the population be within 500,000,000 people of the carrying capacity of planet Earth?

10. Severe acute respiratory syndrome (SARS) is a respiratory illness considered to be the first severe and readily transmissible disease of the twenty-first century (Figure 9-18). The first cases of SARS emerged in November 2002 in China. The virus was carried by an infected doctor to a four-star hotel in Hong Kong, then spread to over 25 countries, principally via airline travel. In March 2003, the World Health Organization advised that all SARS carriers be quarantined, and SARS in Hong Kong was contained over the next 10 weeks. The cumulative cases of SARS over the 10-week period starting March 17, 2003, are displayed in Figure 9-19.[12]

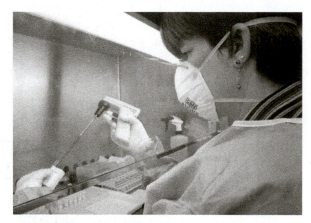

Figure 9-18 A specialist from the Centers for Disease Control processing SARS specimens. *Photo: CDC/Anthony Sanchez.*

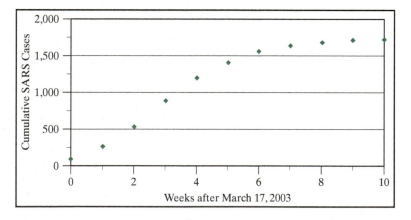

Figure 9-19 SARS in Hong Kong, China

a. Estimate from the graph the cumulative number of SARS cases in the tenth week after March 17, 2003.

b. One estimate for the unrestricted growth rate in SARS cases is 97% per week. Use this rate, and your answer from part (a), to write a logistic difference equation that models the cumulative cases of SARS. Let c represent the cases of SARS and n the number of weeks after March 17.

c. The number of reported SARS cases in Hong Kong on March 17 was 95. Using this number as the initial condition, create a table that lists the cumulative cases of SARS over the 10-week period.

d. The World Health Organization reported that there were 1,683 cumulative cases of SARS in Hong Kong in the eighth week following March 17. What is the error (difference) between your equation's value and the actual value? Express the error as a relative error (as a percent of the actual value).

11. In 1934, G. F. Gause published a landmark text titled *The Struggle for Existence*. (Recall that we mentioned Gause in the introduction to this chapter.) Gause described laboratory investigations concerning the growth of populations of yeast and other microorganisms, and the competition among species. In one experiment, Gause measured the growth of the combined population of the single-celled *Paramecium caudatum* and *Paramecium aurelia* over a 25-day period (Table 9-1).[13] The population counts are from a 0.5 milliliter sample of nutritive medium. Note that the population value on day 1 was never published.

TABLE 9-1

Day	Population	Day	Population	Day	Population
0	2	10	194	19	209
2	10	11	217	20	196
3	10	12	199	21	195
4	11	13	201	22	234
5	21	14	182	23	210
6	56	15	192	24	210
7	104	16	179	25	180
8	137	17	190		
9	165	18	206		

Source: *The Struggle for Existence*

a. Make a graph of the bacteria population data over the 25-day period.

b. With regard to the data, Gause commented that, "*At a certain moment the possibility of growth in a given microcosm is apparently exhausted, and with a continuously maintained level of nutritive resources a certain equilibrium of population is established.*" According to your graph, what is that equilibrium?

c. One estimate for the unrestricted growth rate is 80% daily. Use that value, and your answer to part (b), to write a logistic difference equation.

d. Gause went on to state, "*The oscillations of population around this state of equilibrium are not governed by any apparent law, and depend on various accidental causes.*" What part of your graph was Gause referring to? What type of "accidental causes" could make the population oscillate?

e. After each day of the experiment, Gause changed the nutritive medium to ensure that the population had a constant food supply, and he washed the bacteria to remove waste products. What do you think would have happened to the population if these measures were not taken? Explain in a few sentences and with a diagram.

12. Northern elephant seals (*Mirounga angustirostris*) live in the coastal habitat of California and Baja California (Figure 9-20). Hunted primarily for blubber oil, their population was reduced to about 100 individuals by the year 1910, and they were believed to be doomed to extinction. Fortunately, the species became protected under Mexican and U.S. laws in the early twentieth century. Since then, the population has made a remarkable comeback, totaling about 150,000 seals in the year 2000.[14]

Figure 9-20 Elephant seals bask in the sun. *U.S. Fish and Wildlife Service/Photo by Joe Martin.*

a. A logistic difference equation that models the seal population is given by $p(n) = 1.17p(n-1) - \frac{0.17}{150,000} p(n-1)^2$, where p is the population and n is the number of years. What is the unrestricted growth rate and carrying capacity in this model?

b. Show that between 1910 and 2000 the population of northern elephant seals grew by about 8.5% annually. Then write the *exponential* difference equation that models the growth of the seal population.

c. For each difference equation in parts (a) and (b), estimate the population in year 2020. Use an initial condition that corresponds to the population in year 1910. Comment on the results.

13. Since the Industrial Revolution commenced in the early 1800s, population growth rates in England and Wales have declined as the population has increased.[15] Figure 9-21 shows (population, growth rate) points for each decade from 1811 to 1991.

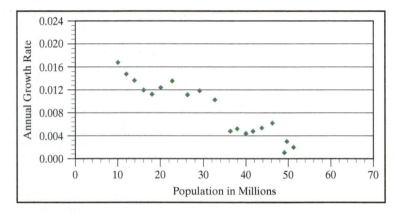

Figure 9-21 Growth rate versus population, England and Wales

a. With a straightedge, draw a best-fitting line through the data presented in the graph. Use the two intercepts to estimate an unrestricted growth rate and carrying capacity for the population of England and Wales.

b. Use your estimates from part (a) to create a logistic difference equation that models the growth of the population from 1811 to 1991.

c. The census of 1811 in England and Wales determined that the population in that year was 10.2 million. Using this number as the initial condition, graph the logistic difference equation for the years 1811–1991 using technology. Sketch on your homework paper.

d. Give two reasonable explanations for the general decline in population growth rates in England and Wales.

14. A study showed that the weekly growth rate of an individual sunflower decreased as the sunflower's height increased (Figure 9-22).[16]

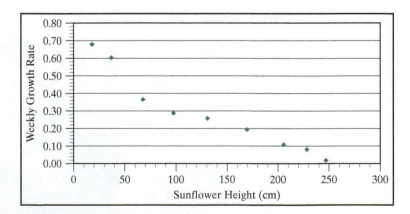

Figure 9-22 Sunflower growth rate versus height

  a. Use the linear trend in the data to estimate an unrestricted growth rate and maximum height (carrying capacity) for the sunflower.

  b. Using your estimates from part (a), write a logistic difference equation that models the growth of the sunflower. Let $n = 0$ refer to the first week.

  c. After the first week of growth, the height of the sunflower was 17.9 cm. Using this number as the initial condition, graph the logistic difference equation for weeks 1-12 using technology. Sketch on your homework paper.

  d. Do you think that the maximum height has anything to do with the carrying capacity of the sunflower's environment? What else could cause the sunflower growth to taper off? Explain.

Equilibrium Values

15. Find the equilibrium values for the following logistic difference equations. *Hint: When solving for E, simplify and set equal to zero before factoring.*

  a. $p(n) = 1.15p(n-1) - \dfrac{0.15}{60}p(n-1)^2$

  b. $p(n) = 1.15p(n-1) - \dfrac{0.15}{60}p(n-1)^2 - 2$

16. Find the equilibrium values for the following logistic difference equations. *Hint: When solving for E, simplify and set equal to zero before factoring.*

  a. $p(n) = 1.16p(n-1) - \dfrac{0.16}{5{,}000}p(n-1)^2$

  b. $p(n) = 1.16p(n-1) - \dfrac{0.16}{5{,}000}p(n-1)^2 - 192$

17. Suppose that a bacteria population in a test tube grows logistically according to the difference equation $p(n) = 1.15p(n-1) - 0.00025p(n-1)^2$, where n is the number of hours.

  a. Graph the difference equation from $n = 0$ to $n = 20$ using the following initial conditions: $p(0) = 200$, $p(0) = 400$, $p(0) = 600$, and $p(0) = 800$. On paper, sketch the four graphs on one set of axes.

  b. What is the carrying capacity of the test tube according to your graph?

  c. Using algebra, find the equilibrium values for the difference equation. Interpret the meaning of the equilibrium values.

18. Suppose that an insect population grows logistically according to the difference equation $p(n) = 1.34p(n-1) - (6.8 \times 10^{-6})p(n-1)^2$, where n is the number of weeks.

  a. Graph the difference equation using the following initial conditions: $p(0) = 5{,}000$; $p(0) = 30{,}000$; $p(0) = 50{,}000$; and $p(0) = 70{,}000$. Make a sketch of the four resulting graphs.

  b. What is the carrying capacity of the environment, according to the graphs?

  c. Using algebra, find the equilibrium values for the difference equation. Interpret the meaning of the equilibrium values.

Harvesting

19. The difference equation found in Example 9-4, concerning the fixed harvest of 50 elk, is $1.30p(n-1) - \dfrac{0.30}{1{,}500}p(n-1)^2 - 50$.

  a. Determine the equilibrium values using algebra. Approximate solutions to the nearest whole elk. *Hint: The resulting equation will not factor so you will need the following information: The solutions to a quadratic equation $Ax^2 + Bx + C = 0$ can be found using the quadratic formula $x = \dfrac{-B \pm \sqrt{B^2 - 4AC}}{2A}$.*

  b. Graph the difference equation from $n = 0$ to $n = 20$ using the following initial conditions: $p(0) = 185$, $p(0) = 300$, $p(0) = 800$, and $p(0) = 1{,}400$. Make a sketch of the four resulting graphs. On the sketch, draw horizontal lines to indicate any equilibrium values.

  c. Use the graph to assess whether the equilibrium values are stable or unstable. Can you think of an interpretation for the lower equilibrium value?

20. The difference equation found in Example 9-5 concerning the proportional harvest of elk is $p(n) = 1.20p(n-1) - \frac{0.30}{1,500}p(n-1)^2$. Determine the equilibrium values using algebra.

21. In Example 9-2, the logistic difference equation modeling the year-to-year growth of the Pacific halibut population is given by $u(n) = 1.71u(n-1) - \frac{0.71}{80}u(n-1)^2$. The initial condition was assumed to equal $u(0) = 20$ million kg.

  a. Suppose that fisheries catch 3 million kg of halibut each year. Write the new difference equation.

  b. Graph the difference equation from part (a); then estimate the long-term maximum mass of the halibut population.

  c. By what percent is the maximum mass below the carrying capacity of the Pacific Ocean?

22. The fin whale (*Balaenoptera physalus*) is second only to the blue whale in size and weight. It is named for the prominent, curved dorsal fin located far back on its body. Called the "greyhounds of the sea," fin whales are capable of bursts of speed of up to 23 mph (37 km/hr).[17] The hunting of fin whales has been prohibited in U.S. waters since the 1972 Marine Mammal Protection Act. In 1986, fin whales became protected worldwide by an International Whaling Commission moratorium. This exercise explores the mathematics of the fin whale population near Antarctica.

  a. The population growth curve of the Antarctic fin whale has been approximated with a logistic model using $K = 400,000$ whales and yearly $r_{max} = 0.08$.[18] Use this information to write the logistic difference equation annual growth model for Antarctic fin whales.

  b. Over the 45 whaling seasons from 1931/32 to 1975/76, an approximate total of 550,000 Antarctic fin whales were caught.[19] Determine the average annual catch during this time period. Then modify the logistic difference equation to include the average annual catch.

  c. Using the carrying capacity as the initial condition in year 1931, graph the modified difference equation over a 70-year period. Record the graph on your homework paper.

  d. According to your model, what was the population of Antarctic fin whales by the time of the 1986 international moratorium on whaling? According to your model, what would have happened to the Antarctic fin whale population had there been no moratorium?

23. For the Pacific halibut population in Example 9-2, the logistic difference equation modeling the year-to-year growth is given by $u(n) = 1.71u(n-1) - \frac{0.71}{80}u(n-1)^2$, and the initial condition was assumed to equal $u(0) = 20$ million kg.

  a. Suppose that each year halibut fisheries catch 15% of the halibut mass. Write the new difference equation, and then simplify it.

  b. What happens to the halibut population over a 20-year period?

  c. Suppose that fisheries around the world agree to a proportional harvest strategy so that, in the long run, there is an equilibrium level of 70 million kg of halibut in the Pacific Ocean. Using a trial-and-error approach, find the correct proportion for the harvest to the nearest 0.5%.

  d. *Challenge*: Solve part (c) using algebra.

24. In a laboratory experiment conducted by T. Carlson in 1913, yeast cells were allowed to multiply in a fixed environment with a steady supply of food. The amount of yeast, initially at 9.6 biomass units, grew logistically toward a maximum of $K = 665$ biomass units. The unrestricted growth rate for the yeast was approximately $r_{max} \approx 60\%$ per hour.[20]

  a. Write the difference equation and initial condition that models the growth of yeast.

  b. Graph the difference equation for hours $n = 0$ to $n = 20$. Sketch the result on your homework paper.

  c. Suppose that after each hour of the experiment, Carlson removed 20% of the yeast cells. Write the new difference equation. Then, using the same initial condition, plot the difference equation on the same graph that you sketched in part (b).

  d. Determine the equilibrium values for the difference equation in part (c). Does your answer agree with your graph?

Periodic and Chaotic Behavior

25. Suppose that an insect population grows logistically according to the values $p(0) = 300$, $K = 1,000$, and $r_{max} = 2.3$ annually.

    a. Write the logistic difference equation; then graph it over a 10-year period using the given initial condition. It may be helpful to set your plotting device to "connected" mode. Record the graph on your homework paper.

    b. Between which values does the population eventually cycle?

26. Suppose that a locust population grows logistically, has an initial population of 750 locusts, and has a carrying capacity of 1,000 locusts.

    a. If the unrestricted growth rate is 2.5 each year, determine the logistic difference equation. Then investigate the long-term behavior of the population using a table of values. Between which values does the population eventually cycle?

    b. Suppose that the unrestricted growth rate is 2.55 each year. Again write the difference equation and investigate the long-term behavior in a table of values. Between which values does the population eventually cycle?

27. Suppose that an insect population grows according to the logistic difference equation with $K = 100$ and $r_{max} = 3.0$ annually.

    a. Write a logistic difference equation that models the annual growth.

    b. Create a table of values from $n = 0$ to $n = 10$ for each initial condition listed in the following table. Record the completed table on your homework paper.

n	Population	Population	Population
0	79	80	81
1			
2			
⋮	⋮	⋮	⋮
10			

    c. Is the population model sensitive to initial conditions? Explain.

28. Suppose that an insect population grows logistically with a carrying capacity of $K = 100$ insects and an initial population of $p(0) = 55$ insects.

    a. Using the annual unrestricted growth rates indicated, create a table similar to the following for years $n = 0$ to $n = 10$.

n	Population ($r_{max} = 2.95$)	Population ($r_{max} = 3.0$)	Population ($r_{max} = 3.05$)
0	55	55	55
1			
2			
⋮	⋮	⋮	⋮
10			

    b. What happens when the unrestricted growth rate is changed just slightly? Are logistic models sensitive to the accuracy of the unrestricted growth rate? Explain.

SCIENCE IN DEPTH

Harvesting and Sustainable Forestry

The definition of sustainable forestry varies from group to group. From a strict **commodity** standpoint, sustainability means that trees can be harvested (i.e., logged) over a very long period of time and that the forest will continue to produce harvestable and profitable trees over that time period. Eventually the timber stand will yield a repeatable "crop" of trees for the sawmill or paper mill. Field crops such as corn are often used as a model for repeatedly cutting down and regrowing trees. This production-oriented definition of sustainability is often referred to as **sustainable yield**.[21]

Figure 9-23 A researcher checks on genetically modified trees in a laboratory. *Photo by Scott Bauer/ USDA Agricultural Research Service.*

For many timber companies and forest parcels, clearcutting of every tree is thought to be the most efficient and sustainable forest practice, even if the clearcut needs some "help" to regenerate and/or flourish. For example, most clearcuts are planted mechanically or by human hand with small but established trees grown in greenhouses or tree farms rather than from naturally wind-borne seed. The planted trees may be **cultivars,** artificially selected over generations for their vitality or other characteristics. One day in the near future, some planted trees will be genetically modified organisms (GMOs) designed to grow fast and repel pests (Figure 9-23). Amendments such as fertilizers and pesticides may be sprayed on the tree plantation to speed up growth and lower mortality. And tree predators such as bear and deer may be killed to prevent tree losses.

A different definition of sustainable forestry is based on the health of the ecosystem, of which the harvestable trees are just one part. In an **ecosystem-based** approach, sustainable forestry preserves both the components (plants, animals, soil, water, etc.) and the functionality (home to organisms, storehouse of carbon, producer of oxygen, etc.) of the ecosystem.[22] Logging of trees is considered to be sustainable if there is little impact on the ecosystem components and functions. The challenge for ecologists is to define and quantify the ecosystem before and after logging so that any changes can be detected.

There are often many subtle connections among the different parts of an ecosystem. For example, the carcasses of spawned-out salmon provide nutrients for organisms living in the streams and rivers but also provide much of the phosphorus and nitrogen found in streamside vegetation growing along salmon-bearing rivers.[23] Fewer salmon results in less nutrients for streamside vegetation, which harms the quality of streams, which in turn harms the salmon. A few years ago, this subtle but important **negative feedback** was not known to biologists. In the Pacific Northwest, logging of old-growth trees has also had a large negative impact on birds such as the marbled murrelet and the spotted owl. The spotted owl needs old-growth trees for nesting, and the favorite food of the spotted owl, the flying squirrel, depends on fungi that grow on the roots of old-growth trees.[24]

 ## CHAPTER PROJECT: TROPICAL FORESTS FOREVER?

The argument has been made that radical harvesting methods such as clearcutting are not sustainable, as the ecosystem is so disturbed that a healthy forest cannot be regenerated. The short time between clearcuts may produce the same overdepletion of nutrients in timberlands that is seen in many cornfields.

In the companion project for Chapter 9 found at the text's Web site (**enviromath.com**), you will investigate different harvest strategies in a tropical forest. You will use the logistic difference equation to model the results of harvesting over time. Which of these harvesting strategies is sustainable, if any? Which strategies lead to complete eradication of the forest?

NOTES

[1]G. F. Gause, *The Struggle for Existence* (1934; repr., New York: Hafner Publishing, 1964).

[2]Colin W. Clark, *Mathematical Bioeconomics: The Optimal Management of Renewable Resources,* 2nd ed. (New York: John Wiley & Sons, 1990), 47–48.

[3]Information about this data set can be found at the QELP Web site: http://www.seattlecentral. edu/qelp/sets/015/015.html

[4]Mike Stark, "Yellowstone's Bison Set Record for Population," *Billings Gazette,* December 7, 2004.

[5]U.S. Dept of the Interior, National Park Service, *Yellowstone, When Bison Leave the Park,* Bison Site Bulletin, October 10, 2002, http://www.nps.gov/yell/publications/pdfs/bisonsb.pdf

[6]Charles J. Krebs, *Ecology: The Experimental Analysis of Distribution and Abundance,* 3rd ed. (New York: Harper and Row, 1985), 351–359.

[7]United Nations News Center, "Desert Locust Swarms Decline in Sahel but Rise in Northwest Africa," October 25, 2004, http://www.un.org/News/

[8]Kari Heliövaara, Rauno Väisänen, and Chris Simon, "Evolutionary Ecology of Periodical Insects," *Trends in Ecology and Evolution* 9, no. 12 (1994): 475–480.

[9]C. Elton, and M. Nicholson, "The Ten Year Cycle in Numbers of Canadian Lynx," *J. Animal Ecology* 11 (1942): 215–244. Data can also be found online at the Time Series Data Library, http://www-personal.buseco.monash.edu.au/~hyndman/TSDL/

[10]James Gleick, *Chaos: Making a New Science* (New York: Viking Penguin, 1987).

[11]United Nations, *The World at Six Billion*, Table 1, 1999, http://www.un.org/esa/population/publications/sixbillion/sixbillion.htm

[12]World Health Organization, "Cumulative Number of Reported Probable Cases of Severe Acute Respiratory Syndrome (SARS)," http://www.who.int/csr/sars/country/en/

[13]Gause, *The Struggle For Existence*, 144–145.

[14]The Marine Mammal Center, "Northern Elephant Seal: *Mirounga angustirostris*," http://www.marinemammalcenter.org/learning/education/pinnipeds/noelephseal.asp

[15]Growth rates based on data from (1) Whitaker's Almanac (1941), as cited by GenDocs: Genealogical Research in England and Wales, "Population Of Great Britain & Ireland 1570–1931," http://www.gendocs.demon.co.uk/pop.html#EW; and (2) U.K. House of Commons, *A Century of Change: Trends in UK Statistics Since 1900*, by Joe Hicks and Grahame Allen, Library Research Paper 99/111 (1999), 6, http://www.parliament.uk/commons/lib/research/rp99/rp99-111.pdf

[16]Graph derived from data presented in H. S. Reed and R. H. Holland, "Growth of Sunflower Seeds," *Proceedings of the National Academy of Sciences* 5 (1919), 140.

[17]American Cetacean Society, "Fin Whale: *Balaenoptera physalus*," http://www.acsonline.org/factpack/finwhl.htm

[18]Clark, *Mathematical Bioeconomics*, 48–49.

[19]Whaling Library, "Catch History of Fin Whales in the Antarctic Since 1931/32 Season," http://luna.pos.to/whale/sta_fin.html

[20]T. Carlson, "Über Geschwindigkeit und Grösse der Hefevermehrung in Würze," *Biochem. Z.* 57 (1913): 313–334. As cited in Krebs, *Ecology*, 216–217.

[21]American Forest Products Association, "Sustainable Forestry Initiative," http://www.afandpa.org/Content/NavigationMenu/Environment_and_Recycling/SFI/SFI.htm

[22]David Suzuki Foundation, "A Cut Above: Ecological Principles for Sustainable Forestry on British Columbia's Coast," http://www.davidsuzuki.org/Forests/Solutions/

[23]James Helfield and Robert Naiman, "Effects of Salmon-Derived Nitrogen on Riparian Forest Growth and Implications for Stream Productivity," *Ecology* 82, no. 9 (2001), 2403–2409.

[24]Pete Taylor, "Major Player," National Wildlife Magazine 38, no. 4, June/July 2000, http://www.nwf.org/nationalwildlife/article.cfm?issueID=30&articleID=277

10

Systems of Difference Equations

Figure 10-1 A child in West Africa receives an immunization. *Photo: CDC/Dr. J. D. Millar.*

Measles is a viral disease that principally infects children and is spread when a contagious child coughs, sneezes, or talks. In the 1960s, a worldwide campaign was launched to eradicate smallpox and measles (Figure 10-1). Although the world was declared smallpox free by 1979, measles was never entirely eliminated. Today, the number of measles cases and the frequency of the outbreaks is much higher in countries with poor environmental conditions and in those areas lacking vaccination programs.

The spread of measles can be modeled with a **system of difference equations**, which encompasses two or more difference equations simultaneously. Before exploring a measles model in the student exercises, we begin by examining several simple population models, in which each age group is given its own difference equation. We then look at the long-term behavior of those populations and learn about stable age distributions. We finish with an investigation modeling the flow of pollution in the Great Lakes. Most of our examples will involve two or three equations, although many real-life models are built upon hundreds or even thousands of equations!

SYSTEMS MODELING

Many of the examples and exercises in this chapter require an understanding of the difference between **discrete** and **continuous** growth. Consider red-legged frogs (*Rana aurora*), which live in the cool, coastal forests of British Columbia, Canada. Female frogs mate and lay their eggs in January or February. Once the embryos hatch in about four weeks, the frog population gets a "pulse" of new growth. Red-legged frog reproduction is an example of discrete growth. Now consider the human (*Homo sapien*) population, whose individuals reproduce throughout the entire year. For humans, new life is added to the population almost every instant of each day. We say that the human population grows continuously (Figure 10-2).

Figure 10-2 The red-legged frog population grows discretely in early spring, while the human population grows year-round in a continuous fashion. *Left photo: U.S. Fish and Wildlife Service/Ryan Hagerty. Right photo: Courtesy of Mark and Carla Ainsworth.*

When setting up difference equations to model populations with discrete growth, we need to make the time step of the difference equations match the reproductive time step of the population. For frogs, that time step is one year. Furthermore, we need to be precise as to when we count the number of individuals in the population—that will make writing the difference equations easier. Let's start with a few simple examples of systems of difference equations.

EXAMPLE 10-1   A Fertile, Short-Lived Organism

Suppose that a population of organisms reproduces at the end of each month and that the maximum age is three months. We divide the population into age groups A, B, and C so that members of group A have ages between 0 and 1 month, those in group B have ages from 1 to 2 months, and group C members have ages from 2 to 3 months. For mathematical convenience, we'll count the number of organisms in each group *at the end of each month.*

Associated with each age group are a monthly **fertility rate** and a **survival rate** as summarized in Table 10-1. The monthly fertility rate for each group indicates the number of offspring *that live for a full month.* In the discussion that follows, though, we'll refer to these surviving offspring simply as "offspring." The monthly survival rates indicate the chance that a member of one age group will live a full month to become a member of the next age group.

TABLE 10-1

Age Group	A	B	C
Age (months)	0–1	1–2	2–3
Monthly fertility rate	0	4	3
Monthly survival rate	0.3	0.1	0.0

A **life-cycle graph** illustrates the survival and fertility information in a concise manner (Figure 10-3). Each age group is displayed as a circle or **node**, and the arrows represent **population flows** from one group to another. The numbers next to the horizontal arrows represent survival rates, whereas those next to the arrows arching back toward group A are fertility rates.

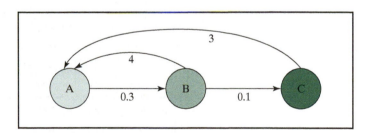

Figure 10-3 A life-cycle graph for a highly fertile but short-lived organism. The numbers next to the arrows indicate monthly survival and fertility rates.

To better understand age group reproduction and survival, suppose that each group starts with 100 individuals in month $n = 0$. How many individuals will be present in each age group in month $n = 1$? As always, it helps to define variables:

Let $a(n)$ represent the population of age group A in month n.

Let $b(n)$ represent the population of age group B in month n.

Let $c(n)$ represent the population of age group C in month n.

Using these variables, the initial populations can be expressed as three separate initial conditions: $a(0) = 100$, $b(0) = 100$, and $c(0) = 100$. As explained earlier, population numbers are always counted at the end of the month. Our task is to find $a(1), b(1),$ and $c(1)$, the populations of groups A, B, and C at the end of month $n = 1$.

The fertility rates indicate the number of offspring per individual per month. The fertility rate for group A is 0 per month, which means that they produce no offspring. The fertility rate of group B is 4 per member, and there are 100 members at the end of month $n = 0$. Thus there will be $4(100) = 400$ offspring from group B that are counted as members of group A at the end of month $n = 1$. (Remember, fertility rates indicate offspring that live one full month!) Using the initial condition and fertility rate for group C, we determine that group C will produce $3(100) = 300$ offspring that are members of group A at the end of month $n = 1$. All total, there will be 700 new members in group A in month $n = 1$.

What happens to the "old" members of group A? There will be $0.3(100) = 30$ members that survive for another month, going from group A in month $n = 0$ to group B in month $n = 1$. Similarly, there will be $0.1(100) = 10$ survivors from group B that end up in group C in month $n = 1$. Because no members of the population live beyond three months, all members of group C in month $n = 0$ will be dead in month $n = 1$. This is a highly fertile but short-lived population!

Putting all the fertility and survival computations together, we get the membership of each group in month $n = 1$:

$a(1) = 4(100) + 3(100) = 700$ ← *New members based on fertility rates*

$b(1) = 0.3(100) = 30$ ← *New members surviving from group A*

$c(1) = 0.1(100) = 10$ ← *New members surviving from group B*

We assume that the same survival and fertility rates in Table 10-1 apply to the three groups in future months. Thus we can compute the membership for each group in month $n = 2$ using population numbers from month $n = 1$. We get

$$a(2) = 4b(1) + 3c(1) \qquad b(2) = 0.3a(1) \qquad c(2) = 0.1b(1)$$
$$= 4(30) + 3(10) \qquad\qquad = 0.3(700) \qquad\quad = 0.1(30)$$
$$= 150 \qquad\qquad\qquad\quad = 210 \qquad\qquad = 3$$

Similarly, the membership of each group in month $n = 3$ can be determined from membership values in month $n = 2$. And those in month $n = 4$ can be determined from month $n = 3$, and so on. We leave those computations for a student exercise but record the results in Table 10-2. Note that for month $n = 4$ our mathematics leads to fractional numbers of individuals. For better accuracy, keep a few decimal places when computing future population values. But when reporting values, round off appropriately: for example, report that $b(4) = 255$ individuals.

TABLE 10-2

Month n	Members in Group A $a(n)$	Members in Group B $b(n)$	Members in Group C $c(n)$
0	100	100	100
1	700	30	10
2	150	210	3
3	849	45	21
4	243	254.7	4.5
5	1,032.3	72.9	25.47

The survival and fertility rates for each age group remain constant as the population "moves" from one month to the next. Each month, new members in group A are the offspring from groups B and C from the previous month. New members in groups B and C are the survivors from groups A and B, respectively. The general formulas (difference equations) giving the membership in month n for each group as a function of the memberships of all groups in month $n - 1$ are as follows:

$$a(n) = 4b(n-1) + 3c(n-1)$$
$$b(n) = 0.3a(n-1)$$
$$c(n) = 0.1b(n-1)$$

The preceding three equations are a **system of difference equations**. With the initial conditions $a(0) = 100$, $b(0) = 100$, and $c(0) = 100$, the system of difference equations can be used to predict the memberships in groups A, B, and C over time.

EXAMPLE 10-2 A Simple System

The following is a simple system consisting of two difference equations.

$$a(n) = 5\,a(n-1) - 2\,b(n-1)$$

$$b(n) = -a(n-1) + 3\,b(n-1)$$

Suppose that $a(0) = 2$ and $b(0) = 1$. Find $a(1)$ through $a(3)$, and $b(1)$ through $b(3)$.

Solution From the difference equations we determine that the amounts at time $n = 1$ are

$$a(1) = 5a(0) - 2b(0) \qquad b(1) = -a(0) + 3b(0)$$
$$= 5(2) - 2(1) \quad \text{and} \quad = -2 + 3(1)$$
$$= 8 \qquad\qquad\qquad = 1$$

Use these last results to compute amounts when $n = 2$:

$$a(2) = 5a(1) - 2b(1) \qquad b(2) = -a(1) + 3b(1)$$
$$= 5(8) - 2(1) \quad \text{and} \quad = -8 + 3(1)$$
$$= 38 \qquad\qquad\qquad = -5$$

The amounts when $n = 3$ are computed in a similar way:

$$a(3) = 5a(2) - 2b(2) \qquad b(3) = -a(2) + 3b(2)$$
$$= 5(38) - 2(-5) \quad \text{and} \quad = -38 + 3(-5)$$
$$= 200 \qquad\qquad\qquad = -53$$

EXAMPLE 10-3 ANWR and the Porcupine Caribou Herd

The Porcupine Caribou Herd (PCH) migrates northward each spring from its wintering grounds in Canada to the coastal arctic plains in and around the Arctic National Wildlife Refuge or ANWR. (For more information about ANWR, see the Chapter 3 "Science in Depth" essay.) The PCH migration, consisting today of approximately 130,000 caribou (*Rangifer tarandus*), constitutes the largest migration of any land animal on Earth.[1] In early June, pregnant caribou cows reach the coastal areas and give birth; births from the PCH typically number in the tens of thousands. The coastal plains are the preferred calving grounds of the caribou, for in late spring and early summer there is an abundance of nutritious plant life, and the numbers of predators and annoying insects have not yet peaked. In mid- to late July, the herd begins its long return trek toward its wintering grounds (Figure 10-4).

Caribou populations can be divided into three **growth stages**: calves, yearlings, and adults. Biologists often use growth stages instead of age groups to model a population, because survival and fertility rates are more a function of size or biological development than age. Annual fertility and survival rates for female caribou in the PCH are given in Table 10-3.[2]

TABLE 10-3

Growth Stage	A = Female Calves	B = Female Yearlings	C = Female Adults
Age in years	0–1	1–2	>2
Yearly fertility rate	0	0	0.41
Yearly survival rate	0.93	0.96	0.84

Source: *Canadian Journal of Zoology*

It is common for biologists to give life statistics for only the females in an animal population. Thus annual fertility rates indicate the total number of *female offspring per*

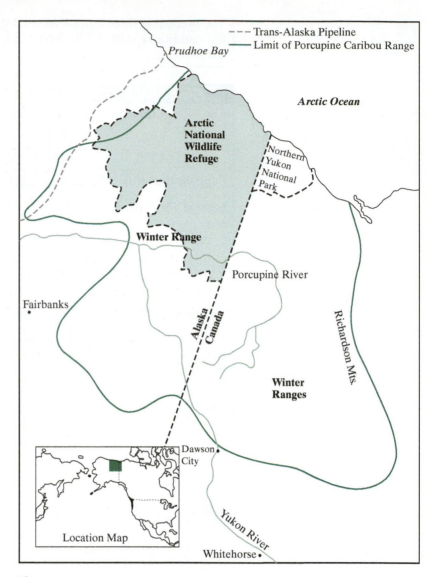

Figure 10-4 The habitat range of the Porcupine Caribou Herd

female each year that survive one full year. In the PCH, female adults produce, on average, 0.41 female offspring that survive to become calves the following year (about 41 surviving female offspring per 100 female adults).

We interpret the yearly survival rates for growth stages as we did with age groups: Calves have a 93% chance of becoming yearlings, and yearlings have a 96% chance of becoming adults. Members in the last stage (adults) can survive for another year and remain in the same stage (adults). The adult survival rate of 0.84 indicates that a typical adult female caribou has an 84% chance of living one additional year. The life-cycle graph in Figure 10-5 indicates this "self-looping" property of the adult stage.

Assume that the PCH population currently has 40 thousand calves, 20 thousand yearlings, and 70 thousand adults. Also assume that 50% of the PCH are female. Based on the rates provided, write a system of difference equations that describes the population dynamics of the female caribou, and predict the total female population five years from now.

Solution Let $a(n)$, $b(n)$, and $c(n)$ represent the female populations in thousands in year n, for growth stages A, B, and C, respectively. The difference equations are

$$a(n) = 0.41\, c(n-1)$$
$$b(n) = 0.93\, a(n-1)$$
$$c(n) = 0.96\, b(n-1) + 0.84\, c(n-1)$$

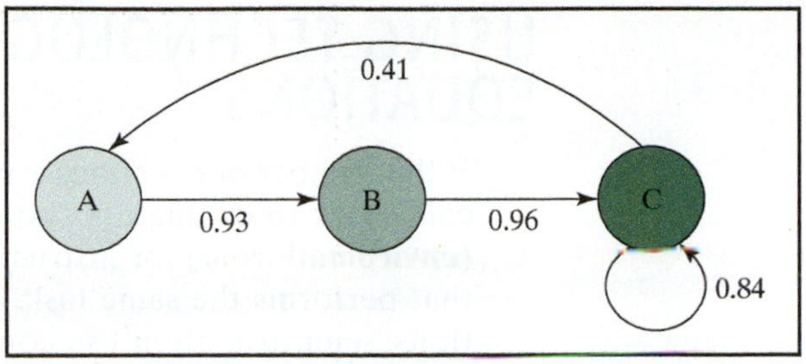

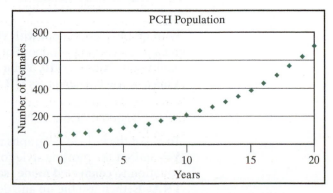

Figure 10-5 A life-cycle graph showing fertility and survival rates for the three growth stages of female caribou. The self-looping arrow for stage C indicates that 84% of the adult caribou population survive to the next year.

With 50% female, the initial caribou populations (in thousands) are $a(0) = 20$, $b(0) = 10$, and $c(0) = 35$. The populations for years $n = 0$ through $n = 5$ are listed in Table 10-4. The last column, total females, was determined by adding the three growth stage populations. Our model predicts that the total female population in five years will equal 116,000 caribou. With the assumption that females make up 50% of the population, the combined female and male population will be twice the values presented in the table.

TABLE 10-4

n	A = Female Calves	B = Female Yearlings	C = Female Adults	Total Females
0	20	10	35	65
1	14	19	39	72
2	16	13	51	80
3	21	15	55	91
4	23	19	61	103
5	25	21	70	116

   A follow-up note to this example is in order. A graph of the total number of PCH females over a 20-year time period shows an exponential-like increase (Figure 10-6). These population values are, of course, based upon a mathematical model, and we must recognize the limitations of extrapolating too far into the future using such a tool. Is the predicted growth really exponential? The answer is "yes!," as we'll see later in this chapter.

Figure 10-6 Graph of the projected female caribou population over a 20-year period. The population appears to increase exponentially.

USING TECHNOLOGY: SYSTEMS OF DIFFERENCE EQUATIONS

If the number of equations is no more than three, we can use the TI-83/84 graphing calculator to evaluate a system of difference equations. Visit the text Web site (**enviromath.com**) for instructions on how to use an online computer application that performs the same task. For systems with more than three difference equations, scientists often use computer software programs based upon matrices or spreadsheet software—those applications will not be covered in this text.

Systems of Difference Equations on the TI-83/84

In *sequence* mode, press **Y=**. Set *n*Min = 0. The calculator allows the use of the variables **u** (**2nd 7**), **v** (**2nd 8**), **w** (**2nd 9**), and **n**. Using these variables, the equations and initial conditions in Example 10-1 are rewritten as follows:

$$u(n) = 4\,v(n-1) + 3\,w(n-1); \quad u(0) = 100$$

$$v(n) = 0.3\,u(n-1); \quad v(0) = 100$$

$$w(n) = 0.1\,v(n-1); \quad w(0) = 100$$

Enter these equations as shown to the right. It's easy to make a mistake when entering an equation or initial condition, *so go slow, and check your entries when finished!*

Table and graph keystrokes for systems are similar to those for a single difference equation. Set *n* values for your table in **TBLSET**; then press **TABLE**.

When inspecting a table of values, you will need to use the right arrow to scroll over to the $w(n)$ column.

With graphs, experiment with **WINDOW** settings to get a good viewing window. For example, set **Xmin = 0**, **Xmax = 10**, **Ymin = 0**, and **Ymax = 1600**. Then press **GRAPH**.

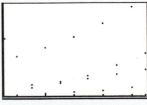

To make the individual graphs more clear, go to **Y=** and set the plotting style for each difference equation to **connected mode** (see Chapter 7). Press **TRACE**, then use the up and down arrows to move between graphs.

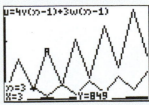

EXPONENTIAL CHANGE AND STABLE AGE DISTRIBUTIONS

We've now looked at several examples in which populations were modeled by systems of difference equations. By using technology, we can study the long-term behavior of these populations. Recall that the female caribou population, consisting of three growth stages, was modeled in Example 10-3 by the following difference equations and initial conditions:

Stage A (calves): $a(n) = 0.41\,c(n{-}1);$ $a(0) = 20$

Stage B (yearlings): $b(n) = 0.93\,a(n{-}1);$ $b(0) = 10$

Stage C (adults): $c(n) = 0.96\,b(n{-}1) + 0.84\,c(n{-}1);$ $c(0) = 35$

To examine the long-term behavior of the population, we'll graph all three equations over a 20-year period. Before entering these equations and initial conditions into a graphing calculator, it is helpful to rewrite them in terms of the variables u, v, and w:

$$u(n) = 0.41\,w(n{-}1);\quad u(0) = 20$$

$$v(n) = 0.93\,u(n{-}1);\quad v(0) = 10$$

$$w(n) = 0.96\,v(n{-}1) + 0.84\,w(n{-}1);\quad w(0) = 35$$

The window settings that we use, and our resulting graphs, are displayed next. We've set the plot style to "connected mode" (see Chapter 7 for directions).

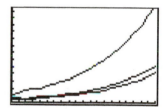

```
WINDOW
 nMin=0
 nMax=20
 PlotStart=1
 PlotStep=1
 Xmin=0
 Xmax=20
↓Xscl=1
```

```
WINDOW
↑PlotStep=1
 Xmin=0
 Xmax=20
 Xscl=1
 Ymin=0
 Ymax=400
 Yscl=30
```

Notice that the three graphs fluctuate slightly in the first few years, but then increase in what appears to be an exponential manner. Although we do not prove it in this text, the female populations of each stage *will* increase exponentially in the long term and all at the same rate. It follows that the total female population will also increase exponentially in the long term at the same rate. We saw exponential-like growth in the plot of the total female population (Figure 10-6).

We can estimate the long-term growth multiplier M for each stage by working with actual population values, but keep in mind that *long term* means different n values for different populations. For the female caribou population, we'll jump ahead 15 years and see what we get. The difference equations predict that the population of calves will equal 84.131 when $n = 15$, and 94.884 when $n = 16$. Assuming exponential change, we write

$$84.131M = 94.884$$

which means that the annual growth multiplier for calves is

$$M = \frac{94.884}{84.131} \approx 1.128$$

Using the formula $M = 1 + r$, we determine that the annual growth rate for calves is $r = 0.128 = 12.8\%$.

The growth multipliers and growth rates for yearlings and adults are computed in the same manner; the results are summarized in Table 10-5. We find that the growth rates between years $n = 15$ and $n = 16$ are the same for calves, yearlings, and adults. In future years, the growth rates for the difference equations will remain fixed at

$r = 12.8\%$. (We'll have you explore this in the student exercises.) Because each of the three caribou stages increases by 12.8% annually in the long term, so will the total female caribou population.

TABLE 10-5 Long-term multipliers are found by dividing population values from two successive years.

	Calves $u(n)$	Yearlings $v(n)$	Adults $w(n)$
$n = 15$	84.131	69.379	231.423
$n = 16$	94.884	78.242	260.999
Growth multiplier	$M = \dfrac{94.884}{84.131} \approx 1.128$	$M = \dfrac{78.242}{69.379} \approx 1.128$	$M = \dfrac{260.999}{231.423} \approx 1.128$
Growth rate	$r = 12.8\%$	$r = 12.8\%$	$r = 12.8\%$

When populations are modeled by difference equations that resemble those for caribou calves, yearlings, and adults, the results will be similar. In the long term, each age group or growth stage, as well as the total population, will grow or decay by the same exponential rate. (By resemble, we mean no higher powers of $u(n-1)$, $v(n-1)$, or $w(n-1)$, and no constants such as those in affine models.) An outcome of this fact is that the *percentage* of the population determined by each age group or stage will eventually be constant. We can see evidence of this by computing the percentages of calves, yearlings, and adults from $n = 0$ to $n = 15$ (Table 10-6). In the long term, the percentage of calves approaches 21.9%, while the percentages of yearlings and adults approach 18.0% and 60.1%, respectively. According to our model, the long-term population will be distributed among calves, yearlings, and adults according to the percentages 21.9%, 18.0%, and 60.1%. These percentages make up what is called a **stable age distribution**. In Table 10-6 entries in the last four columns are computed using unrounded values.

TABLE 10-6 In the long term, population percentages approach fixed levels

n	Female Caribou Population				Percent of Total		
	Calves	Yearlings	Adults	Total	Calves	Yearlings	Adults
0	20	10	35	65	30.8%	15.4%	53.8%
1	14	19	39	72	19.9%	25.9%	54.2%
2	16	13	51	80	20.0%	16.7%	63.3%
3	21	15	55	91	22.8%	16.3%	60.8%
4	23	19	61	103	22.1%	18.8%	59.1%
5	25	21	70	116	21.6%	18.3%	60.2%
6	29	23	79	130	21.9%	17.8%	60.4%
7	32	27	88	147	21.9%	18.0%	60.0%
8	36	30	100	166	21.8%	18.1%	60.1%
9	41	34	113	187	21.8%	18.0%	60.2%
10	46	38	127	211	21.9%	18.0%	60.1%
11	52	43	143	238	21.9%	18.0%	60.1%
12	59	48	161	268	21.9%	18.0%	60.1%
13	66	55	182	303	21.9%	18.0%	60.1%
14	75	62	205	341	21.9%	18.0%	60.1%
15	84	69	231	385	21.9%	18.0%	60.1%

Using the fertility and survival rates estimated by biologists, we created a systems model which forecasts the populations of female calves, yearlings, and adults.

We predict that each of these population subgroups, as well as the total population, will increase 12.8% annually in the long term. Furthermore, our model predicts that the percentages of calves, yearlings, and adults will eventually become constant. Models, such as this one, are great for predicting the future of a population, for at least a few years out. But the models can also be used to help *quantify* the consequences of human actions that impact the environment. For example, if oil drilling in ANWR destroys some caribou calving grounds and lowers birth rates by 25%, how will the long-term growth rate of the PCH population change? If global warming lengthens the growing season for arctic vegetation, making caribou adult survival rates increase from 84% to 94%, what will be the new stable age distribution for caribou? These are just a few of the types of questions we'll let you wrestle with in the end-of-chapter exercises.

WHAT ELSE BESIDES POPULATIONS?

Up to this point, we have used systems of difference equations to model animal populations, writing separate equations for each age group or growth stage of the population. Systems of difference equations are used to model many other types of environmental problems, and the key to those models is breaking down the problem into component parts. Each component part then gets its own difference equation. When the difference equations are pulled together and analyzed as a system, an amazingly powerful model evolves! Let's see how this approach can be used to examine pollution flows in the Great Lakes.

EXAMPLE 10-4 Pollution in the Great Lakes

Lake Erie and Lake Ontario are two immense bodies of fresh water that lie on the border of the United States and Canada (Figure 10-7). The two lakes are connected by the Niagara River, a 35-mile watercourse that drops 200 feet at Niagara Falls. Over the centuries, the quality of these and many other lakes has diminished through human actions: Toxic chemicals and heavy metals are dumped by factories, pesticides and sediment run off agricultural fields and urban streets, sewage is discharged from municipal sources, and so on (Figure 10-8). Some of these pollutants settle to the lake bottoms; others are taken up by plants and animals.

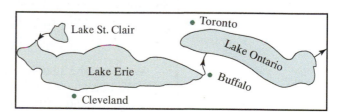

Figure 10-7 A map showing the geography of Lakes Erie and Ontario. Lake Erie flows into Lake Ontario via the 35-mile-long Niagara River.

Figure 10-8 Water pollution from a point source (left) and a non–point source (right). For more information, read the "Science in Depth" essay at the end of this chapter. *Left photo: USFWS. Right photo by Lynn Betts, USDA National Resource Conservation.*

Another way in which pollutants are removed is through a natural flushing process—when clean water enters a lake, polluted water exits. About every three years the water in Lake Erie is replaced through natural flushing, which means that during any one-year period about 33% of the water is flushed. For Lake Ontario, the flushing time is about six years, so each year about one-sixth or 17% of the water is flushed.[3] Suppose that we stopped polluting Lakes Erie and Ontario. How would the pollution level change over time if pollution is removed only through the natural flushing process of each lake?

To make the problem simpler, several assumptions need to be made. First, assume that 100% of the water from Lake Ontario comes from Lake Erie. Second, assume that both lakes have equal amounts of pollution at the present time. Third, assume that the amount of pollution that Lake Ontario picks up in any given year is the amount of pollution flushed from Lake Erie in the previous year. Fourth, assume that the pollution is well mixed within the water, so that when a certain percentage of water is removed, the same percentage of pollution tags along.

Because Lake Erie has 33% of its water flushed each year, and new pollution has ceased, we conclude that its pollution should drop by 33% each year. This is equivalent to having 67% of the pollution remain each year. Let $E(n)$ represent the amount of pollution in Lake Erie in year n. The difference equation that describes the year-to-year change in pollution is

$$E(n) = 0.67E(n-1)$$

Now let $O(n)$ represent Lake Ontario's pollution amount in year n. If Lake Ontario were *not* connected to Lake Erie, its 17% flushing rate would allow 83% of the pollution to remain from year to year, and its difference equation would be

$$O(n) = 0.83\,O(n-1)$$

Because Lake Ontario is fed by Lake Erie, it picks up pollution that Lake Erie flushes out. We assumed that the amount of pollution that Lake Ontario picks up in year n is the amount flushed from Lake Erie in year $n-1$, which is $0.33\,E(n-1)$. Thus the difference equation describing the pollution amount in Lake Ontario is

$$O(n) = 0.83\,O(n-1) + 0.33\,E(n-1)$$

We now have two equations describing the pollution dynamics of this system of lakes:

$$\text{Lake Erie:} \quad E(n) = 0.67\,E(n-1)$$

$$\text{Lake Ontario:} \quad O(n) = 0.83\,O(n-1) + 0.33\,E(n-1)$$

We made the assumption that both lakes have equal amounts of pollution initially. We don't have an exact number, so we make both initial conditions equal to one: $E(0) = 1$ and $O(0) = 1$. This is a common approach when we are more concerned about the *change* in pollution level, not the absolute amounts. Using these initial conditions and the difference equations for the two lakes, we can generate the pollution levels over the course of several years (Table 10-7).

TABLE 10-7

n	$E(n)$	$O(n)$	n	$E(n)$	$O(n)$
0	1	1	7	0.06061	0.70604
1	0.67	1.16	8	0.04061	0.60601
2	0.4489	1.1839	9	0.02721	0.51639
3	0.30076	1.13077	10	0.01823	0.43758
4	0.20151	1.03779	11	0.01221	0.36921
5	0.13501	0.92787	12	0.00818	0.31047
6	0.09046	0.81468	13	0.00548	0.26039

The table indicates that the amount of pollution in Lake Erie decreases rather quickly at first, and then more slowly later. We should not be surprised at this result, for Lake Erie's difference equation is the familiar exponential decay model. Note that after five years there is $E(5) = 0.13501$ amount of pollution remaining—about 13.5% of the original amount.

For Lake Ontario, the pollution amount actually *increases* at first, reaching a high point of 118.39% of the original amount in year $n = 2$. Afterward the amount decreases slowly. After five years, there is about 92.8% of the original amount remaining.

How can we explain the fact that the pollution amount would actually rise in Lake Ontario in the first two years? The answer lies with the fact that more pollution is flushed into Lake Ontario in those years than is flushed out—its pollution level must rise! Eventually the amount flushing into Lake Ontario is less than the amount flushing out—at that point, Lake Ontario's pollution level begins to fall. We can see from this example how strangely systems can behave!

CHAPTER SUMMARY

Simultaneous modeling of different age groups of a population, or a series of lakes in an ecosystem, can be accomplished with **systems of difference equations**. Populations are best modeled by making the time step of the difference equations match the reproductive time step of the population. A difference equation, based upon fertility and survival rates, can be written for each age group to account for the number of individuals entering and leaving that age group each time period.

Systems of difference equations can also be used to model populations broken into growth stages. Consider, for example, the following three difference equations, which describe female calves (stage A), yearlings (stage B), and adults (stage C) in terms of year n. The initial condition for each stage is 100 individuals.

$$a(n) = 0.9\,c(n-1); \quad a(0) = 100$$

$$b(n) = 0.8\,a(n-1); \quad b(0) = 100$$

$$c(n) = 0.7\,b(n-1) + 0.6\,c(n-1); \quad c(0) = 100$$

The system of difference equations can be diagrammed with a **life-cycle graph** (Figure 10-9).

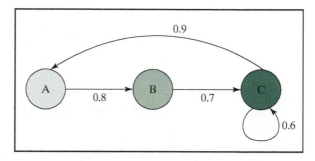

Figure 10-9 A life-cycle graph modeling a population composed of three stages.

The first equation indicates that each female adult gives birth to 0.9 female newborns each year (on average) that will survive to be calves the following year. The second equation tells us that 80% of the female calves survive to become yearlings. The third equation shows that the adult female population in any year consists of 70% of the surviving female yearling population, plus 60% of the surviving female adult population, from the previous year.

In the long term, the annual growth rate of each stage of the population described previously stabilizes at approximately 5.38%. The long-term growth rate can be found by working with consecutive population values for large values of n (Table 10-8).

TABLE 10-8

	Calves $a(n)$	Yearlings $b(n)$	Adults $c(n)$
$n = 24$	350.18	265.83	410.03
$n = 25$	369.03	280.14	432.10
Growth multiplier	$M = \dfrac{369.03}{350.18} \approx 1.0538$	$M = \dfrac{280.14}{265.83} \approx 1.0538$	$M = \dfrac{432.10}{410.03} \approx 1.0538$
Growth rate	$r = 5.38\%$	$r = 5.38\%$	$r = 5.38\%$

With all growth stages (and the total population as well) growing at 5.38% in the long term, the population approaches a **stable age distribution**. For example, by the year $n = 25$, the total population is about 1,081 individuals. The distribution of calves, year-lings, and adults is about $\frac{369}{1,081} \approx 34\%$, $\frac{280}{1,081} \approx 26\%$, and $\frac{432}{1,081} \approx 40\%$, respectively.

END *of* CHAPTER EXERCISES

Systems Warm-Up

1. In Example 10-1, the number of members in groups A, B, and C in month $n = 2$ is $a(2) = 150, b(2) = 210$, and $c(2) = 3$.

   a. Find the membership of each group in months $n = 3, 4$, and 5 to verify that Table 10-2 is correct.

   b. Find the total population of all groups combined, from $n = 0$ to $n = 5$.

2. Continue with the computations in Example 10-2 to determine the amounts when $n = 4$ and 5. Show your work at each step.

3. A system of three difference equations is given, along with a table listing initial conditions. Copy the table onto your homework paper; then complete the missing entries *by hand*.

$$a(n) = a(n-1) + b(n-1)$$
$$b(n) = b(n-1) + c(n-1)$$
$$c(n) = c(n-1) + a(n-1)$$

n	$a(n)$	$b(n)$	$c(n)$
0	20	10	30
1			
2			
3			

4. A system of three difference equations is given, along with a table listing initial conditions. Copy the table onto your homework paper; then complete the missing entries *by hand*.

$$a(n) = 2\,b(n-1)$$
$$b(n) = 3\,c(n-1)$$
$$c(n) = a(n-1)$$

n	$a(n)$	$b(n)$	$c(n)$
0	50	80	25
1			
2			
3			

5. A system of four difference equations is given, along with a table listing initial conditions. Copy the table onto your homework paper; then complete the missing entries *by hand*.

$$a(n) = 2\,c(n-1) + 1$$
$$b(n) = 3\,d(n-1) + 1$$
$$c(n) = 2\,a(n-1) - 2$$
$$d(n) = 3\,b(n-1) - 2$$

n	$a(n)$	$b(n)$	$c(n)$	$d(n)$
0	2	5	3	4
1				
2				
3				

6. A system of four difference equations is given, along with a table listing initial conditions. Copy the table onto your homework paper; then complete the missing entries *by hand*.

$$a(n) = 0.5\,b(n-1) + a(n-1)$$
$$b(n) = 0.2\,d(n-1) + a(n-1)$$
$$c(n) = 0.2\,c(n-1) + a(n-1)$$
$$d(n) = 0.5\,b(n-1) - a(n-1)$$

n	$a(n)$	$b(n)$	$c(n)$	$d(n)$
0	100	100	100	100
1				
2				
3				

7. A population has three age groups: A, B, and C.
   a. Suppose that the yearly fertility rate is 2 for group A, 3 for group B, and 0 for group C. Explain in your own words what these rates mean.
   b. Suppose that the yearly survival rate is 0.6 for group A, 0.25 for group B, and 0 for group C. Interpret these rates in your own words.
   c. Make a table summarizing the information from parts (a) and (b); then draw a life-cycle graph for this population.
   d. Write a system of difference equations modeling the population.

8. Let $a(n), b(n)$, and $c(n)$ represent the populations of age groups A, B, and C, in that order. Suppose that the difference equations governing these groups is given by the following system:

$$a(n) = 5\,b(n-1) + 2\,c(n-1)$$
$$b(n) = 0.6\,a(n-1)$$
$$c(n) = 0.8\,b(n-1)$$

   a. Describe the overall population in a few sentences, making specific references to the three groups. Include a table with fertility and survival rates.
   b. Draw a life-cycle graph for this population.
   c. What percentage of group A *dies* each time period? And for group B?

9. For the given life-cycle graph, write a system of difference equations.

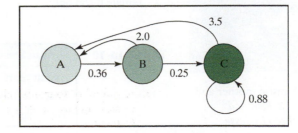

10. For the given life-cycle graph, write a system of difference equations.

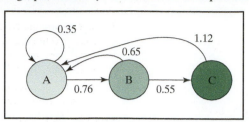

Applications

11. A study was conducted on the biology of the sand smelt (*Atherina boyeri*) living in the English Channel, near the Fawley power station in Southampton, United Kingdom. The study found that sand smelt rarely survive to four years of age. Age-specific survival and fertility rates for females are given in Table 10-9.[4]

TABLE 10-9

Age Group	A	B	C
Age (years)	0–1	1–2	2–3
Total eggs per female	1,423	4,567	8,718
Annual survival rate for females	0.1082	0.1082	0

Source: *Journal of Fish Biology*

a. The male:female sex ratio in the sand smelt population was found to be 1:1, and the percentage of sand smelt that survived from the egg stage to exactly one year of age was 0.16%. Determine the annual fertility rate for *female* sand smelt in each age group. Round answers to two decimal places. (The annual fertility rate will equal the number of eggs that survive one year and result in female smelt.)

b. Draw a life-cycle graph, and then write a system of difference equations for the population of female English Channel sand smelt.

c. Suppose that there are 1,000 female sand smelt initially in each of the groups A, B, and C. Create a table listing the number of female fish in each age group for years $n = 1$ through $n = 7$. Sum the individual age group values to determine the total population of female sand smelt for each year.

d. Use the total population values from years $n = 5$ and $n = 6$ to determine the rate of growth (expressed as a percent) for the overall population.

e. Repeat part (d) for years $n = 6$ and $n = 7$. What can you conclude?

f. In the long term, what is the stable age distribution?

g. Small sand smelt that swim near the Fawley power station are often sucked into the cooling-water intake pipes. Suppose that this action reduces the fertility rates that you found in part (a) by 40%. According to your revised system of difference equations, what is the new stable age distribution?

12. Perhaps the greatest symbol of the western United States is the American bison (*Bison bison*, commonly called a buffalo). Once numbering in the tens of millions, the American bison population was reduced significantly in the mid to late 1800s. Today the American bison are primarily confined to Yellowstone National Park and number in the thousands. (See Chapter 9 for more details.) Similar to the PCH caribou, the American bison have three principal growth stages: calves, yearlings, and adults (Figure 10-10). Estimates for the survival and fertility rates of these stages are given in Table 10-10.[5] These statistics are for female bison only.

Figure 10-10 A bison mother and calf. *Courtesy of U.S. Fish and Wildlife Service/photo by Jesse Achtenberg.*

TABLE 10-10

Growth Stage	A = Calves	B = Yearlings	C = Adults
Age in years	0–1	1–2	2+
Yearly fertility rate	0	0	0.42
Yearly survival rate	0.6	0.75	0.95

Source: *Discrete Dynamical Systems*

a. Draw a life-cycle graph; then write a system of difference equations describing the populations of the three growth stages of female American bison.

b. In Yellowstone National Park there are approximately 4,000 bison roaming today (both male and female). Assume that 20% are calves, 10% are yearlings, and 70% are adults. Also assume that 50% are females. Determine the initial conditions for the three female bison growth stages.

c. Find the total female population in years $n = 5$ and $n = 6$; then determine the percent growth rate for the total population.

d. Repeat part (c) for years $n = 9$ and $n = 10$. What can you conclude?

e. In the long term, what percentage of the total population do the individual growth stages attain? Do these percentages change much from $n = 0$?

13. Revisit Example 10-3 on the Porcupine Caribou Herd (PCH). Determine the growth rates for female PCH calves, yearlings, and adults between the years $n = 20$ and $n = 21$. Repeat for the years $n = 30$ and $n = 31$. See Table 10-5 for assistance. Describe what you find.

14. At the end of the discussion on stable age distributions, two questions were posed concerning the human impact on the arctic environment and the potential effects on the Porcupine Caribou Herd (PCH). See Figure 10-11. Modify the system of difference equations for the PCH to answer these questions.

    Question 1: If oil drilling in the Arctic National Wildlife Refuge destroys some caribou calving grounds and lowers birth rates by 25%, how will the long-term growth rate of the PCH population change? Justify your answer.

    Question 2: If global warming lengthens the growing season for arctic vegetation, making caribou adult survival rates increase from 84% to 94%, what will be the new stable age distribution for the PCH? Justify your answer.

15. The Umfolozi National Park is located in KwaZulu-Natal, a province in eastern South Africa. This tropical wilderness is home to a variety of large mammal populations, including the waterbuck (*Kobus ellipsiprymnus*). Waterbuck are in the same animal family as deer and antelope and are primarily a foraging animal living on grasslands. As their name might indicate, waterbuck also live in areas that have lots of water, as they need to drink every few days and use small water bodies to hide from predators such as lions. Helicopter-based surveys from 1970 to 1975 indicated that the waterbuck population in Umfolozi National Park declined from 1,098 to 494. In 1976, biologists launched a three-year study to analyze waterbuck ecology and evaluate management actions that would lead to increased waterbuck numbers. The study determined annual survival and fertility rates for female waterbuck calves, yearlings, and adults. Those numbers are given in the life-cycle graph in Figure 10-12.[6]

Figure 10-11 An adult male caribou displaying a full set of antlers. *Photo courtesy of U.S. Fish and Wildlife Service/photo by Jon Nickles.*

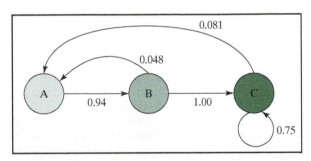

Figure 10-12 A life-cycle graph for the three growth stages of female waterbuck

a. Write a system of difference equations modeling the female waterbuck population.

b. Ground surveys during the three-year study indicated that the male:female sex ratio was 1:2.24. What percent of the population was female? Use your answer to this question to estimate the number of females sighted in the helicopter survey in 1975.

c. Assume that the number of females sighted by helicopter in 1975 was evenly split among calves, yearlings, and adults. With $n = 0$ referring to 1975, write the initial conditions.

d. By what percent will the female waterbuck population grow or shrink in the long term? Show your work. If your forecast is true, what is the fate of the waterbuck in the park?

16. Follow-up to the previous exercise. One reason that park biologists conducted the waterbuck study in the Umfolozi National Park was to enable them to *better manage* the

population. So what strategies can park biologists take to boost population numbers? Consider the following options and questions:

a. Lions (*Panthera leo*) are one of the predators of waterbuck and commonly feed on newborn offspring and calves. Suppose that wildlife biologists in the park reduce the number of lions so that the fertility rates for yearlings and adults are both quadrupled. Write a new system of difference equations. Then determine the long-term effect on the total population growth rate.

b. In times of drought, grasses and reeds that are a principal part of the waterbuck's diet can be in short supply. Suppose that park managers provide supplementary food for waterbuck during these droughts, so that adult survival rates increase to 96%. Write a new system of difference equations. Then determine the long-term effect on the total population growth rate. Show your work.

c. Assuming that the strategies given in parts (a) and (b) have the same monetary cost, which is the most effective?

d. Which strategy do *you* think is better? Consider variables other than monetary costs.

17. Refer to Example 10-4 about pollution in Lake Erie and Lake Ontario. Answer the following questions *using all decimal places that your technology device provides*.

a. What is Lake Erie's pollution level in year $n = 2$? What quantity of pollution will be flushed away in that year?

b. What is Lake Ontario's pollution level in year $n = 2$? What quantity of pollution will be flushed away in that year?

c. Use answers from part (a) to compute the pollution amount for Lake Erie in year $n = 3$.

d. Use your answers from parts (a) and (b) to compute the pollution level for Lake Ontario in year $n = 3$.

e. Use technology to create a table of values from $n = 0$ to $n = 3$, based on the difference equations and initial conditions. Verify that your answers to parts (c) and (d) are correct. Are they?

18. Suppose that the flushing rates for Lake Erie and Lake Ontario are modified over time to equal 29% and 19%, respectively.

a. List two environmental factors that might modify flushing rates.

b. What are the new difference equations for Lakes Erie and Ontario?

c. Make a table listing the pollution levels for years $n = 0$ to $n = 6$.

d. Lake Erie's pollution level drops immediately. What kind of function describes that drop? Is there another formula that can be used to model Lake Erie's annual pollution level?

e. Life-cycle graphs are used to describe flows from one age group to the next in populations. Create a similar diagram to display the flow of water (and pollution) from Lake Erie to Lake Ontario. Make one diagram for the original flushing rates, and one for the modified flushing rates.

Extended Exercises

19. Human exposure to **lead**, a heavy metal, can have grave health effects. In high enough concentrations, lead can damage the central nervous system, decrease the production of red blood cells, and impair vital organs such as the liver and kidneys. (See the "Science in Depth" article in Chapter 8 for more information on lead poisoning.) In the 1970s, doctors examined a 53-year-old male patient from Los Angeles (LA) to determine the rates at which lead moves among the body's systems. A flow chart illustrating those findings is shown in Figure 10-13.[7]

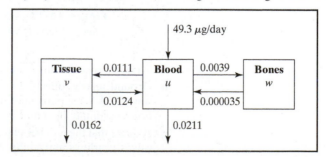

Figure 10-13 Lead is stored in the body in three principal regions: tissue, blood, and bones. Lead moves between tissue and blood, and between blood and bones. Lead exits the body through tissue and blood.

Lead enters the stomach and lungs via contaminated food, air and water, and then moves into the bloodstream (Figure 10-15). For the LA patient, the amount of lead that entered the bloodstream from the stomach and lungs was 49.3 micrograms (μg) each day. Once in the blood, lead can move to the tissue and bones. For the LA patient, 0.0111 or 1.11% of the lead in the blood moved to the tissue each day, and 0.0039 or 0.39% of the lead in the blood moved to the bones each day.

Lead can also move from the tissue and bones back into the blood. For the LA patient, 1.24% of the lead in the tissue moved to the blood each day, and 0.0035% of the lead in the bones moved to the blood each day. This last percentage indicates that lead does not exit the bones very quickly—we'll see shortly what effect this has on overall lead storage in the body.

Lead exits the body through tissue material such as feces, hair, nails, and perspiration. The LA patient lost 1.62% of the tissue lead each day in this manner. Lead can exit the bloodstream as the kidneys purify the blood; the lead is eventually passed out of the body through the urinary system. The LA patient removed 2.11% of his blood lead each day through this process.

To model the flow of lead through the body, define variables: Let $u(n)$, $v(n)$, and $w(n)$ represent the amount of lead in the blood, tissue, and bones, respectively, on day n. In all three equations the lead amounts are measured in micrograms.

To write the difference equation for the lead in the blood, $u(n)$, focus only on the flows of lead into and out of the bloodstream (Figure 10-14). The blood loses $0.0111 + 0.0039 + 0.0211 = 0.0361$ or 3.61% of its lead each day; this means that it retains $1 - 0.0361 = 0.9639$ or 96.39% of its lead from one day to the next. We start building the difference equation for blood lead as follows:

$$u(n) = 0.9639\,u(n-1)$$

The blood is also gaining lead: 0.0124 of the tissue's lead, and 0.000035 of the bone's lead, as well as the fixed amount from the environment of 49.3μg/day. We add these amounts onto the previous difference equation to get

$$u(n) = 0.9639\,u(n-1) + 0.0124\,v(n-1) + 0.000035\,w(n-1) + 49.3$$

a. Draw a flow diagram (similar to Figure 10-14) that illustrates the lead flows *only for the tissue*. Then write the difference equation, $v(n)$, for the tissue's lead amount.

b. Repeat part (a) to find the difference equation for the bone's lead amount, $w(n)$.

c. Assume that the LA patient has no lead in his body at time $n = 0$. Write the initial conditions. Then using technology, graph the three difference equations from $n = 0$ to $n = 360$ days. *Hint: Set the maximum height of your graph window to 2,500.* Record the graph on your homework paper.

d. Describe the general trend in lead accumulation in each of the three reservoirs: blood, tissue, and bones. Can you explain why one of these reservoirs stores so much more lead in the long term?

e. The California Occupational Lead Poisoning Prevention Program considers a blood lead level of 25 μg/dL (micrograms of lead per deciliter of blood) or higher to be "dangerous."[8] Did the blood lead level of the LA patient exceed this dangerous level after 360 days? *Note: An average adult body contains about 5 liters of blood.*

Figure 10-14 A diagram of the daily lead flows into and out of the bloodstream.

49.3 μg/day

0.0111   **Blood** u 0.0039

0.0124   0.000035

0.0211

Figure 10-15 Common sources of lead poisoning include solder and pipes (left) and old, peeling paint (right). *Photo credit: Langkamp/Hull.*

20. Follow-up to the previous exercise.

a. Find the lead amounts in the blood, tissue, and bones after 360 and 361 days. Organize the results as in the following table. Record your answers using one decimal place. *Note: A graphing calculator may take several minutes to compute the desired output.*

n	Blood = $u(n)$	Tissue = $v(n)$	Bones = $w(n)$
360			
361			

b. Your table indicates that the amounts of lead in the blood and tissue are not rising significantly after the 360th day. By what amount does the lead in the bones increase from day 360 to day 361?

c. On the 360th day, the LA patient ingested 49.3 μg of lead. Considering your answer to part (b), account for the missing lead on the 361st day.

d. Suppose the lead in the environment were cleaned up, so that the LA patient had no more exposure to lead after the 360th day. Make a graph of the lead amounts in the blood, tissue, and bones after 360 additional days. Record the graph on your paper. *Hint: Set the initial conditions to equal the lead amounts on the 360th day, and modify the blood lead difference equation to account for no additional lead intake. Graph the equations using the same window as in the previous exercise.*

e. Describe the effects on the body of having no more exposure to lead. Give a reasonable explanation for what you find.

21. Measles is similar to HIV in that it compromises the immune system and makes those infected more susceptible to other diseases such as pneumonia and diarrhea. Measles can be prevented with vaccinations, although in many parts of the world vaccination programs are lacking (Figure 10-16). The Measles Initiative, an international program dedicated to eradicating measles in Africa, reports that

> *Each year, a disease barely remembered by most Americans kills one million children, nearly a half million of those in Africa alone. This fact makes measles the single leading cause of death among children in Africa—more than HIV, more than tuberculosis, and more than malnutrition. In a place where health conditions are extremely poor, living conditions are more than difficult, and access to health care is minimal, measles can be easily prevented with a simple vaccination.[9]*

The time sequence in which a child contracts measles is quite fascinating. A child who has not yet caught the measles is referred to as a **susceptible**. Once a child catches the disease, there is a **latent period** of about one week before the child becomes contagious. The **contagious period**, the time when a child can spread the measles to other children, also lasts about one week. Once the child is past the contagious period, he or she is **immune** from catching the measles in the future (Figure 10-17).[10] Because the latent and contagious periods each last one week, it is natural to build a difference equation model that uses a time step of one week. Start by defining the following two variables:

$$S(n) = \text{number of susceptible children at the start of week } n$$

$$C(n) = \text{number of contagious children at the start of week } n$$

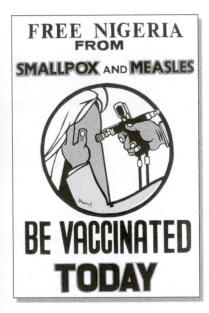

Figure 10-16 A pre-1979 poster from Nigeria promoting vaccinations for smallpox and measles. *Source: CDC.*

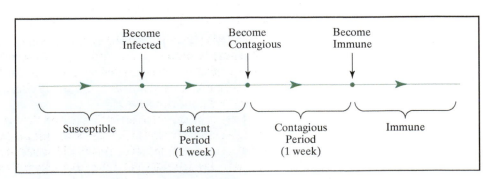

Figure 10-17 A time sequence illustrating the four stages of measles.

To determine the number of susceptible children that become infected and eventually contagious, we need information on how "efficiently" a contagious child spreads the measles. Suppose that a contagious child spreads the measles to 0.04% of the susceptible children in a city. If there are 5,000 susceptible children, one contagious child will infect $(0.0004)(5,000) = 2$ children. Similarly, if 25 contagious children have the same efficiency of 0.04%, and each can spread the measles to 5,000 susceptible children, then 25 contagious children would infect $(0.0004)(5,000)(25) = 50$ susceptible children. One week later these 50 newly infected children will be contagious.

a. In a 1982 study of measles epidemics in Nigeria, each contagious child in an urban area infected 0.003% of the susceptible children.[11] At the start of week $n-1$ there are $S(n-1)$ susceptible children and $C(n-1)$ contagious children. Write the difference equation for $C(n)$, the number of children that are contagious one week later. *Hint: Consider the computations in the preceding paragraph.*

b. The number of susceptible children at the start of week n is equal to the number of susceptible children at the start of week $n-1$, minus the number that became infected at the start of week $n-1$, plus the number that are born in week $n-1$. For the Nigerian city investigated, researchers estimated that the number of births each week was 360 and that the number of childhood deaths was negligible. Write the difference equation for $S(n)$. *Hint: The number that became infected at the start of week $n-1$ is equal to the number that became contagious at the start of week n.*

c. The same 1982 study assumed that the Nigerian city had 20 contagious children and 30,000 susceptible children initially. Write the initial conditions.

d. Make a graph of the number of susceptible children and a graph of the number of contagious children (separately) over a 400-week time period. Record each separately on your homework paper. *Suggestion: Use a window height of 50,000 for the graph of $S(n)$, and 4,000 for the graph of $C(n)$.*

e. Based on the model, how many children become contagious at the peak of each outbreak in the Nigerian city? How often do the peaks occur, or, in other words, what is the frequency of the measles epidemics?

22. Follow-up to previous exercise.

a. The 1982 study mentioned in the previous exercise also investigated measles epidemics in Britain. Similar difference equations and initial conditions were obtained for a British city, except that the number of births was estimated at 120 each week. Write the new system of difference equations.

b. Graph each difference equation over a 400-week period. Use graph windows similar to those in the previous exercise. Record on your homework paper.

c. Measles epidemics are observed in Britain every two to three years. Does your model agree with this fact?

d. How are the British measles epidemics predicted by your model similar to those predicted for Nigeria? How are they different?

SCIENCE IN DEPTH

A River Runs through Europe

Old, abandoned gold mines with their large piles of rock tailings are the target of modern gold companies trying to squeeze the last bit of precious metal out of an old mine. The rock tailings, left over from crushing and extraction of the easily recoverable metal, are processed again using a technique called **heap leaching**.[12] Cyanide acid is poured over the heap of finely crushed tailings, tiny bits of gold and other metals are leached out of the pile, and the gold is recovered using other chemicals. An undesirable result of the heap leach process is a large, artificial reservoir filled with poisonous cyanide and metal-rich waste.

One such heap leach project is the Baia Mare gold mine in northern Romania, once owned and operated by Esmeralda Exploration of Australia (now bankrupt). Following heavy rains at the end of January 2000, the artificial dam holding back the reprocessed tailings at the Baia Mare mine gave way, and about 100,000 cubic meters (over 3 million cubic feet) of cyanide and metal-rich waste flowed into the Lapus River, and then into the Somes (Szamos) River, and then into the Tizsa River, and finally into the Danube River (Figure 10-18).[13] Fish kills were extensive along this river system, despite the diluting effect of downstream transport of the wastes. While cyanide breaks down rather quickly, metals, cyanide by-products, and other chemicals used in the heap leach process remain in the rivers today.

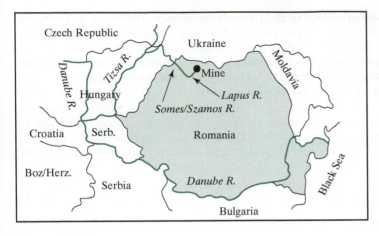

Figure 10-18 Map of the flow of the metal-rich wastewater from the Baia Mare mine to the Black Sea.

Downstream from the Baia Mare mine, the Danube River flows through northern Serbia (the former Yugoslavia). In 1999, NATO forces bombed various chemical industries along the Danube River in Serbia, targeting petrochemical facilities, fertilizer manufacturing plants, pharmaceutical companies, and other industries. Large quantities of pollutants flowed into the Danube during and after the bombing. Fresh water intake at a Bulgarian nuclear power plant on the Danube River was threatened due to oil slicks originating from damaged petroleum facilities in Serbia.

The Danube flows into the Black Sea, the largest oxygen-poor sea on planet Earth.[14] The Black Sea is fed by fresh water from rivers throughout central Europe; 60% of the fresh water discharge into the sea comes from the Danube. The Black Sea is also fed by salt water from the Mediterranean Sea. The fresh water floats on top of the denser salt water; little mixing takes place between these two layers. Increased amounts of nitrogen and phosphorus in the streams and rivers feeding the Black Sea have caused blooms of algae and plankton in the upper layer. Fish stocks have declined as a result of the algal blooms in the Black Sea; overfishing and the introduction of exotic species have also contributed to the decline.

The nitrogen and phosphorus originate from both **point sources** and **non–point sources**. Point sources are discrete, well-defined origins of pollutants. For example, human sewage is collected by a system of pipes, and then the untreated or poorly treated sewage is pumped into the Danube River and its tributaries. Non–point sources of pollution are distributed over a wide area. For example, farmers in central Europe apply fertilizers on their fields to boost crop yields. Excess fertilizer is washed into the intricate stream network that covers all of central Europe.

CHAPTER PROJECT:
POLLUTION IN A CHAIN OF LAKES

It took about three weeks for the plume of heavy metals and chemicals from the Baia Mare mine disaster to reach the Black Sea. The plume entering the Black Sea was much diluted compared to the toxic levels of metals and cyanide at the source. Is dilution the solution to pollution?

In the companion project to Chapter 10 found at the text Web site (**enviromath.com**), you will use difference equations to model a system of lakes connected by rivers. You'll track pollution levels as water moves downstream from one lake to another, and evaluate the dilution effect. And you'll modify the connecting rivers, Army Corps of Engineers style, while analyzing the impacts of this pollution bypass.

NOTES

[1] The Caribou Commons Project, http://www.cariboucommons.com/issue/issue.html

[2] Values derived from data presented in S. G. Fancy, K. R. Whitten, and D. E. Russell, "Demography of the Porcupine Caribou Herd, 1983–1992," *Canadian Journal of Zoology* 72 (1994): 840–846.

[3] University of Wisconsin Sea Grant Institute, Great Lakes Online, "Gifts of the Glaciers," http://www.seagrant.wisc.edu/communications/greatlakes/GlacialGift/index.html

[4] P. A. Henderson and R. N. Bamber, "On the Reproductive Biology of the Sand Smelt *Atherina boyeri* Risso (Pisces: Atherinidae) and Its Evolutionary Potential," *Biological Journal of the Linnean Society* 32 (1987): 395–415. Based on data from A. W. H. Turnpenny, R. N. Bamber, and P. A. Henderson, "Biology of the Sand-Smelt (*Atherina presbyter* Valenciennes) Around Fawley Power Station," *Journal of Fish Biology* 18 (1981): 417–427.

[5] James T. Sandefur, *Discrete Dynamical Systems: Theory and Applications* (Oxford: Claredon Press, 1990), 282.

[6]Values derived from data presented in D. A. Melton, "Population Dynamics of Waterbuck (*Kobus ellisiprymnus*) in the Umfolozi Game Reserve," *African Journal of Ecology* 21 (1983): 77–91.

[7]Flow parameters based upon model values presented in E. Batschelet, L. Brand, and A. Steiner, "On the Kinetics of Lead in the Human Body," *Journal of Mathematical Biology* 8 (1979): 15–23. As cited in ODE Architect mathematics software (John Wiley and Sons, 1999).

[8]California Dept of Health Services, Occupational Health Branch, *Lead in the Workplace*, April 2002, http://www.dhs.ca.gov/ohb/OLPPP/regshort.pdf

[9]The Measles Initiative, http://www.measlesinitiative.org/

[10]Glenn Fulford, Peter Forrester, and Arthur Jones, *Modelling with Differential and Difference Equations, Australian Mathematical Society Lecture Series* 10 (Cambridge: Cambridge University Press, 1997), 394–397.

[11]Roy Anderson and Robert May, "The Logic of Vaccination," *New Scientist*, November, 1982, 410–415.

[12]Earthworks, "Cyanide Heap Leach Packet," http://www.earthworksaction.org/pubs/Cyanide_Leach_Packet.pdf

[13]Anonymous, "Death of a River," *The Bulletin (Quarterly Magazine for the Regional Environmental Center for Central and Eastern Europe)* 9, no. 2, April 2000, http://bulletin.rec.org/Bull92/Default.html

[14]University of Delaware College of Marine Studies, "A Black Sea Journey," http://www.ocean.udel.edu/blacksea/

PART FOUR

Elementary Statistics

11

Fundamentals of Statistics

L andfills are becoming packed with many of the 270 million automobile tires that U.S. residents discard each year.[1] Efforts to decrease this waste stream are quite innovative: burning shredded tires to generate electricity, and using whole tires to construct artificial shorelines (Figure 11-1). In another pioneering program, tires are shredded to create rural septic fields. But what happens to these old tires when exposed to the environment? Are they really all that safe? Samples taken from those same septic fields show that heavy metals can be leached from tires into the groundwater. You'll get to work with that data in the end-of-chapter exercises using **statistics**. Statistics is the *process* by which we organize, summarize, examine, and interpret numbers.

This chapter introduces a wide collection of statistical tools for exploring data sets and discusses when, why, and how to use those tools effectively. We will examine various methods to measure the center or average value of data, as well as the spread or variability in data. We will also introduce the display of data with boxplots, and look at their connection to histograms and skew. This chapter will conclude with a few words on the important topic of sampling—without sampling, we would not have discovered the hazards of burying shredded tires.

Figure 11-1 An artificial coast in Port Wing, Wisconsin. *Photo: Minnesota Sea Grant, Dale R. Baker.*

MEASURES OF CENTER AND OTHER DESCRIPTIVE STATISTICS

Chapter 3 illustrated how to display data with pie charts, bar charts, histograms and other visual aids. Data can also be summarized and described with numbers called **descriptive statistics**. We introduce the most commonly used descriptive statistics in the following two examples.

EXAMPLE 11-1 Ponderosa Pine Snags and Logs

Nutrients stored in snags (dead trees still standing) and logs (dead trees which have fallen) are vital to the health of forest ecosystems (Figure 11-2). A study was conducted by scientists at the H. J. Andrews Long Term Ecological Research Center in the Pacific Northwest to measure the amount of nutrients in snags and logs in a sample forest. In one plot of the forest, the lengths of several ponderosa pine (*Pinus ponderosa*) snags and logs were measured.[2] The results are given in units of meters.

17.1 20.8 3.0 18.0 31.0 13.0 9.1 5.0 10.0 9.1

The **size** of the data set, the number of numbers, is $n = 10$. The **minimum** value is the smallest length in the data set (3.0 meters), and the **maximum** is the largest

Figure 11-2 A pine snag can stand for decades after the tree has died. The snag provides valuable forest habitat, and the decomposing wood releases nutrients back into the soil. *Photo: Langkamp/Hull.*

(31.0 meters). The **range**, the difference between the maximum and the minimum, is $31.0 - 3.0 = 28.0$ meters. It is common to compute a **center value** (often called an average value) for a data set; however, there are three common definitions of center used in statistics—mean, median, and mode. The **mean** is computed by summing all the values and dividing by the number of values. In this example, the mean is computed as

$$\bar{x} = \frac{17.1 + 20.8 + 3.0 + 18.0 + 31.0 + 13.0 + 9.1 + 5.0 + 10.0 + 9.1}{10}$$

$$= 13.61 \text{ meters}$$

The symbol $\bar{x}$ (pronounced "x bar") is often used to represent the mean. Note that it's widespread practice to state the mean using *one more decimal place than in the original data*. When using the mean in other computations, more decimal places may be required.

The **median** is the middle number when the data are arranged in ascending or descending order. The lengths are shown arranged in ascending order.

3.0 5.0 9.1 9.1 10.0 13.0 17.1 18.0 20.8 31.0

Because this data set has an even number of values ($n = 10$), the median is computed by taking the mean of the two middle numbers:

$$\text{median} = \frac{10.0 + 13.0}{2} = 11.5 \text{ m}$$

If a data set has an odd number of values, then the median is equal to the middle number (as we'll see in the next example). The **mode** is the most frequently occurring value. In this data set the mode is 9.1 m. In this text we focus more on the mean and median because the mode is not used very often in analyzing environmental data. It is also common to state the **proportion** of a data set with some characteristic. By *proportion* we mean "fraction or percentage of the total" (which is different than how we defined proportion in Chapter 2). For example, the proportion of snags and logs that have lengths less than 15.0 meters is 6/10 or 60%.

EXAMPLE 11-2 Central American Fertility Rates

Statistics on death, fertility, and migration are essential in forecasting population change, which in turn is critical for planning social services. Table 11-1 shows total fertility rates for Central American countries in the year 2003.[3] The total fertility rate measures the number of live births for an average woman during her lifetime. Costa Rica had the minimum total fertility rate (2.38), while Guatemala had the maximum rate (4.67).

TABLE 11-1

Country	2003 Total Fertility Rate (per woman)
Belize	3.86
Costa Rica	2.38
El Salvador	3.25
Guatemala	4.67
Honduras	4.07
Nicaragua	3.00
Panama	2.53

Source: U.S. Census Bureau

The mean total fertility rate for the Central American countries in 2003 was

$$\bar{x} = \frac{3.86 + 2.38 + 3.25 + 4.67 + 4.07 + 3.00 + 2.53}{7} = 3.394$$

When the total fertility rate data are arranged in ascending order, the middle value is 3.25, which is the median. There is no number in the data that occurs more than once, so there is no mode. We can find the proportion of the countries that have total fertility rates less than, for example, 4.00. That proportion is 71% (5 out of 7).

WEIGHTED MEANS

Care must be taken when interpreting the significance of any descriptive statistic, whether environmental or not. The mean of the total fertility rates for the seven Central American countries is 3.394. This number is not the mean for a typical Central American *woman*; rather, it's the mean for a typical Central American *country*. The mean of the seven country values and the mean for all women in Central America are not the same because the countries have different numbers of women.

How do we compute the mean total fertility rate of all Central American women? The answer is with a **weighted mean**, which factors in population size to "weight" each country's fertility rate. To compute the weighted mean, first multiply each country's total fertility rate by that country's population of women *expressed as a percent of the total number of Central American women.*[4] For each country we obtain a "weighted" total fertility rate (Table 11-2). Second, sum the weighted total fertility rates to arrive at the mean total fertility rate for Central America, which is 3.72. (We use the symbol Σ, the Greek capital letter sigma, to represent "sum.") Notice that Guatemala, with both the highest total fertility rate and largest population of women, contributes the most toward Central America's total fertility rate.

TABLE 11-2

Country	Total Fertility Rate (per woman)	Population Weight (percent of total number of women in Central America)	Total Fertility Rate (weighted)
Belize	3.86	0.7% = 0.007	(3.86)(0.007) = 0.02702
Costa Rica	2.38	9.8% = 0.098	(2.38)(0.098) = 0.23324
El Salvador	3.25	16.9% = 0.169	(3.25)(0.169) = 0.54925
Guatemala	4.67	35.0% = 0.350	(4.67)(0.350) = 1.63450
Honduras	4.07	17.0% = 0.170	(4.07)(0.170) = 0.69190
Nicaragua	3.00	13.1% = 0.131	(3.00)(0.131) = 0.39300
Panama	2.53	7.5% = 0.075	(2.53)(0.075) = 0.18975
			Sum of values $\Sigma \approx 3.72$

EXAMPLE 11-3 Mean Rainfall

Suppose that four adjacent watersheds have areas and annual rainfall amounts as given in Table 11-3. The mean rainfall for the region encompassing all four watersheds can be determined using a weighted mean.

TABLE 11-3

Watershed	Area (hectares)	Mean Rainfall (cm)
A	200	30
B	50	75
C	150	115
D	100	50

The total area for all watersheds is 500 hectares, so the weights for watersheds A, B, C, and D are 40%, 10%, 30%, and 20%, respectively. The mean annual rainfall is

$$(40\%)(30\,\text{cm}) + (10\%)(75\,\text{cm}) + (30\%)(115\,\text{cm}) + (20\%)(50\,\text{cm}) = 64\,\text{cm}$$

In Example 11-1 we computed the mean of the snag and log lengths using a familiar process: add up the lengths and divide by 10. That method of computing the mean appears quite different than using a weighted mean, but they are actually the same. We computed the mean length as follows:

$$\bar{x} = \frac{17.1 + 20.8 + 3.0 + 18.0 + 31.0 + 13.0 + 9.1 + 5.0 + 10.0 + 9.1}{10}$$

But dividing by 10 is the same as multiplying by 1/10 or 0.1. Thus the preceding fraction can be rewritten as

$$\bar{x} = 0.1\,(17.1 + 20.8 + 3.0 + 18.0 + 31.0 + 13.0 + 9.1 + 5.0 + 10.0 + 9.1)$$

Distributing 0.1 across the parentheses results in

$$\bar{x} = 0.1(17.1) + 0.1(20.8) + 0.1(3.0) + 0.1(18.0) + 0.1(31.0)$$
$$+ 0.1(13.0) + 0.1(9.1) + 0.1(5.0) + 0.1(10.0) + 0.1(9.1)$$

In this last expression, all log and snag lengths are multiplied by 0.1; that is, all snags and logs have the same "weight," equal to 0.1 or 10%. In general, when we compute the mean of a set of values, we compute a weighted mean with all weights being the same.

QUARTILES AND THE 5-NUMBER SUMMARY

Recall that the median divides a data set into upper and lower halves. When a data set is divided into quarters, the **1st quartile** (denoted by Q1) is the number that separates the first quarter of the data set from the second. Similarly, the **3rd quartile** (denoted Q3) divides the third quarter of the data from the fourth. See Figure 11-3.

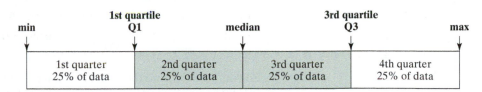

Figure 11-3 A diagram of the 5-number summary

All together, the five numbers consisting of the minimum, Q1, the median, Q3, and the maximum are called the **5-number summary**. The 5-number summary contains information on the center value (the median) as well as how "spread out" each quarter of data is from the center value.

As with the median, the numbers Q1 and Q3 may or may not be actual numbers from the data set. Determining Q1 and Q3 can be a little challenging, but there is a simple rule that is easy to follow:

- **Q1 is the median of the lower half of the data (all data values less than the median).**

- **Q3 is the median of the upper half of the data (all data values greater than the median).**

It should be noted that Q1 and Q3 do not always perfectly split a data set into equal-sized quarters, especially with small data sets. Also, rules for computing Q1 and Q3 vary somewhat with different calculators and computer software packages, although the values should be close. Let's look at a few examples.

EXAMPLE 11-4 Wisconsin Forests

Estimates for the amounts of forest in 13 counties in central Wisconsin in 1996 are given in Table 11-4.[5] Let's examine the percent forest values for the 13 counties, listed in ascending order in the last column of the table. Clearly the minimum is 31.9% (Waushara County) and the maximum is 61.0% (Adams County). The middle number of 39.8% is the median (Eau Claire County).

TABLE 11-4

No.	County	Total Forest (×1,000 acres)	Total Land (×1,000 acres)	Percent Forest
1	Waushara	127.7	400.7	31.9%
2	Marquette	95.8	291.6	32.9%
3	Portage	173.8	516.1	33.7%
4	Chippewa	232.4	646.7	35.9%
5	Marathon	376.1	988.9	38.0%
6	Waupaca	183.4	480.7	38.2%
7	Eau Claire	162.4	408.1	39.8%
8	Clark	321.0	778.0	41.3%
9	Wood	215.4	507.4	42.5%
10	Monroe	273.4	576.6	47.4%
11	Juneau	271.7	491.3	55.3%
12	Jackson	366.8	631.9	58.0%
13	Adams	253.1	414.6	61.0%

Source: U.S. Dept of Agriculture

To determine Q1, we find the median of the percent forest values that lie below 39.8%; thus Q1 is the mean of the values for the 3rd and 4th ranked counties:

$$Q1 = \frac{33.7\% + 35.9\%}{2} = 34.8\%$$

To compute Q3, find the median of the percent forest values greater than 39.8%. Do this by computing the mean of the values for the 10th and 11th ranked counties:

$$Q3 = \frac{47.4\% + 55.3\%}{2} = 51.35\%$$

The 5-number summary is 31.9%, 34.8%, 39.8%, 51.35%, 61.0%. From these numbers we can tell that the middle 50% of all central Wisconsin counties are between 34.8% and 51.4% forested.

EXAMPLE 11-5 E-Waste

There is growing concern about the amount of electronic products disposed of in landfills. These discarded products are collectively referred to as **e-waste**. Not only does e-waste occupy scarce landfill space, it often contains toxic substances (Figure 11-4). For example, e-waste consisting of cathode ray tubes (CRTs) from computer monitors and televisions contains lead and other heavy metals that can leach into groundwater.

In a University of Florida hazardous waste study, 36 models of CRTs were tested for the presence of **leachable lead**.[6] Leachable lead is a form of lead that readily dissolves into groundwater. The results of the study are given in Table 11-5.

The lead concentration data (in milligrams per liter or mg/L) are sorted in ascending order. Clearly the minimum is 1.0 mg/L and the maximum is 85.6 mg/L. Because there are 36 data values, an even number, the median is the mean of the 18th and 19th values:

$$\text{median} = \frac{9.1 + 9.4}{2} = 9.25 \text{ mg/L}$$

Figure 11-4 E-waste is often shipped to less-wealthy countries, although the Basel Ban prohibits the transboundary movement of hazardous wastes. This photo shows migrant workers in Guiyu, China breaking down old computers for parts and heavy metals, exposing themselves to toxic materials. *Photo courtesy of the Basel Action Network: www.ban.org.*

TABLE 11-5

Year Made	CRT Maker	Lead Concentration (mg/L)	Year Made	CRT Maker	Lead Conentration (mg/L)
90	Clinton	1.0	89	Panasonic	9.4
84	Matsushita	1.0	97	Toshiba	10.6
85	Matsushita	1.0	87	NEC	10.7
87	Matsushita	1.0	98	Samsung	15.4
89	Samsung	1.0	92	Chunghwa	19.3
86	Phillips	1.0	97	Chunghwa	21.3
84	Goldstar	1.5	94	Zenith	21.5
94	Sharp	1.5	77	Zenith	21.9
94	Zenith	1.6	87	NEC	26.6
97	Toshiba	2.2	96	Orion	33.1
97	KCH	2.3	85	Sharp	35.2
91	Chunghwa	2.8	92	Phillips	41.5
93	Toshiba	3.2	84	Quasar	43.5
84	Matsushita	3.5	92	Toshiba	54.1
84	Sharp	4.4	85	Toshiba	54.5
98	Samsung	6.1	93	Panasonic	57.2
95	Samsung	6.9	89	Samsung	60.8
98	Chunghwa	9.1	89	Hitachi	85.6

Source: Florida Center for Solid and Hazardous Waste Management

With 18 data values below 9.25 mg/L, Q1 is the mean of the 9th and 10th numbers:

$$Q1 = \frac{1.6 + 2.2}{2} = 1.9 \text{ mg/L}$$

Similarly, Q3 is the mean of the 27th and 28th numbers:

$$Q3 = \frac{26.6 + 33.1}{2} = 29.85 \text{ mg/L.}$$

The 5-number summary is therefore

$$1.0 \text{ mg/L}, 1.9 \text{ mg/L}, 9.25 \text{ mg/L}, 29.85 \text{ mg/L}, 85.6 \text{ mg/L}$$

We can tell from this summary that the lowest 25% of CRT models have lead concentrations from 1.0 mg/L to 1.9 mg/L, while the highest 25% vary from 29.85 mg/L to 85.6 mg/L. Thus the CRT models with the largest lead concentrations have values much more spread out than those with the lowest concentrations.

BOXPLOTS

A quick way to display graphically the 5-number summary is with a **boxplot**, also referred to as a **box-and-whisker plot**. The boxplot of the 5-number summary for the e-waste data in Example 11-5 is illustrated in Figure 11-5. The boxplot consists of a box drawn from Q1 to Q3, with a vertical line through the box at the median. From the left and right sides of the box, "whiskers" are drawn out to the minimum and maximum values. In the boxplot, is it easy to see the small spread of the lowest 25% of the data, and the large spread of the highest 25%.

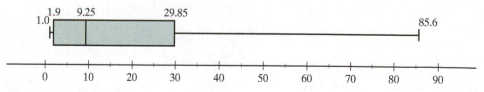

Figure 11-5 Boxplot of the e-waste data

Boxplots are great visual aids for comparing the relative positions and spread of several data sets. Figure 11-6 shows boxplots illustrating percent forested land for all counties in Wisconsin, divided into five forestry sectors.[7] We see that, in general, the northern counties have a greater amount of forestry cover, while those in southern Wisconsin have the least. Central Wisconsin counties (those listed in Example 11-4) have the smallest spread in percent forest values (the range is $61.0\% - 31.9\% = 29.1\%$), indicating that those counties are more similar in forest coverage as compared to counties in other sectors. What other patterns do you notice?

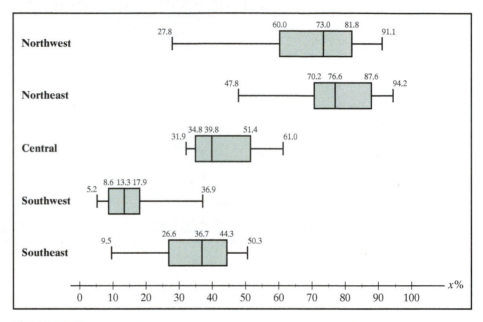

Figure 11-6 Boxplots for five forestry sectors in Wisconsin, with numbers indicating percent forest cover for counties in those sectors.

USING TECHNOLOGY: FINDING DESCRIPTIVE STATISTICS

Technology makes computing descriptive statistics, such as the mean and 5-number summary, quite easy and is an essential tool when working with large data sets. In this section we summarize the procedures for finding descriptive statistics on the TI-83/84 calculator. The text Web site (**enviromath.com**) provides instructions on how to use online computer applications that perform the same tasks. *Note: When using these other applications, keep in mind that the values of Q1 and Q3 are often computed with alternate formulas, providing slightly different results.*

Entering Univariate Data on the TI-83/84

Select **STAT** > **1: Edit** to view the list editor. You should see the window displayed to the right showing lists L1, L2, and L3.

*Note 1: To clear a list, use the arrow keys to highlight the list name (e.g. L1), then press **CLEAR** > **ENTER**. Do not press **DEL** or you will delete the list from the list editor.*

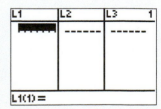

*Note 2: If any of the lists L1 through L6 are not present in the list editor, select **STAT** > 5:SetUpEditor, then press **ENTER**.*

Enter the 36 CRT lead concentration values from Example 11-5. Enter these data (in any order) into one list, such as L1. Scan the list to ensure you've entered the data correctly.

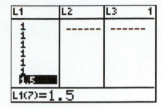

Descriptive Statistics on the TI-83/84

Select **STAT** > **CALC** > **1-VAR Stats**. After **1-VAR Stats** is pasted on the homescreen, enter **L1**.

Press **ENTER**. There are two screens of information. Use the up and down arrows to scroll between screens.

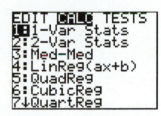

The first screen displays the mean, $\bar{x}$ = 18.70277778, and the number of data points, n = 36. You may ignore the other symbols on the first screen for now. The second screen displays the 5-number summary.

SHAPE OF A DATA SET

A boxplot gives us a crude picture of the amount of spread in a data set, but for a more refined picture, we often turn to histograms. Histograms display the frequency, or relative frequency, of univariate data over a set of intervals or bins. We discussed how to construct histograms in Chapter 3; in case you're a bit rusty at this, we'll go through the construction process again in the following example.

EXAMPLE 11-6   Weights of Black Bears

Table 11-6 gives weights (in pounds) of 54 black bears (*Ursus americanus*) that were captured in the wild.[8] The data are arranged in ascending order, from the lowest bear weight of 26 pounds to the highest of 514 pounds.

TABLE 11-6

26	62	90	132	154	204	270	365
29	64	94	140	166	204	316	416
34	65	105	140	166	212	332	436
40	76	114	144	180	220	344	446
46	79	116	148	182	220	348	514
48	80	120	150	202	236	356	
60	86	125	150	202	262	360	

Source: Gary Alt

We pick (somewhat arbitrarily) the bins 0 to 100 lb, 100 to 200 lb, and so on, and tally the number of bears that fall into each bin to determine frequency (Table 11-7). We use the convention that if a data value falls on a bin boundary, the value is tallied in the higher bin. We also compute the percent frequency for each bin in order to construct a relative frequency histogram.

TABLE 11-7

Bin	Frequency	% Frequency
0–100 lb	16	$16/54 \approx 29.6\%$
100–200 lb	17	$17/54 \approx 31.5\%$
200–300 lb	10	$10/54 \approx 18.5\%$
300–400 lb	7	$7/54 \approx 13.0\%$
400–500 lb	3	$3/54 \approx 5.6\%$
500–600 lb	1	$1/54 \approx 1.9\%$

To construct the frequency and relative frequency histograms, draw vertical bars to indicate the frequency or relative frequency of each bin (Figure 11-7). Recall that there is a fundamental difference between bar graphs and histograms. In bar graphs, the bins represent discrete (disconnected) categories; thus the bars are often drawn without touching each other. In histograms, the horizontal axis is continuous, and because the bins touch each other, so must the bars.

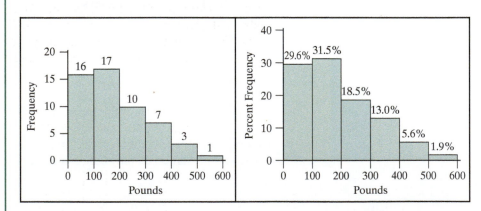

Figure 11-7 A frequency histogram (left) and relative frequency histogram (right) for the bear weight data. Both histograms have the same overall shape; only the vertical scales are different.

Now that we have the histograms of the bear weights in front of us, we can observe some characteristics of the data. In a histogram, a bin with the highest frequency is called a **modal bin**. For the bear data, the modal bin is 100–200 pounds. We can also see that over 60% of the weights fall between 0 and 200 pounds. Because the median weight will divide the data into two equal-sized sets, with 50% of the data above and 50% below, we know that the median must lie between 100 and 200 pounds. By computing one-variable statistics for the bear weights using technology, we find that this is true. See the following screens.

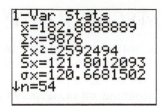 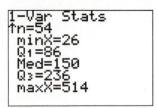

With a median weight of 150 pounds, the lower 50% of the data are "squished" into the first and second histogram bars, whereas the upper 50% of the data are "stretched" over the second through sixth bars. This makes the histograms asymmetric (unbalanced), giving an appearance of having a "tail to the right."

In general, a histogram that is not symmetric is called **skewed**. Some basic shapes of histograms are displayed in Figure 11-8. Those that have negative skew will have a tail to the left, those with zero skew are perfectly symmetric, and those with positive skew will have a tail to the right.

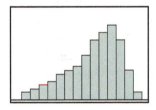 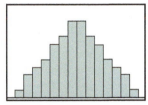 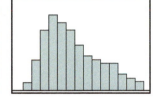

Figure 11-8 Data sets with (left to right) negative, zero, and positive skew

Each histogram in Figure 11-8 has a single, principal "peak"; histograms with this characteristic are called **unimodal**. But histograms come in many other shapes, often with multiple peaks and valleys. In some histograms the peaks and valleys are inexplicable; in others, they might clue us in to some underlying truth. For example, a histogram of total fertility rates for all countries in the world (Figure 11-9) is **bimodal**, meaning that it has two principal peaks.[9] Perhaps there are socioeconomic or environmental conditions that put countries into two classes—those with women that have between 1.5 and 2.2 children, and those that have between 4.3 and 5.7 children.

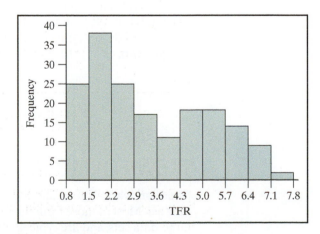

Figure 11-9 A histogram of the total fertility rates (TFRs) for 177 countries in the year 2000. The histogram shows two distinct peaks—one centered at 1.85 children, the other at 5.0 children.

USING TECHNOLOGY: HISTOGRAMS

Creating histograms often involves sorting and binning large data sets; technology can greatly reduce the time involved. Furthermore, it's easy to manipulate the appearance of histograms by adjusting the bin width automatically. Instructions for creating histograms on the TI-83/84 graphing calculator are given next. Visit the text Web site (**enviromath.com**) for instructions on how to use online computer applications that perform the same task.

Plotting a Histogram on the TI-83/84

Select **STAT** > **1: Edit** and enter the 54 bear weights (Table 11-6) into a single list, such as L1.

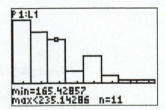

Select **STATPLOT** (**2nd Y=**), and choose **Plot 1**. In Plot 1, select **ON**, choose the histogram symbol under **Type**, and enter **L1** under **Xlist**. Set **Freq** to **1**.

To create a histogram automatically, select **ZOOM** > **9:ZoomStat**. Press **TRACE** and use the left and right arrows to view the bin boundaries and frequency of each bar.

Select **WINDOW** to adjust manually the size of the graph and the bin width. To duplicate the frequency histogram in Figure 11-7, set **Xmin = 0**, **Xmax = 600**, and **Xscl = 100** (bin width of 100 lb). Set **Ymin = 0**, **Ymax = 20**, and **Yscl = 5**. Leave **Xres = 1**. Press **GRAPH** or **TRACE** to view the histogram. *Note: The TI-83/84 does not create relative frequency histograms.*

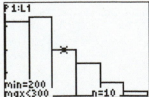

A SKEW FORMULA

There are various formulas available that can be used to assign a skew number to a data set. Most of those formulas are cumbersome to work with by hand, especially with large data sets, and are typically not programmed into calculators. Fortunately, there is a skew formula that is fairly simple, called **Bowley's skew formula**:

$$\text{Bowley skew} = \frac{Q3 - 2 \cdot \text{median} + Q1}{Q3 - Q1}$$

The Bowley skew will always fall between −1 and 1:

$$-1 \le \text{Bowley skew} \le 1$$

As you can see, the Bowley formula uses only Q1, the median, and Q3—essentially the middle 50% of a data set. More complex formulas factor in all data values, from the minimum to the maximum, when computing a skew value. For any data set, if the distance from Q1 to the median is greater than the distance from the median to Q3, the Bowley skew will be negative. The opposite case will make the Bowley skew positive.

Next we compute the Bowley skew for the bear weight data, using the 5-number summary provided by the graphing calculator.

$$
\begin{aligned}
\text{Bowley skew} &= \frac{Q3 - 2 \cdot \text{median} + Q1}{Q3 - Q1} \\
&= \frac{236 - 2(150) + 86}{236 - 86} \\
&= 0.15
\end{aligned}
$$

```
1-Var Stats
↑n=54
 minX=26
 Q1=86
 Med=150
 Q3=236
 maxX=514
```

The skew value of 0.15 is positive, agreeing with the fact that the distance from Q3 to the median is greater than the distance from Q1 to the median. The sign of the skew value also agrees with the fact that the bear weight frequency histogram has a tail to the right. The magnitude (size) of the skew value is most useful when comparing two data sets, to determine which has greater or lesser skew. *Note: With other technology devices, the values of Q1 and Q3 may be slightly different; thus the Bowley skew may not equal 0.15. This will also hold true for other examples and exercises in this text.*

Let's investigate another example to gain more experience with skew and shapes of data sets.

Figure 11-10 A reservoir near Dillon, Colorado is filled in spring by snowmelt. *Photo: Scott Bauer/USDA Agricultural Research Service.*

EXAMPLE 11-7   Western Water

The National Water and Climate Center is part of the Natural Resources Conservation Service—an agency of the U.S. Department of Agriculture. The climate center monitors water levels in western U.S. basins (lakes and reservoirs) which are used for irrigation and hydropower (Figure 11-10). A histogram of the water levels in 300 basins measured on December 31, 1998 is given in Figure 11-11.[10] Note that

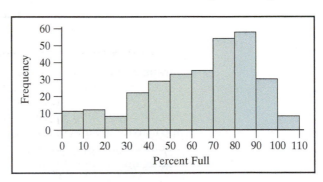

Figure 11-11 A frequency histogram of the water levels in the 300 basins

water levels are expressed as a percentage of total basin capacity, or as "Percent Full." The 5-number summary and boxplot are displayed in Figure 11-12, with all values given in percent form.

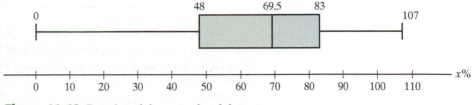

Figure 11-12 Boxplot of the water level data

The value of the Bowley skew is

$$\text{Bowley skew} = \frac{Q3 - 2 \cdot \text{median} + Q1}{Q3 - Q1}$$

$$= \frac{83 - 2(69.5) + 48}{83 - 48}$$

$$= -0.23$$

The sign of the Bowley skew (negative) agrees with the fact that the middle 50% of the data are negatively skewed (Q1 lies further from the median than does Q3).

COMPARING THE MEAN AND MEDIAN

The mean and median, both considered to be "averages," measure the location of the center of a data set. But these center values are not always the same. When data sets have negative skew (tail to the left), the mean will lie to the left of the median. For positively skewed data, the mean lies to the right of the median. We say "the mean follows the extreme," indicating that the mean will lie closer to the tail end (extreme values) in the histogram (Figure 11-13). For example, the bear weights in Example 11-6 have a median of 150 pounds and a mean of 182.9 pounds. In that data set, the mean lies to the right of the median, reflected in the fact that the histogram has a positive skew (tail to the right).

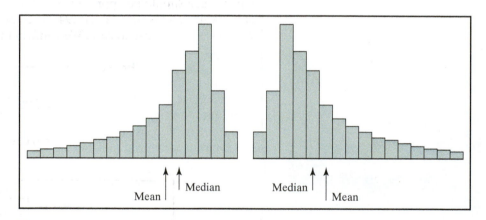

Figure 11-13 The mean follows the extreme

To understand why the mean gets "pulled" toward extreme values, suppose that the heaviest bear in our previous example increases its weight from 514 pounds to 600 pounds, while all other bear weights stay the same. We make this change in the data, and again compute the one-variable statistics. See the following screens.

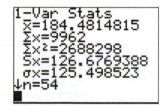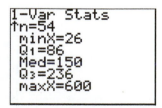

Notice how the mean has increased from $\bar{x} = 182.9$ lb to $\bar{x} = 184.5$ lb, and yet the median has stayed the same at 150 lb. You might have predicted this because the mean is the sum of all 54 bear weights divided by 54, so the mean will increase when any weight increases. The median, on the other hand, is just the middle number, and the numbers near the middle have not changed. We say that the mean is **sensitive** to extreme (high or low) values whereas the median is **resistant** to extreme values.

When extreme values on the left- and right-hand sides of a histogram "balance out" (i.e., when the histogram is symmetric), the mean and median will have approximately the same value. Let's investigate a data set that shows this balance.

EXAMPLE 11-8 Wheat Yields

The Rothamsted Experimental Station in Great Britain is one of the oldest agricultural research stations in the world. In 1843 the station began a series of long-term experiments to measure the effects of inorganic and organic fertilizers on crop yields. In 1910, a wheat yield experiment was conducted on a 20 × 25 grid of equal-sized plots to measure the regularity of wheat growth.[11] Summary statistics of yield measurements are displayed next. Note that the original data were measured in bushels.

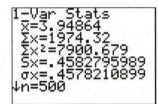
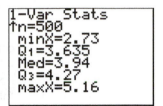

A boxplot and histogram of the 500 wheat yields (Figure 11-14) show that the shape of the data is highly symmetric. The high symmetry indicates that the mean and median should be approximately equal, and this is exactly the case (the mean is 3.94864 while the median is 3.94). The high symmetry also suggests that the skew value should be close to zero. We confirm this by computing Bowley's skew:

$$\text{Bowley skew} = \frac{Q3 - 2 \cdot \text{median} + Q1}{Q3 - Q1}$$

$$= \frac{4.27 - 2(3.94) + 3.635}{4.27 - 3.635}$$

$$= 0.039$$

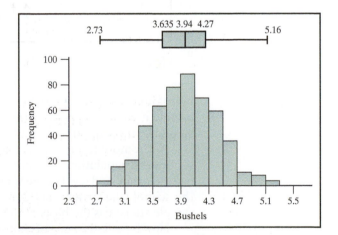

Figure 11-14 A nearly-symmetric boxplot and histogram of wheat yields. Symmetric, bell-shaped histograms are examined more thoroughly in Chapter 12 and Chapter 13.

Recall that the reason we compute the mean or median of a data set is to find a single number that best represents all the data; that is, an average or center value. Knowing that the mean is more sensitive than the median to extreme values helps us choose whether the mean or median best represents the center. This choice, of course, depends upon the information that we want to convey. For example, suppose rainfall totals for seven days in Kansas City were

$$0 \text{ cm} \qquad 0 \text{ cm} \qquad 0 \text{ cm} \qquad 0 \text{ cm} \qquad 0 \text{ cm} \qquad 5 \text{ cm} \qquad 9 \text{ cm}$$

We might use the daily median of 0 cm to indicate that it did not rain "on average," or at least on 50% of the days, and ignore the highly rainy days. On the other hand, we might use the daily rainfall mean of 2.0 cm as the typical value in forecasting long-term reservoir levels—a computation requiring rainfall data for all days, extreme or not.

SAMPLING

In Example 11-2, we examined fertility rates for the seven countries that make up Central America. Basic measures such as the mean, median, and range were used to summarize and describe the fertility data. In Example 11-4, forest levels were given for the counties that make up central Wisconsin, and we used the 5-number summary to measure the center and spread of those levels. In both of these examples, the original collection of data was represented with just a few numbers; as we mentioned earlier, those numbers are collectively called descriptive statistics.

The field of statistics involves much more than simply describing data sets. By using statistical methods, scientists are able to explore more sophisticated environmental questions. What is the mean lead level in the drinking water piped into our school? Is forest biodiversity impaired because of certain types of logging practices? Does exposure to DDT cause breast cancer in women? These types of questions can only be answered accurately by adhering to strict protocols involving project design, data analysis, and interpretation. In the concluding paragraphs of this chapter, we give a brief overview of some key elements of project design.

Suppose that we try to answer our first question posed: What is the mean lead level in the drinking water piped into our school? To answer this question, we decide to measure the amount of lead in 1 liter of water. We cannot just collect 1 liter of water from one faucet, because lead levels may vary tremendously for a number of reasons: water temperature, time of day, type of metal plating on the faucet, age of the nearby pipes, whether the pipes are made of lead, or whether lead solder was used to fasten the pipes together. The important thing to understand is that lead levels in our school's water system will have some *intrinsic variability*.

Using statistical terminology, we say that all liters of water flowing through our school's water system make up the **population**. We need to measure the lead level of all individual members of the population (each liter of water in the water system), and then compute the mean. For most schools this task would be highly impractical because of limitations in time and money! Instead, we collect a **sample**—a smaller number of liters of water—and compute the mean of the lead levels in these liters. If the sample is collected properly (more on this later), the mean lead level in the sample should be a good estimate of the mean lead level in the population.

A few more terms are necessary at this point. A **parameter** is a number that describes the population, whereas a **statistic** is a number that describes the sample. We use the mean lead level of the sample (the statistic) to predict the mean lead level for the school (the parameter). Inferring (predicting) a population parameter from a sample statistic is the basis of the field called **inferential statistics**.

Now back to the issue of choosing a sample. If our sample liters were collected from just one drinking faucet at 8 A.M. on Monday, after the water had sat over the weekend, they would most likely give an erroneous estimate for the overall lead levels. We want our sample liters of water to be representative of all liters of water in the entire school's water system. There are several ways to proceed. First, we could collect a **systematic sample**, in which, for example, we collect 1 liter of water from one wash basin and one drinking fountain on each floor of each building. In this way, no sector of our school is left without representation. Alternatively, we could collect a systematic sample in which we gather water every hour of the day, ensuring that no times are unrepresented.

We could also collect a **simple random sample** of, for example, 25 separate liters of water. In a simple random sample, every sample of size n has an equal chance of being selected. In our example, every possible collection of $n = 25$ liters of water running through the school's pipes would have an equal chance of being chosen. Random sampling often involves assigning each individual in the population a number, then using a random number generator (like those found on a calculator) to choose the individuals that make up the sample. Selecting liters of water through random sampling would be quite difficult, requiring us to keep track of the flow through all water pipes. Another complication with random sampling is the following. If the sample size is small, as is often the case with expensive environmental monitoring, we are left with the possibility that, by pure chance alone, we did not measure water lead levels in one sector of our school, or at one time of the day. There are methods to combine systematic sampling and random sampling that avoid the pitfalls of each—those methods are left for more advanced texts.

Even when we use one of the aforementioned sampling methods, sources of error can be introduced while collecting data that will give false results. For example, we might measure lead levels using equipment that is not sensitive enough (not enough precision), or the equipment may be defective and produce wildly different readings (not enough accuracy). We could also introduce human errors by failing to operate the equipment correctly, failing to write down the correct results, or simply failing to be careful! Another typical source of human error involves not sampling correctly—perhaps we missed a floor in our systematic sampling routine, or we favored drinking fountains over wash basins while we sampled. Or perhaps we were short on time and money and conducted a **convenience sample**, in which we simply collected water from the first two fountains that we walked up to.

No matter which sampling method chosen, and even if there is no error in data collecting, we must understand that samples have inherent variability. Let's explore this more deeply. Suppose that we have enough money to bring in three experts to estimate the lead level in our school's water. They all agree to use a random sampling method and test 25 separate liters of water. Each works independently on choosing his or her simple random sample, and when finished measuring and calculating, they report the sample mean lead levels (Table 11-8):

TABLE 11-8

Expert	Sample Mean Concentration of Lead
A	14 ppb
B	16 ppb
C	13 ppb

Because the sample means are all different, should we conclude that something went wrong? Certainly not. If fact, we might be surprised if all sample means were the same! Sample means and other sample statistics will differ from one sample to another—that's **sample variability**. Fortunately, we can often predict the degree to which sample statistics vary. That prediction of variability in samples helps us determine how certain we are in estimating the population (entire school) parameter. These topics will be further explored in Chapter 13.

CHAPTER SUMMARY

Univariate (one-variable) data can be summarized with **descriptive statistics**. Three statistics that measure the center of a data set are the **mean**, **median**, and **mode**. The mean, $\bar{x}$, is the sum of the data values divided by the number of data values. The median is the middle value when the data are sorted in ascending or descending order. The mode is the most frequently occurring number in the data set. The **range** of a data set is the difference between the maximum and minimum values and measures the **spread** in the data. A **weighted mean** is used to find a center value of a data set when individual data values have different "weights" or "impacts" on the total.

The **5-number summary** consists of the following values: minimum, Q1, median, Q3, and maximum. Q1 is the **1st quartile** and divides the lower 25% of the data set from the upper 75%. Q3 is the **3rd quartile** and divides the upper 25% of the data set from the lower 75%. A **boxplot** is a graph that uses the 5-number summary to illustrate both the center and spread of a data set.

A data set can be tallied into **bins** to view the distribution of data values. The **frequency** of a bin is the number of values that fall into that bin; the **relative frequency** is the percentage of values in that bin. Table 11-9 illustrates how a sample of 24 water

TABLE 11-9

Bin	Frequency	Relative Frequency
0–2 ppm	3	12.5%
2–4 ppm	8	33.3%
4–6 ppm	6	25.0%
6–8 ppm	4	16.7%
8–10 ppm	3	12.5%

pollution values might be binned. Values in Table 11-9 can be displayed with a **frequency histogram** or **relative frequency histogram** (Figure 11-15). In a histogram, the heights of the bars indicate the frequency or relative frequency for each bin.

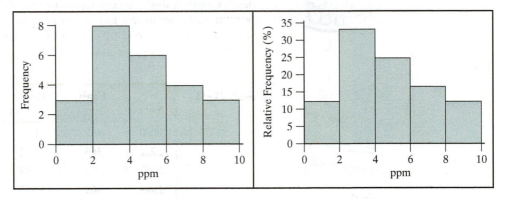

Figure 11-15 Histograms illustrating frequency (left) and relative frequency (right)

The histograms in Figure 11-15 have **positive skew**, with a tail to the right. A data set with **negative skew** has a tail to the left. A data set with **zero skew** is symmetric, and the mean and median have equal values. In a skewed data set, the mean lies closer to the tail of the histogram than does the median. The mean is **sensitive** to extreme values, while the median is **resistant**. The skew of a data set can be measured with Bowley's formula:

$$\text{Bowley skew} = \frac{Q3 - 2 \cdot \text{median} + Q1}{Q3 - Q1}$$

A **population** is a collection of measurements under study, and a **sample** is a subset of the population. Samples are used to make predictions about a population. Samples should be representative of the population, but even good samples have some variability. Common sampling procedures include **systematic**, **random**, and **convenience** sampling.

END *of* CHAPTER EXERCISES

Measures of Center and Other Basic Statistics

1. One use for the millions of tires that are discarded each year in the United States is in rural septic systems. Instead of gravel, shredded tires are buried underground to create septic drain fields. Tire chips submerged in these fields, though, can leach heavy metals into the groundwater. Researchers conducted a laboratory study to determine just how much metal would leach from tire chips. Five samples were taken, with the resulting concentration data given in Table 11-10, in units of mg/L.[12]

TABLE 11-10

Sample Number	Iron (Fe)	Manganese (Mn)	Aluminum (Al)	Chromium (Cr)	Copper (Cu)	Zinc (Zn)
1	7.10	0.78	0.27	0.19	0.13	1.06
2	6.82	0.77	0.15	0.07	0.15	1.06
3	7.12	0.70	0.42	0.05	0.20	1.13
4	7.42	0.71	0.48	0.02	0.27	1.11
5	6.46	0.70	0.36	0.14	0.16	· 0.70

Source: Chelsea Center for Recycling and Economic Development

    a. For each of the dissolved metals, determine the concentration mean and median. Include units of measure.

    b. Make a table listing the six metals along with their concentration ranges. Sort the metals from lowest to highest concentration range. Explain in a short sentence what the ranges indicate.

    c. What percent of the chromium samples were below 0.10 mg/L?

2. Dungeness crabs (*Cancer magister*) are one of the most important commercial species of crab harvested in the eastern Pacific Ocean. In British Columbia, Canada, size limits are used as the primary conservation measure to sustain the crab population. Only Dungeness males with shell widths exceeding 165 mm can be kept; undersized and female crabs must be returned to the water immediately upon capture (Figure 11-16). Table 11-11 lists British Columbia crab boat and harvest data for the years 1980 to 1999.[13]

Figure 11-16 To sustain the Dungeness crab population in waters off British Columbia, fishermen keep only the largest male crabs. The crab in this photo exceeds the minimum width. *Courtesy of Area "A" Crab Association and Archipelago Marine Research Ltd; photograph by Jason Scherr.*

TABLE 11-11

Year	Crab Boats	Harvest (tonnes)	Year	Crab Boats	Harvest (tonnes)
1980	343	1,701	1990	497	2,129
1981	366	1,315	1991	223	1,858
1982	352	999	1992	224	3,334
1983	369	957	1993	224	6,289
1984	387	1,156	1994	223	5,995
1985	362	1,164	1995	223	4,539
1986	386	1,320	1996	221	4,931
1987	363	1,631	1997	220	3,943
1988	331	1,508	1998	220	2,918
1989	330	1,519	1999	220	2,783

Source: Fisheries and Oceans Canada

    a. Determine the mean and median harvests from 1980 to 1999. First express your answers in tonnes; then convert to kilograms and report using scientific notation. *Note*: 1 tonne = 1,000 kg.

    b. What are the minimum and maximum harvests for this time period? What is the range in harvest values? What percent of the harvests exceeded 4,000 tonnes?

    c. For each year of data, determine the harvest per crab boat. Give results to the nearest 0.1 tonne per boat. Examine the resulting numbers, and describe any patterns or irregularities that you find.

Weighted Means

3. Seven different categories of land are given in Table 11-12, along with the percentage of global land area and **plant biomass** for each land type. Plant biomass is the mass or weight of the vegetation growing on the land type, expressed in kilograms of carbon per square meter (kg C/m^2). Calculate the mean plant biomass (in kg C/m^2) for all of Earth's land area.[14]

TABLE 11-12

Land Type	Percentage of World Land Area	Plant Biomass (kg C/m^2)
Forest	29%	10.0
Grassland	27%	2.3
Desert	13%	0.3
Tundra	8%	0.8
Wetland	2%	2.7
Cultivated	11%	1.4
Rock and Ice	10%	0.0

Source: *Biogeochemistry: An Analysis of Global Change*

4. Northern regions of the United States and much of Canada apply salts to roads in winter to keep water on the roads from freezing. Some of these salts eventually make their way to the Great Lakes, polluting waters with chloride (salt) ions. Use the information in Table 11-13 to compute the mean chloride concentration of all Great Lakes water combined.[15]

TABLE 11-13

Great Lake	Volume of Lake (miles^3)	Chloride Concentration (mg/liter)
Erie	116	17
Huron	850	6
Michigan	1,180	11
Ontario	395	22
Superior	3,000	2

Source: U.S. Environmental Protection Agency

5. In its definition of *poverty*, the U.S. Census Bureau uses a set of family income levels that vary by family size and composition. If a family's total income is below the poverty level, then all members in the family are considered poor. Table 11-14 shows Census Bureau poverty data, averaged over years 2000 and 2001. The poverty rate for each race is defined as the number in poverty divided by the total number of that race.[16]

TABLE 11-14

Race	Number in Poverty ($\times 1,000$)	Poverty Rate
White (Non-Hispanic)	14,819	7.6%
Black	8,059	22.6%
American Indian and Alaska Native	726	22.5%
Asian and Pacific Islander	1,266	10.1%
Hispanic	7,872	21.5%

Source: U.S. Census Bureau

a. Determine the total number of people of each race in the United States from the data in the table. Sum these totals to find an estimate for the U.S. population. Does your last answer seem reasonable?

b. Express the population of each race as a percent of the total U.S. population.

c. Determine the mean poverty rate for all people in the United States by using a weighted mean.

d. The poverty rate for blacks and the poverty rate for American Indian and Alaskan Natives are nearly the same. Which of these rates plays a larger factor when computing the overall U.S. rate? Explain.

e. Divide the total number in poverty for all races by the U.S. population that you computed in part (a). Interpret the result.

6. In the United States each year there is a great amount of public debate about gasoline prices. Some argue that gasoline should be taxed at a higher rate as an incentive to reduce consumption of a limited natural resource and reduce greenhouse gas emissions. Opponents to this argument claim that gasoline taxes are regressive (impact the poor more than the rich), and that additional taxes will take away important jobs dependent upon

transportation. Table 11-15 lists the means of all gasoline prices advertised at service stations for five geographic regions of the United States on November 4, 2002, as well as the human population for each region in July 2000.[17]

TABLE 11-15 Region	Advertised Gasoline Prices ($ per gallon)	Human Population July 2000 (millions)
East Coast	1.426	102.341
Midwest	1.469	77.707
Gulf Coast	1.389	37.216
Rocky Mountains	1.462	9.261
West Coast	1.499	52.361

Source: U.S. Dept of Energy and U.S. Census Bureau

a. Determine the mean of the five regional mean prices listed in the table.

b. Determine the mean of all service station gasoline prices across the United States. To answer this question, you will have to estimate the number (or relative number) of service stations in each region. Explain how you accomplish this.

c. Which U.S. region do you live in? What is the typical price of gasoline in that region today? Use these facts to estimate the mean of all service station gasoline prices across the United States. You may assume that today's population of each region is distributed as it was in July 2000. You may also assume that gas prices have changed *by the same percentage* for each region.

5-Number Summary, Boxplots, and Skew

7. For each data set, determine the 5-number summary then draw a boxplot (try doing these *by hand*).

a. 26   32   40   44   50   59   60   69

b. 15   15   20   21   30   40   60   62   69

c. 3.4   3.4   3.8   3.9   4.0   4.0   4.2   4.6   5.0   5.2

d. 100   120   120   131   140   150   150   152   160   160   160

8. A data set of size $n = 18$ is given.

12.2   15.6   18.0   8.0   6.2   4.4   2.3   4.5   21.0

16.0   12.2   16.5   8.2   2.5   5.3   2.2   10.0   4.1

a. Sort the data in ascending order, determine the 5-number summary, and draw a boxplot (try these tasks *by hand*).

b. What is the range of the data set? Is there a mode?

c. How many of the data values lie below Q1? What percent of 18 is your answer?

d. How many of the data values lie below Q3? What percent of 18 is your answer?

9. Boxplots of residential fresh water use for all counties in Maryland, Louisiana, and Ohio are displayed in Figure 11-17. Values have units of gallons per person per day. For example, in the most water-frugal county in Maryland, residents use 11 gallons per person per day on average.[18]

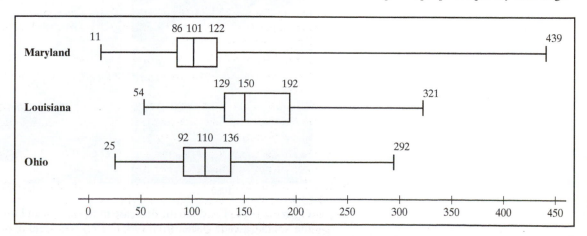

Figure 11-17 Boxplots of fresh water use in Maryland, Louisiana, and Ohio

a. Suppose that you live in Louisiana, and your county ranks at the 1st quartile in water use. How much water, on average, does each person in your county use each day?

b. In which state does the median county use the least water, on a per capita basis?

c. In which state is the spread in county water use the least, on a per capita basis?

d. For these three states, is county per capita water use positively or negatively skewed? Answer this by first looking at the boxplots and then using the Bowley skew formula.

10. Examine the 36 CRT lead concentrations given in Table 11.5 (see Example 11-5).

a. Divide the data into two subsets, one for early-model CRTs (years 1990 and earlier), and one for late-model CRTs (years 1991 and later).

b. Compute the 5-number summary for the lead concentrations in each data set. Then draw boxplots, one above the other, for comparison.

c. Compare the central values (medians) and spreads (ranges).

d. Compute the Bowley skew value for each data set. Do the *signs* of the skews make sense? Do the *sizes* of the skews make sense? Explain.

e. According to the boxplots, would an early-model or a late-model CRT be more likely to have a high lead concentration?

11. Soil density samples were taken randomly before and after a mature conifer forest was logged. Density is mass per volume and is a measure of soil compaction. The sample area was relatively small; only the upper portion of the soil was sampled. Table 11-16 lists the density measurements in grams per cubic centimeter (g/cm^3).[19] For comparison, the density of water is 1 g/cm^3. Note that there were more samples taken after the logging operation.

TABLE 11-16

Before Logging						
0.15	0.14	0.34	0.64	0.13	0.50	0.81
0.10	0.46	0.19	0.21	0.19		
After Logging						
0.61	0.38	0.77	0.65	0.60	0.34	0.15
0.87	0.56	0.78	0.73	0.85	1.13	0.96
1.09	0.90	1.10	0.20	1.04		

Source: Michael D. Purser

a. For each data set, find the 5-number summary. Organize your work in a table.

b. Draw a boxplot for each data set (one above the other).

c. Is soil more compact (more dense) before or after logging? How can you tell? Which number(s) best support your answer?

d. In which data set is variation in soil density greatest? How do you know?

12. Most glaciers around the world have been shrinking over the last 30 years from global warming (Figure 11-18). Among the many consequences are lower volumes of glacial melt water that support valley ecosystems. Table 11-17 lists the decrease in length of a sample of 12 glaciers from the North Cascade mountains in Washington State between 1980 and 2003.[20]

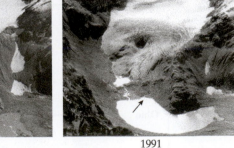

1960   1991

Figure 11-18 The dramatic decrease in the length and thickness of the South Cascade glacier from 1960 to 1991 is evident in these aerial photographs. *Courtesy of the U.S. Geological Survey.*

TABLE 11-17

Glacier	Decrease in Length (meters)	Glacier	Decrease in Length (meters)
Boulder	430	Lynch	115
Columbia	85	Mazama	210
Easton	225	Quien Sabe	156
Hinman	243	South Cascade	160
Honeycomb	150	Squak	250
Ice Worm	38	Sulphide	95

Source: North Cascade Glacier Climate Project

a. Sort the 12 numbers in ascending order and find the 5-number summary. Be sure to include units for each number. Draw a boxplot.

b. What is the range for the full data set?

c. What is the range for the first quarter of the data? For the fourth quarter? Write a sentence interpreting the meaning of these ranges.

d. Suppose glaciers shrink even further. How will the boxplot that you drew in part (a) change?

Histograms, Skew, Mean versus Median

13. A frequency histogram for a set of data is shown in Figure 11-19.

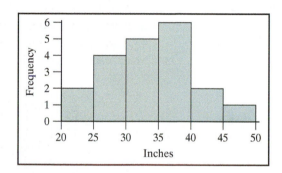

Figure 11-19

a. What is the size of the data set?

b. Create a table that lists bins, frequencies, and relative frequencies.

c. What is the modal bin for the histogram?

d. Briefly describe the approximate position of each of the following: Q1, median, and Q3.

14. A frequency histogram for a set of data is shown in Figure 11-20.

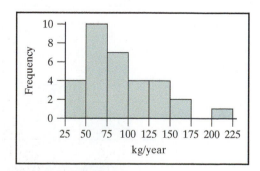

Figure 11-20

a. What is the size of the data set?

b. Create a table that lists bins, frequencies, and relative frequencies.

c. Is the histogram unimodal or bimodal? Explain.

d. For this data set, which is larger: the mean or the median?

15. Consider the following four histograms shown in Figure 11-21.

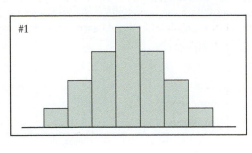
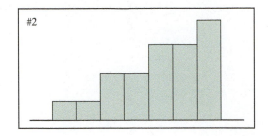
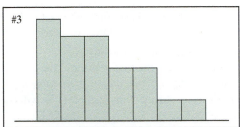
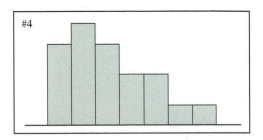

a. Which of the data sets are most symmetric?

b. In which data sets are the skews negative?

c. In which data sets are the skews positive?

d. In which data sets will the mean lie left of the median?

e. In which data sets will the mean and median be approximately equal?

16. Match each of the four histograms to its correct boxplot.

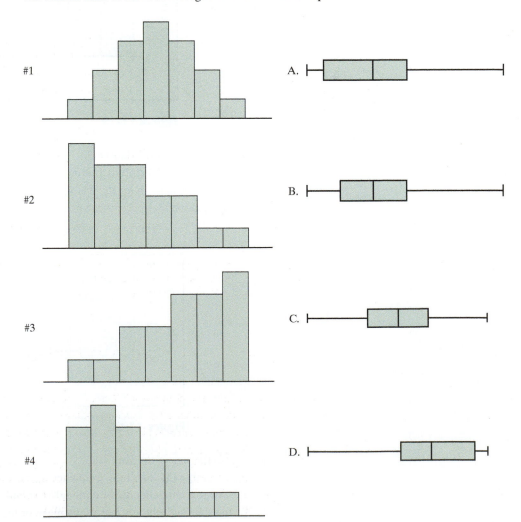

Figure 11-21 A forest consisting primarily of Pacific silver fir. *Photo: Langkamp/Hull.*

17. The Pacific silver fir tree (*Abies amabilis*) grows in mountain forests from Alaska to northwestern California in areas that receive heavy precipitation. Although silver fir often grow in mixed stands with other conifers, they can also grow alone in pure stands (Figure 11-21). The diameters of 52 silver fir from a pure stand are given in Table 11-18 in inches, with all measurements taken at "belt height."[21]

TABLE 11-18

24	15	15	9	17	13	19	21	10	23	9	17	12
15	14	15	12	14	20	20	27	15	13	9	12	18
22	19	22	9	16	12	16	7	14	20	15	17	15
21	14	16	20	11	11	9	18	13	15	11	10	12

Source: Greg Langkamp and Joseph Hull

a. Using technology, find the mean, the 5-number summary, and the range. Be sure to include units.

b. Construct a frequency histogram for the data, using at least six bins. On your homework paper list the bins and frequencies, and sketch the corresponding histogram.

c. Discuss the skew of the data, using both the histogram's shape and Bowley's skew value.

d. What is the modal bin for your histogram? What does that bin indicate?

18. The industrial use of **cadmium**, a metal, is primarily in nickel-cadmium (Ni-Cad) batteries, although it can also be found in semiconductors and ceramic glazes. Workers employed in industries that use cadmium face the risk of exposure to air-borne cadmium dust that can cause respiratory and kidney damage. Cigarette smoke also contains cadmium and is the biggest source of cadmium exposure for the general population. In Table 11-19 is a sample of 35 air quality readings for cadmium, measured in units of milligrams per cubic meter of air.[22]

TABLE 11-19

0.044	0.020	0.040	0.057	0.055	0.061	0.047
0.030	0.066	0.045	0.050	0.037	0.061	0.051
0.052	0.052	0.039	0.056	0.062	0.058	0.054
0.044	0.049	0.039	0.061	0.062	0.053	0.042
0.046	0.030	0.039	0.042	0.070	0.060	0.051

Source: *Statistical Methods in the Biological and Health Sciences*

a. Use technology to find the mean, the 5-number summary, and the range. Be sure to include units.

b. Set up bins (at least six) and find the relative frequency for each bin. Then construct a relative frequency histogram.

c. Does the histogram have negative, zero, or positive skew?

d. How are the relative sizes of the mean and median related to the shape of the cadmium histogram? Explain.

Sampling

19. Suppose that researchers wish to estimate the mean width of the Atlantic surf clam (*Spisula solidissima*). They collect a simple random sample of 80 clams, measure each, and calculate a mean width of 10.2 cm.

a. What is the population? What is the sample?

b. Is the number 10.2 cm a parameter or a statistic?

c. Do the researchers know the mean width for the population?

d. Researchers might select clams that burrow just under the surface rather than clams that have burrowed deeply. If, for example, smaller clams reside closer to the surface, the mean clam width of the sample would be smaller than that of the population. List another possible source of sampling error, and describe its impact on the accuracy of predicting the population's mean width.

20. Sulfur dioxide (SO_2), the main culprit in acid rain, is a product of the combustion of gasoline in automobiles. Suppose that you wish to determine the mean concentration of sulfur dioxide emitted from automobiles in your city. After obtaining expensive measuring equipment, you consider how best to sample the automobile exhaust.

    a. Describe how you would gather a systematic sample to ensure that SO_2 automobile emissions from all times of the day would be computed in the sample mean.

    b. Describe how you would collect a simple random sample to estimate the mean SO_2 concentration emitted from automobiles in your city.

    c. Suppose that the mean of your sample is $\bar{x} = 0.30$ ppm. What can you conclude about the population mean?

21. The brightness of light bulbs is measured in lumens, while the energy consumption is measured in watts. The **energy efficiency** of a light bulb is found by dividing the number of lumens by the number of watts. Suppose that a simple random sample of $n = 10$ light bulbs is collected from a hardware store, with luminosity and wattage values as listed in Table 11-20.

TABLE 11-20

Lumens	Watts	Lumens	Watts
210	25	1,200	75
220	25	1,600	100
820	60	1,500	100
830	60	1,750	120
1,170	75	1,800	120

    a. Determine the energy efficiency of each bulb in the sample. Organize your work in a table, use one decimal place of precision, and include units.

    b. Estimate the mean energy efficiency for all light bulbs in the hardware store.

    c. Estimate the percentage of light bulbs in the hardware store that have energy efficiencies above 14.0 lumens per watt.

    d. Scan the data presented in the table. What information would make you doubt that the data were collected randomly?

22. Suppose that researchers at FLEET (Florida Lakes Ecosystem Evaluation Team) pull out the map for the environs of Lake Wales, Florida and determine that the total area of the more than 100 lakes on the map is 2.05×10^7 square meters.

    a. The rectangular map represents an area of 13.8 km by 9.7 km. Convert this area into square meters.

    b. What proportion of the map area is composed of lakes?

    c. FLEET members conjecture that the proportion of the entire state of Florida covered by lakes is approximately the same as your answer to part (b). Is this true? Answer yes or no, and support your answer.

SCIENCE IN DEPTH
Energy Makeover

Imagine a two-bedroom apartment with a single bathroom, a living room, a kitchen, some hall closets, and a balcony that doubles as a back porch. The apartment is shared by two individuals who want to cut their enormous (and growing) energy bills for electricity and natural gas. The apartment is heated by natural gas and the oven is gas powered, except for the broiler unit. The roommates call in the Green Team (the authors Greg and Joe) for a consultation.

The group starts at the front door. Greg: "The outer hallway seems unheated, so you are probably losing a lot of heat at the front door. You don't have a **door sweep**; have your landlord install one along the bottom outside edge of the door, or get permission to do it yourself. You can also buy a **draft guard**, which is a cloth tube filled with sand or other materials, to put against the inside of the door at night." Joe: "I can see some space between the wall and the door frame. You should block that space with thin beads of caulking. And the weather stripping around the frame is old; your landlord should have it replaced. Or you can volunteer to replace the weather stripping in exchange for some rent relief!"

The group turns into the kitchen. Greg: "You have a lot of stuff piled on top of the refrigerator, between the refrigerator and an over-head cabinet. That means heat from the compressor can't escape easily. Remove that stuff to insure good airflow around the refrigerator." Joe: "You have an electric toaster-oven that you're using to reheat leftovers. This appliance is a **resistance heater** that generates heat by forcing electrons through wires. Resistance heaters are electricity hogs. I'd replace the toaster oven with a microwave oven, which uses much less electricity."

The group crowds into one of the bathrooms, which also houses the hot water heater. Greg: "Your hot water heater needs an insulating blanket to reduce heat loss. The hot water pipe coming out of the heater could use some wrapping insulation too. You also need a low-flow flexible showerhead with an on-off switch to reduce hot water consumption in the shower. And you should turn your thermostat down on the hot water heater until the temperature of the hot water at the kitchen faucet is 120°F." Joe: "You're heating the bathroom with a small electric space heater. This is another inefficient resistance heater, turning high-quality electricity into low-quality heat. The resistance heater can be replaced with a wall- or ceiling-mounted radiant heat panel. Radiant heat warms the body and surfaces without warming the air. Or you can tough it out!"

The group migrates to the bedroom. Greg: "In the winter, you need to insulate your windows as much as possible. Cut out pieces of transparent plastic to size, and tape to the window frame with removable tape. You won't mind the small loss of light in your bedroom. Heavy curtains help cut heat loss as well." Joe: "In the summertime, you can use energy efficient fans to help circulate the air, which makes the room seem cooler. Your window-mounted room air conditioner is rather old and has a low energy efficiency rating. Room air conditioners can use a lot of electricity, so you might think about replacing this older unit."

The group enters the living room. Greg: "The floor register against this wall is somewhat inconvenient for heating the living room. You can buy an **air deflector** that secures to the register and redirects the warm air stream out into the room. And one of your registers is partially blocked by this end table. All of the warm air registers should be unobstructed." Joe: "During cool months, turn down the thermostat to 65°F when you're home, 60°F at night, and 50°F when everyone leaves. In addition, the lighting needs to be redone in the living room, as there are too many traditional incandescent bulbs. Compact fluorescents are much more energy efficient and generate little waste heat."

They exit onto the balcony. Greg: "During the summertime, it's important to block the sun from entering the apartment, to keep the rooms cool. Your landlord should install a retractable awning over this balcony, to keep the sun out during the summer. A retractable awning can also let sun in during the winter." Joe: "I would install heavy draperies over the sliding glass doors to the balcony, to help with insulation in the winter. In order for draperies to work well, they need to shut off the space around the doors. The draperies should just touch the floor, and should be closed off on top with a cornice or box. In addition, the sliding glass doors are rather old. I'd lean on the landlord to have them replaced with glass that has a higher insulating value."

CHAPTER PROJECT:
ELECTRIC BILLS AND SUPER BULBS

Can the home modifications suggested in the "Science in Depth" article really save on energy? And if so, how much money could you save? In the companion project to Chapter 11 found at the text's Web site (**enviromath.com**), you and your classmates will provide electric bills for your apartments or homes. You'll compile and analyze the data to gain a picture of energy consumption habits. Are you an extreme energy user? Finally, you'll investigate the energy and monetary savings when switching to compact fluorescent light bulbs in your own home.

NOTES

[1] U.S. Environmental Protection Agency, "10 Fast Facts on Reycling," http://www.epa.gov/reg3wcmd/solidwasterecyclingfacts.htm

[2] H. J. Andrews Long Term Ecological Research Center, *Log and Snag Dimensions*, by Mark Harmon (PI), TD12, plot 1, 1991, http://www.fsl.orst.edu./lter/

[3] U.S. Census Bureau International Database, http://www.census.gov/ipc/www/idbnew.html

[4] Ibid.

[5] U.S. Dept of Agriculture, *Wisconsin Forest Statistics* 1996, by Thomas L. Schmidt, U.S. Forest Service Resource Bulletin NC-183 (1997).

[6] State University System of Florida, Florida Center For Solid And Hazardous Waste Management, *Characterization of Lead Leachability from Cathode Ray Tubes Using the Toxicity Characteristic Leaching Procedure*, by Timothy G. Townsend et al., Report #99-5 (1999).

[7] *Wisconsin Forest Statistics* 1996.

[8] Data form Gary Alt and Minitab, Inc. As cited in Mario F. Triola, *Elementary Statistics*, 9th ed. (Reading, MA: Addison-Wesley, 2005), 760.

[9] United Nations Dept of Economic and Social Affairs Population Division, *World Feritility Report 2003*, March 12, 2004, http://www.un.org/esa/population/publications/worldfertility/World_Fertility_Report.htm

[10] U.S. Dept of Agriculture, National Water and Climate Center, "Reservoir Reports Individual Sites and Basin Summaries, for the end of December 1998," http://www.wcc.nrcs.usda.gov/wsf/reservoir/resv_rpt.html

[11] W. B. Mercer and A. D. Hall, "The Experimental Error of Field Trials," *Journal of Agricultural Science (Cambridge)* 4 (1911): 107–132. More information about this data set can be found in Data Set #059 at the QELP Web site: http//www.seattlecentral.edu/qelp/sets/059/059.html

[12] University of Massachusetts, Chelsea Center for Recycling and Economic Development *Preliminary Investigation of Tire Shreds for Use in Residential Subsurface Leaching Fields Systems*, by Sukalyan Sengupta and Heather J. Miller, Technical Report no. 12, http://www.chelseacenter.org/pdfs/TechReport12.PDF

[13] Fisheries and Oceans Canada, Pacific Region, "Stats on 2000 Crab Fishery," http://www.pac.dfo-mpo.gc.ca/ops/fm/shellfish/crab/default_e.htm

[14] Values summarized from data presented in W. H. Schlesinger, *Biogeochemistry: An Analysis of Global Change*, 2nd ed.(San Diego: Academic Press, 1997).

[15] Year 2002 chloride concentrations estimated from U.S. Environmetal Protection Agency, Great Lakes Monitoring, "Limnology Program Overview of Results," Table 2, http://www.epa.gov/glnpno/monitoring/limnology/

[16] U.S. Census Bureau, *Poverty in the United States: 2001*, by Bernadette D. Proctor and Joseph Dalaker, Publication no. P60-219 (2002), http://www.census.gov/prod/2002pubs/p60-219.pdf

[17] U.S. Dept of Energy, "Gasoline and Diesel Fuel Prices," http://tonto.eia.doe.gov/oog/info/gdu/gasdiesel.asp; U.S. Census Bureau, "Times Series of State Population Estimates," Table ST-2001EST-01, http://www.census.gov/popest/archives/2000s/vintage_2001/ST-2001EST-01.html

[18] U.S. Dept of the Interior, *Estimated Use of Water in the United States in 2000*, by Susan S. Hutson et al., U.S. Geological Survey Circular 1268 (March 2004), http://water.usgs.gov/pubs/circ/2004/circ1268/

[19] Michael D. Purser, "The Impact of Clearcut Logging with High-Lead Yarding on Spatial Distribution and Variability of Infiltration Capacities on a Forest Hillslope" (Master's thesis, Univ. of Washington, 1988), 92–93.

[20] The North Cascade Glacier Climate Project, "Glacier Changes Through Time Photographic Series," http://www.nichols.edu/departments/glacier/

[21] G. Langkamp and J. Hull, July 2004, unpublished data.

[22] J. Susan Milton, *Statistical Methods in the Biological and Health Sciences*, 3rd ed. (New York: McGraw-Hill, 1999), 35.

12

Standard Deviation

There is an old saying, "Don't wade across a stream with an average depth of two feet." Those words are kept in mind by the student pictured in Figure 12-1 as she conducts a stream transect. An average value is important, but equally important is information about the spread or variability in a set of numbers.

A common way to measure spread is with **standard deviation**, which measures the "average distance" of all data values from their mean. In this chapter we explore what is meant by standard deviation and explain how it is computed. Then we investigate the connection between standard deviation and z-scores, outliers and Chebychev's Rule. This chapter concludes with a discussion about bell-shaped or normal data sets, and the Empirical Rule that governs them.

STANDARD DEVIATION

Chapter 11 introduced our first measure of spread—the **range**. Recall that the range is the difference between the maximum and minimum values of a data set. While range is a useful measure of spread, it is only based upon the lowest and highest data values. Consequently, the range fails to measure the spread of *all* data values. To better understand this shortcoming, consider the following two data sets. Both data sets have a range of $500 - 300 = 200$, but Data Set 2 has interior values that are more clustered towards the center value of 400.

Data Set 1:	300	301	301	400	499	499	500
Data Set 2:	300	399	399	400	401	401	500

Our second measure of spread—**standard deviation**—measures the "average distance" of all data values away from the mean. Without doing any computations, we should expect the standard deviation of Data Set 1 to be larger than that of Data Set 2, because its values lie farther, on average, from the mean.

The steps involved in computing standard deviation are best understood in a practical example, and so we focus on compact fluorescent lights (CFLs). CFLs are considered to be more environmentally friendly than incandescent lights (the standard bulb) because they consume much less electricity for the same amount of light output (Figure 12-2). For example, the 275 million CFLs used in North America in the year 2000 not only saved energy, but also lowered power plant carbon emissions by 3.5 million tons and sulfur emissions by 69,000 tons.[1] That's certainly good news!

A downside to CFLs is that they contain mercury, while standard incandescent lights do not. After CFLs are discarded into landfills, the mercury can leach into groundwater and enter the food chain (Figure 12-3). People or wildlife exposed to mercury risk serious adverse health effects, especially to the central nervous system and kidneys. CFLs should be treated as hazardous waste and disposed of accordingly!

Figure 12-1 A student measuring stream discharge. *Photo by Langkamp/Hull.*

Figure 12-2 A typical incandescent bulb (left) and a compact fluorescent light or CFL (right). *Photo: Langkamp/Hull.*

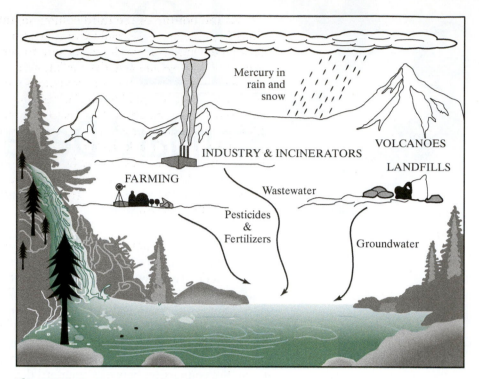

Figure 12-3 The environmental sources of mercury are numerous, although the burning of coal is the number one culprit. Once in the environment, mercury works its way up the food chain.

Suppose we collect a simple random sample of nine CFLs (all 15 watt), and determine with laboratory testing that the mercury weights of each are as follows:

4.6 mg   4.8 mg   5.1 mg   5.0 mg   5.1 mg   5.2 mg   5.4 mg   5.1 mg   4.7 mg

The mean mercury weight is

$$\bar{x} = \frac{4.6 + 4.8 + 5.1 + 5.0 + 5.1 + 5.2 + 5.4 + 5.1 + 4.7}{9} = 5.00 \text{ mg}$$

From our sample we determine that a typical CFL contains 5.00 milligrams of mercury. But what about the spread of mercury weights—how much, on average, do CFL mercury weights differ from the mean? To answer this question, start by computing the **deviation from the mean** for each weight. If x denotes the mercury weight and $\bar{x}$ the mean, then the deviation from the mean is represented by $x - \bar{x}$. See Table 12-1 for these results.

TABLE 12-1

Original Data x	Mean $\bar{x}$	Deviation from the Mean $x - \bar{x}$	Square of Deviation from the Mean $(x - \bar{x})^2$
4.6 mg	5.00 mg	−0.4 mg	0.16 mg^2
4.8 mg	5.00 mg	−0.2 mg	0.04 mg^2
5.1 mg	5.00 mg	0.1 mg	0.01 mg^2
5.0 mg	5.00 mg	0.0 mg	0.00 mg^2
5.1 mg	5.00 mg	0.1 mg	0.01 mg^2
5.2 mg	5.00 mg	0.2 mg	0.04 mg^2
5.4 mg	5.00 mg	0.4 mg	0.16 mg^2
5.1 mg	5.00 mg	0.1 mg	0.01 mg^2
4.7 mg	5.00 mg	−0.3 mg	0.09 mg^2
		$\sum (x - \bar{x}) = 0.0$ mg	$\sum (x - \bar{x})^2 = 0.52$ mg^2

At the bottom of the third column in Table 12-1 we find the sum of all deviations from the mean. (Recall that the symbol Σ, or sigma, represents "sum.") Notice how all deviations from the mean sum to zero. This happens to be true for *all* data sets—the positive and negative deviations from the mean always cancel out! But this very fact makes the sum of the deviations a poor choice to measure spread (all data sets would have a spread of zero!). We'd like to use the deviations so that they don't sum to zero. The preferred choice is to compute the **squares of the deviations** as in the fourth column of Table 12-1.

The next step in calculating standard deviation is to compute the **variance** of the sample, which is the mean of the $(x - \overline{x})^2$ values. To compute variance, we would expect to sum these values and divide by $n = 9$. This is *not* the case. Instead we sum the values and divide by $n - 1 = 8$ (the explanation for this will follow shortly). We obtain the following result:

$$\text{sample variance} = \frac{\text{sum of squared deviations}}{n - 1}$$

$$= \frac{\Sigma(x - \overline{x})^2}{n - 1}$$

$$= \frac{0.52 \text{ mg}^2}{8}$$

$$= 0.065 \text{ mg}^2$$

The sample variance is 0.065 mg². Variance can be used as a measure of spread but has one big drawback—the units are squared. In this example, the original CFL data have units of milligrams (mg), but the variance has units of milligrams squared (mg²). We prefer that our measure of spread has the same units as the original data. To convert variance into a measure with non-squared units, take its square root. The result is the **sample standard deviation**, denoted by *s*.

$$s = \text{sample standard deviation} = \sqrt{\text{sample variance}}$$

$$= \sqrt{\frac{\Sigma(x - \overline{x})^2}{n - 1}}$$

$$= \sqrt{0.065 \text{ mg}^2}$$

$$= 0.2549509757 \text{ mg}$$

The number of decimal places used when reporting standard deviation is typically one more than in the original data, just as with the mean. We would report that the sample mercury weights have a standard deviation of $s = 0.25$ mg. This value indicates that the mercury weights in the sample deviate from the mean *on average* by about 0.25 milligrams.

Now let's return to our computation of the sample variance. Why did we divide by $n - 1 = 8$ instead of $n = 9$? To understand the answer to this question, recall that our basic purpose for collecting a sample is to predict some characteristic value of a population. In this example, we want the standard deviation of the sample (the statistic) to predict the standard deviation of the population (the parameter). But statisticians have found, in general, that dividing by n in the computation makes the sample standard deviation *underestimate* the population standard deviation. To remedy this lower estimate, we divide the sum of the $(x - \overline{x})^2$ values by $n - 1$; dividing by a smaller number makes the sample standard deviation a bit larger and gives us a better estimate of the population standard deviation.

Using the sample standard deviation of $s = 0.25$ mg, we predict that the population (mercury weights in *all* 15-watt CFLs) has a standard deviation of 0.25 mg as well. Using more advanced statistical procedures, we could quantify the accuracy of this prediction with some stated probability of success—that topic will be explored in Chapter 13. In general, by taking larger samples, we will obtain more accurate estimates of any population parameter. But in the case of testing for mercury in CFLs, larger samples have higher laboratory costs and possibly greater health risks to those analyzing the lights.

You might be wondering about the relative size of the standard deviation. Is a value such as 0.25 mg large or small? Just as with the mean, the size of the standard deviation can only be interpreted in context. If a company is trying to produce CFLs so that the standard deviation in mercury weights is no more than 0.1 mg, then 0.25 mg is a big number—the spread in mercury levels is higher than what is acceptable. On the other hand, when comparing mercury weights in CFLs produced by two different companies, and one's standard deviation is 1.75 mg while the other's is 0.25 mg, then the number 0.25 mg seems pretty low. In this case, the company producing bulbs with less mercury variability might have more precise manufacturing methods.

We computed the sample standard deviation of mercury levels in a sample of CFLs "by hand" to better understand the background mathematics. When the size of a data set is large, we rely on technology to tackle this extensive computational work. The mean and standard deviation are typically listed as part of the summary statistics produced by a calculator or software package. Chapter 11 provided instructions on how to compute one-variable descriptive statistics on the TI-83/84 graphing calculator. (Review that information now if necessary.) Following are two screens of summary statistics for the CFL data.

```
1-Var Stats          1-Var Stats
x̄=5                  ↑n=9
Σx=45                 minX=4.6
Σx²=225.52            Q₁=4.75
Sx=.2549509757        Med=5.1
σx=.240370085         Q₃=5.15
↓n=9                  maxX=5.4
```

The first screen shows that the mean of the nine mercury weights is $\bar{x} = 5$, but be sure to use two decimal places and units and write $\bar{x} = 5.00$ mg. The first screen also displays the standard deviation, using the symbol Sx:

$$Sx = s = 0.2549509757 \approx 0.25 \text{ mg}$$

Note that some calculators and computer programs also display the standard deviation for a population. For example, the TI-83/84 graphing calculator displays the population standard deviation as

$$\sigma x = 0.240370085 \approx 0.24 \text{ mg}$$

The standard deviation of a population is computed in the same way as for a sample, except that the sum of the squared deviations is divided by n rather than $n - 1$. In this text, we'll assume that all of our data sets are sample data sets, as it is rare to measure *every* object in a population. So when we work with standard deviation, we'll always compute the sample standard deviation, denoted by s.

EXAMPLE 12-1 Flame Retardants

Polybrominated diphenyl ethers (PBDEs) are chemicals used as flame retardants in many home products such as furniture, carpeting, televisions, computers, and coffee makers. Although PBDEs save lives, they also leach into home environments and accumulate in human serum, tissue, and breast milk. Studies have shown that North American individuals have PBDE concentrations 17 times higher than those in Europe, where PBDEs have mostly been banned. The concentrations of PBDEs in 17 dust samples collected from homes in Maryland and Washington D.C. are listed in Table 12-2, in parts per billion (ppb).[2]

TABLE 12-2

30,100	5,780	2,030	7,080	14,990	4,590
2,730	4,390	5,480	4,250	3,030	5,470
2,580	2,440	780	1,700	3,800	

Source: *Environmental Science and Technology*

We enter the data into our calculator (in any order) and compute the one-variable statistics.

```
1-Var Stats
 x̄=5954.117647
 Σx=101220
 Σx²=1383902000
 Sx=6987.606045
 σx=6778.973598
↓n=17
```

The PBDE concentrations have a sample mean of $\bar{x} = 5{,}954.1$ ppb and a sample standard deviation of $s = 6{,}987.6$ ppb. This is a large standard deviation, as expected from a data set with a large spread in values.

CALCULATING POSITION USING Z-SCORES

We introduced standard deviation as an *overall* measure of the spread of a data set away from its mean. We now use standard deviation to help describe the position of a single data point. The position is given by the **z-score**, which measures the number of standard deviations a data point lies above or below the mean.

To better understand the meaning of a z-score, let's return to our example of mercury in compact fluorescent lights (CFLs). Our sample of $n = 9$ CFLs has mercury weights with a mean of $\bar{x} = 5.00$ mg and a standard deviation of $s = 0.25$ mg. Using these values, we compute the mercury weights that are one, two, and three standard deviations from the mean. See Table 12-3. Note that we would obtain more precise estimates for these numbers using additional decimal places in the standard deviation.

TABLE 12-3

$\bar{x} - 3s$	$\bar{x} - 2s$	$\bar{x} - s$	$\bar{x}$	$\bar{x} + s$	$\bar{x} + 2s$	$\bar{x} + 3s$
4.25 mg	4.50 mg	4.75 mg	5.00 mg	5.25 mg	5.50 mg	5.75 mg

Now suppose that we purchase a single CFL and determine that its mercury content equals 5.25 mg. We give this CFL a z-score of 1 ($z = 1$), indicating that its mercury content lies one standard deviation above the mean. If we measure another CFL and its mercury content equals 4.50 mg, two standard deviations below the mean, it gets a z-score of $z = -2$. Thus for each of the numbers in Table 12-3, we can assign a z-score indicating the number of standard deviations that the number lies above or below the mean. See Table 12-4.

TABLE 12-4

$\bar{x} - 3s$	$\bar{x} - 2s$	$\bar{x} - s$	$\bar{x}$	$\bar{x} + s$	$\bar{x} + 2s$	$\bar{x} + 3s$
4.25 mg	4.50 mg	4.75 mg	5.00 mg	5.25 mg	5.50 mg	5.75 mg
$z = -3$	$z = -2$	$z = -1$	$z = 0$	$z = 1$	$z = 2$	$z = 3$

If we examine another CFL, odds are quite good that the number of standard deviations above or below the mean will not have an integer value. For example, if the mercury content of a CFL weighed 4.55 mg, Table 12-4 indicates that the z-score lies between $z = -2$ and $z = -1$ (actually a bit closer to $z = -2$). To compute the actual z-score, we determine that the weight is 0.45 mg less than the mean, which is equivalent to 0.45 mg/0.25 mg $= 1.8$ standard deviations below the mean. Thus the z-score is $z = -1.8$.

Our last example shows that computing z-scores can be a bit complicated. Fortunately, we can simplify the process by using a formula. For a data point with value x, taken from a sample with mean $\bar{x}$ and standard deviation s, the z-score can be computed with the following formula:

z-score formula: $\quad z = \dfrac{x - \bar{x}}{s}$

Now back to the CFL with mercury weighing $x = 4.55$ mg. Using the z-score formula, we get

$$z = \frac{4.55 \text{ mg} - 5.00 \text{ mg}}{0.25 \text{ mg}} = \frac{-0.45 \text{ mg}}{0.25 \text{ mg}} = -1.8$$

Notice that the formula automatically produces a negative z-score when the actual value lies below the mean. Using more decimal places in the standard deviation would give more precise results in the z-score.

EXAMPLE 12-2 Nuclear Energy Production

Nuclear power plants produce far less greenhouse gases than coal-based plants but generate high-level radioactive waste that remains toxic for thousands of years. Furthermore, many nuclear power plants have serious design flaws that pose an imminent threat of radiation release into the environment (Figure 12-4).

Figure 12-4 A nuclear power plant in Dukovany, Czech Republic. Some nuclear power plants lack adequate containment structures that could prevent the release of radiation. *Photo: U.S. Dept of Energy International Nuclear Safety Program.*

In 1990, 32 states in the United States produced energy from nuclear-powered plants. The total energy produced was 576,650 gigawatt-hours (GWh), with state-by-state annual amounts listed in Table 12-5.[3]

TABLE 12-5 Nuclear energy production in the United States, 1990

State	Production (GWh)	State	Production (GWh)	State	Production (GWh)
Alabama	13,050	Maine	4,900	Ohio	10,700
Arizona	20,600	Maryland	1,300	Oregon	6,100
Arkansas	11,300	Massachusetts	5,100	Pennsylvania	57,800
California	32,700	Michigan	21,600	South Carolina	42,900
Connecticut	19,800	Minnesota	12,100	Tennessee	14,000
Florida	21,800	Missouri	8,000	Texas	15,900
Georgia	24,800	Nebraska	7,500	Vermont	3,600
Illinois	71,900	New Hampshire	4,100	Virginia	23,800
Iowa	3,000	New Jersey	23,800	Washington	5,700
Kansas	7,900	New York	23,600	Wisconsin	11,200
Louisiana	14,200	North Carolina	25,900		

Source: U.S. Dept of Energy

The graphing calculator's descriptive statistics are displayed next:

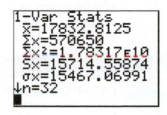

We find that the mean energy produced is $\overline{x}$ = 17,832.8 GWh and the standard deviation is s = 15,714.6 GWh.

Now let's investigate the state producing the most nuclear energy: Illinois. Its z-score is

$$z = \frac{71,900 \text{ GWh} - 17,832.8 \text{ GWh}}{15,714.6 \text{ GWh}} = 3.4$$

indicating that its nuclear energy production is 3.4 standard deviations above the mean. The state producing the smallest amount of nuclear energy (of the 32 that produce some nuclear energy) is Maryland; its z-score is

$$z = \frac{1,300 \text{ GWh} - 17,832.8 \text{ GWh}}{15,714.6 \text{ GWh}} = -1.1$$

indicating that its production is slightly more than one standard deviation below the mean. The z-score for Nebraska is

$$z = \frac{7,500 \text{ GWh} - 17,832.8 \text{ GWh}}{15,714.6 \text{ GWh}} = -0.66$$

which indicates that Nebraska's nuclear energy production is about two-thirds of a standard deviation below the mean.

OUTLIERS

Data values that are extremely large or extremely small, as compared with other values in the data set, are called **outliers**. In this text, we define an outlier as a number that lies three or more standard deviations from the mean. This definition is a bit subjective, and there may be times when you have to use another cutoff level, such as two standard deviations.

Sometimes outliers occur in data sets because of error. Suppose that we collect air samples from 50 locations in an urban area. We measure the dust concentration of each air sample and determine that the mean is 25 μg/m^3 and the standard deviation is 3 μg/m^3. If one of those samples has a dust concentration of 43 μg/m^3, its z-score would equal

$$z = \frac{43 \ \mu\text{g/m}^3 - 25 \ \mu\text{g/m}^3}{3 \ \mu\text{g/m}^3} = 6$$

This high of a z-score indicates that the dust concentration of 43 μg/m^3 is an outlier. We might suspect that something went wrong in the measuring or recording stage of our data collection process. If we knew which sample this number referred to, we could go back and measure it again. But if remeasuring were not an option, it might be better to throw out this data value. Discarding outliers is common practice when they are thought to be caused by error.

Now let's return to the nuclear production data in Example 12-2. The z-score for Illinois is z = 3.4, indicating that Illinois produces an extremely large amount of nuclear energy, compared to other states. Should we discard this data point because we suspect there was some error? This is a difficult question to answer if we know nothing about the *reliability* of the data. Also remember that extremely large and small values can occur without any error involved. Sometimes there is a good explanation for the extreme value—perhaps there is some environmental or economic incentive for Illinois to generate so much nuclear energy. Working with outliers is a double-edged sword. Leave them in and the data set may be seriously flawed. Throw them out and you may be dumping some important information!

EXAMPLE 12-3 Hydroenergy Production

Hydropower plants produce no greenhouse gases or toxic emissions and rely principally on water and gravity—two resources that are plentiful. But many hydropower plants are constructed using artificial dams which greatly impact the environment. Dams prevent fish migration, are constructed out of enormous amounts of concrete, and trap river sediment that limits their life span to 100 to 200 years. Furthermore, reservoirs behind dams flood once-fertile river valley bottoms, and their large surface area causes increased evaporation that alters the natural hydrologic cycle. Hydropower facilities that make use of natural falls, such as the one pictured in Figure 12-5, cause less damage to the environment.

In 1990, 48 states in the United States produced hydroenergy (energy from hydropower facilities) totaling 299,000 gigawatt-hours (GWh). State-by-state annual values are listed in Table 12-6.[4] The mean amount of hydroenergy produced by these 48 states is $\bar{x} = 6{,}225.6$ GWh with a standard deviation of $s = 14{,}272.9$ GWh.

In Example 12-2, we found that Illinois produced the most nuclear energy. We see in Table 12-6 that the state of Washington produces the most hydroenergy. But which state, Illinois or Washington, has a production level that is more extreme?

Figure 12-5 The Rainbow Falls hydropower facility on the Ausable River in New York State. *Photo: Langkamp/Hull.*

TABLE 12-6 Hydroenergy production in the United States, 1990

State	Production (GWh)	State	Production (GWh)	State	Production (GWh)
Alabama	10,300	Louisiana	697	Ohio	173
Alaska	980	Maine	3,960	Oklahoma	2,870
Arizona	8,180	Maryland	2,310	Oregon	40,800
Arkansas	4,890	Massachusetts	1,090	Pennsylvania	3,190
California	23,900	Michigan	3,040	Rhode Island	6
Colorado	1,320	Minnesota	843	South Carolina	3,880
Connecticut	452	Missouri	2,190	South Dakota	4,270
Florida	173	Montana	10,700	Tennessee	11,800
Georgia	4,710	Nebraska	833	Texas	1,570
Hawaii	89	Nevada	1,620	Utah	481
Idaho	7,450	New Hampshire	1,980	Vermont	1,100
Illinois	771	New Jersey	17	Virginia	4,050
Indiana	441	New Mexico	215	Washington	87,300
Iowa	13	New York	29,400	West Virginia	1,330
Kansas	12	North Carolina	7,070	Wisconsin	1,150
Kentucky	2,880	North Dakota	1,720	Wyoming	611

Source: U.S. Geological Survey

We can answer this question by comparing z-scores. The z-score for Washington's hydroenergy production is

$$z = \frac{87{,}300 \text{ GWh} - 6{,}225.6 \text{ GWh}}{14{,}272.9 \text{ GWh}} = 5.7$$

Because Washington's hydroenergy production has a higher z-score than Illinois's nuclear energy production ($z = 5.7$ versus $z = 3.4$), Washington's hydroenergy production is more extreme.

CHEBYCHEV'S RULE

Chapter 11 introduced the 5-number summary, which consists of the minimum, Q1, the median, Q3, and the maximum. Using the 5-number summary, we can state several properties or "rules of thumb" about the spread of data about the median. For example, the middle 50% of the data lie between Q1 and Q3. Are there similar rules of thumb involving the mean and standard deviation? The answer is "yes," and they are given by Chebychev's Rule:

Chebychev's Rule: For *any* set of data, the following are true.

- At least 75% of the data lie within two standard deviations of the mean; that is, between $\overline{x} - 2s$ and $\overline{x} + 2s$, or in the interval $(\overline{x} - 2s, \overline{x} + 2s)$.
- At least 89% of the data lie within three standard deviations of the mean; that is, between $\overline{x} - 3s$ and $\overline{x} + 3s$, or in the interval $(\overline{x} - 3s, \overline{x} + 3s)$.

Figure 12-6 displays Chebychev's Rule graphically. Chebychev's Rule is very conservative and gives lower limits for the percentages of data that lie within two and three standard deviations from the mean. In many data sets, *all* of the data will lie within these bounds.

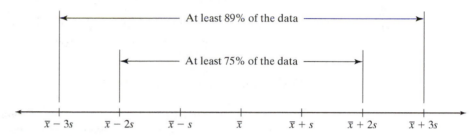

Figure 12-6 Chebychev's Rule applies to any data set.

Let's see how Chebychev's Rule applies to our sample of nine mercury weights from CFLs. Recall that the sample has a mean of $\overline{x} = 5.00$ mg and a standard deviation of $s = 0.25$ mg. Chebychev's Rule says that at least 75% of the sample data will lie between $\overline{x} - 2s$ and $\overline{x} + 2s$. In other words, at least 75% of the lights sampled will have mercury weights between

$$\overline{x} - 2s = 5.00 - 2(0.25) = 4.50 \text{ milligrams}$$

and

$$\overline{x} + 2s = 5.00 + 2(0.25) = 5.50 \text{ milligrams}$$

Similarly, Chebychev's Rule tells us that at least 89% of the sample bulbs will have mercury weights above $\overline{x} - 3s = 4.25$ mg and below $\overline{x} + 3s = 5.75$ mg. By scanning the original nine mercury weights, we find that all of the data lie within two standard deviations of the mean, so Chebychev's Rule is easily satisfied.

Because the CFLs were chosen randomly, approximately the same percentages should hold for *all* CFLs (the population). So if we accidently crack open a CFL in our living room, there is at least a 75% chance that the amount of mercury released is between 4.50 and 5.50 milligrams. Estimates such as these help us quantify risks, but we emphasize that care must be taken when making such predictions—neither the error nor the certainty in these claims is known. Let's consider another example.

EXAMPLE 12-4 Recycled Water

In large cities of the United States, household and industrial sewage water is piped to wastewater treatment plants for purification. The treated waste water is often released into nearby water bodies, but can be used for other purposes where health risks are low. For example, the San Francisco Bay cities of San Jose, Milpitas, and

Santa Clara now use 10% of their treated wastewater (called **recycled water**) for irrigation and industrial needs. In 2004, weekly water quality samples showed that the concentration of total dissolved solids (TDS) in their recycled water had a mean of 721 ppm and a standard deviation of 8 ppm.[5]

So what do these numbers tell us about the weekly measurements? By Chebychev's Rule, at least 75% of the weekly samples have TDS concentrations in the interval

$$(\bar{x} - 2s, \bar{x} + 2s) = (721 - 2(8), 721 + 2(8))$$
$$= (705 \text{ ppm}, 737 \text{ ppm})$$

Also, at least 89% of the weekly TDS concentrations lie in the interval

$$(\bar{x} - 3s, \bar{x} + 3s) = (721 - 3(8), 721 + 3(8))$$
$$= (697 \text{ ppm}, 745 \text{ ppm})$$

Again, the percentages given by Chebychev's Rule are conservative estimates; it is quite possible that 100% of the weekly TDS concentrations fall within three, or even two, standard deviations of the mean.

NORMAL DISTRIBUTIONS

Histograms of many data sets are bell shaped, with a single peak in the middle and tails falling off symmetrically to the left and right. These distributions are called **normal distributions**. A frequency distribution that is bell shaped or normal is displayed in Figure 12-7. Normal distributions of measurements are quite common in environmental data sets.

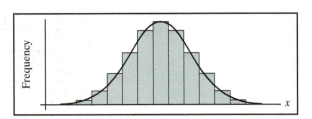

Figure 12-7 The typical bell shape of a normal distribution

A normal distribution is perfectly symmetric about a vertical center line and has a bell shape (Figure 12-8). The x-position of the center line is also the value of the mean, median, and modal bin (the bin with the highest frequency). Using these properties of normal distributions, we can examine a data set to check whether its distribution is approximately normal or not.

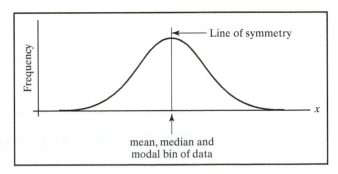

Figure 12-8 Normal distributions are symmetric, with the mean, median, and modal bin all positioned at the line of symmetry.

EXAMPLE 12-5 Columbia River Velocities

Scientists collected data below the Priest Rapids Dam on the Columbia River to measure the variation in water velocity. At 55% of the river depth, they measured velocities every minute for 66 minutes. Results are displayed in Table 12-7 with all values given in meters per second (m/s).[6]

TABLE 12-7 Columbia River velocities in meters per second

2.7	3.2	3.3	3.4	3.5	3.5
2.9	3.2	3.3	3.4	3.5	3.6
3.0	3.2	3.3	3.4	3.5	3.6
3.0	3.2	3.3	3.4	3.5	3.6
3.1	3.2	3.4	3.4	3.5	3.6
3.1	3.3	3.4	3.4	3.5	3.6
3.1	3.3	3.4	3.4	3.5	3.6
3.1	3.3	3.4	3.4	3.5	3.7
3.2	3.3	3.4	3.5	3.5	3.7
3.2	3.3	3.4	3.5	3.5	3.8
3.2	3.3	3.4	3.5	3.5	3.9

Source: U.S. Geological Survey

To examine the normality of these sample values, we bin the data and construct a histogram (Figure 12-9). We also enter the values into the graphing calculator and compute the one-variable statistics, as displayed next.

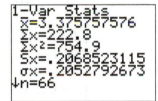
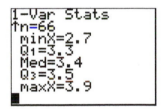

We see that the histogram is fairly symmetric with a single "hump" in the middle—not quite bell shaped, but close. The mean and median of the sample velocities are both about 3.4 m/s and lie close to the modal bin of 3.3–3.5 m/s in the histogram. We can conclude that the velocity data are close to being normally distributed.

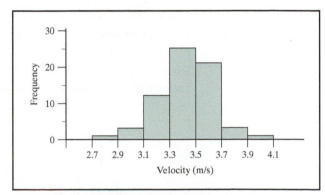

Figure 12-9 The river velocities are close to being normally distributed.

A data set that is normal (a bell-shaped distribution) can be described with just the mean and standard deviation. The mean indicates where the distribution is centered and the standard deviation describes the distribution's spread (how far the tails extend to the sides). We can estimate the size of the standard deviation by approximating the locations of the **inflection points** on the bell curve that "fit" the distribution. Inflection points are located where a curve changes shape, from "opening upward" to "opening downward." For example, Figure 12-10 shows a bell curve that approximates a data set with a mean of $\bar{x} = 8$. The inflection points of the bell curve are marked with open circles. The horizontal distance between the line of symmetry and either inflection point is equal to the standard deviation. This distance is 2 in Figure 12-10, so we estimate that the data represented by this bell curve have a standard deviation of $s = 2$.

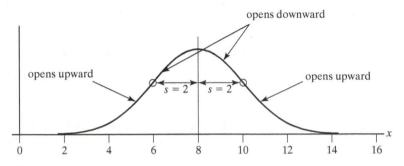

Figure 12-10 The points where the curvature changes are called inflection points (open circles). These points are located one standard deviation from the line of symmetry. For this bell curve, the standard deviation s is equal to 2.

THE EMPIRICAL RULE

Notice that the bell curve in Figure 12-10 tapers off on both sides of the mean, almost touching the x-axis at $x = 2$ (three standard deviations below the mean) and at $x = 14$ (three standard deviations above the mean). The fact that almost 100% of normal data lie within three standard deviations of the mean is an important property of normal distributions. Further properties of normal data sets, and their relationship to the mean and standard deviation, are described by the Empirical Rule.

Empirical Rule: For normal data sets, the following are true.

- About 68% of the data lie within one standard deviation of the mean; that is, in the interval $(\bar{x} - s, \bar{x} + s)$.

- About 95% of the data lie within two standard deviations of the mean; that is, in the interval $(\bar{x} - 2s, \bar{x} + 2s)$.

- About 99.7% of the data lie within three standard deviations of the mean; that is, in the interval $(\bar{x} - 3s, \bar{x} + 3s)$.

Figure 12-11 illustrates the percentages and intervals stated in the Empirical Rule, which is sometimes called the **68-95-99.7 Rule**. Be careful! The wording in the Empirical Rule can be easily confused with that in Chebychev's Rule. The Empirical

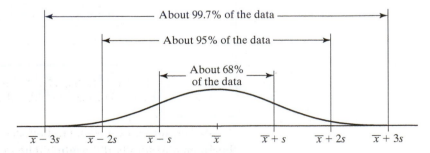

Figure 12-11 Percentages of bell-shaped distributions

Rule states "*about* xx%" whereas Chebychev's Rule says "*at least* xx%." Also, the Empirical Rule only applies to *normal* data sets, whereas Chebychev's Rule applies to *all* data sets.

Data sets from real-world examples are rarely perfectly normal (perfectly bell shaped); often the data have positive or negative skew, or are multimodal (two or more peaks). But if a data set is close to normal, it can be modeled as if it were normal. We concluded that the river velocity data in Example 12-5 are fairly normal because the mean and median lie close to the modal bin, and the histogram is fairly symmetric and bell shaped. We can further test the normality of the data by checking if the Empirical Rule holds for the data set. Using the mean of $\bar{x} = 3.38$ m/s and standard deviation of $s = 0.21$ m/s, we compute the seven values listed in Table 12-8:

TABLE 12-8

$\bar{x} - 3s$	$\bar{x} - 2s$	$\bar{x} - s$	$\bar{x}$	$\bar{x} + s$	$\bar{x} + 2s$	$\bar{x} + 3s$
2.75 m/s	2.96 m/s	3.17 m/s	3.38 m/s	3.59 m/s	3.80 m/s	4.01 m/s

By tallying stream velocity data from Table 12-7, we observe the following:

- $48/66 \approx 73\%$ of the data lie within one standard deviation of the mean (between 3.17 m/s and 3.59 m/s.)
- $62/66 \approx 94\%$ of the data lie within two standard deviations of the mean (between 2.96 m/s and 3.80 m/s.)
- $65/66 \approx 98.5\%$ of the data lie within three standard deviations of the mean (between 2.75 m/s and 4.01 m/s.)

These observed percentages (73%, 94%, 98.5%) are fairly close to the theoretical percentages given by the Empirical Rule (68%, 95%, 99.7%). This is further evidence that the Columbia River velocities are almost normally distributed. Let's consider one additional example.

EXAMPLE 12-6 Wheat Yields

Recall our example in Chapter 11 of wheat yields from plots at the Rothamsted Experimental Station in Great Britain. For the 500 equal-area plots involved in the study, the mean yield was 3.95 bushels and the median was 3.94 bushels. The bin from 3.9 to 4.1 bushels has the highest frequency (Figure 12-12) and therefore is the modal bin. The fact that the mean, median, and modal bin all lie close together, combined with the obvious bell-shaped nature of the histogram, leads us to think that the data are normally distributed.

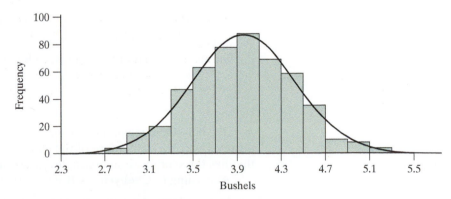

Figure 12-12 The bell-shaped distribution of wheat yields from 500 equal-area plots. The mean and median lie within the modal bin

Using the mean of $\bar{x} = 3.949$ bushels, and the standard deviation of $s = 0.458$ bushels, we compute the seven numbers listed in Table 12-9, with all measurements given in bushels.

TABLE 12-9	$\bar{x} - 3s$	$\bar{x} - 2s$	$\bar{x} - s$	$\bar{x}$	$\bar{x} + s$	$\bar{x} + 2s$	$\bar{x} + 3s$
	2.575	3.033	3.491	3.949	4.407	4.865	5.323

According to the Empirical Rule, about 340 (68% of 500) plots in the sample have yields between $\bar{x} - s = 3.491$ bushels and $\bar{x} + s = 4.407$ bushels. We could also say that 340 plot yields fall in the interval $\bar{x} \pm s = 3.949 \pm 0.458$ bushels. Or we could say that 340 plot yields fall within about a half bushel of the mean.

Because of the symmetric nature of the data set, the middle 68% of the data can be divided in two, giving about 34% of the yields between $\bar{x} = 3.949$ and $\bar{x} + s = 4.407$ bushels. Likewise, another 34% of the yields fall in the interval $\bar{x} - s = 3.491$ to $\bar{x} = 3.949$ bushels.

The Empirical Rule also states that about 95% of the plots have yields from 3.033 bushels ($\bar{x} - 2s$) to 4.865 bushels ($\bar{x} + 2s$). In actuality, 8 plots had yields below 3.033 bushels and 12 had yields above 4.865 bushels. This means that $480/500 = 96\%$ of plot yields were between $\bar{x} - 2s$ and $\bar{x} + 2s$, very close to what the Empirical Rule predicts.

Columbia River velocities and wheat yields on equal-area plots are just two examples of real-world data that are almost normally distributed. Data that are the product of many independent factors are often normal. Let's think about why this might be true with wheat yields. The yield (volume of plant material) for each plot depends on many factors: sunlight, precipitation, wind, air and soil temperatures, air and soil nutrients, plant genetics, and so on. Taken individually, some factors will be above average and some below as they contribute to plant growth. Taken as a whole, these factors even out to produce an "average-sized" plant, which makes equal-area plots have about the same yield. These average plots create the peak in the normal distribution. But some plots, by random chance, will have yields that deviate from the mean. Larger and larger deviations are less and less likely. This decreasing likelihood produces the *decreasing tails* on each side of the mean.

CHAPTER SUMMARY

A common way to measure spread is with **standard deviation**, which measures the spread or "average distance" of all data values x from their mean $\bar{x}$. Sample standard deviation, s, is computed with the formula

$$s = \sqrt{\frac{\sum(x - \bar{x})^2}{n - 1}}$$

An individual data point can be assigned a **z-score**, which is the number of standard deviations the data point lies above or below the mean. The z-score formula is

$$z = \frac{x - \bar{x}}{s}$$

A value below the mean has a negative z-score, while a value above the mean has a positive z-score. The z-score of the mean always equals 0. An **outlier** is a data point that lies three or more standard deviations from the mean ($z \le -3$ or $z \ge 3$).

According to **Chebychev's Rule**, every data set has the following properties:

- *At least* 75% of the data lie within two standard deviations of the mean.
- *At least* 89% of the data lie within three standard deviations of the mean.

Histograms of **normal distributions** are bell shaped and symmetric about a vertical line whose x-position is the same as the mean, median, and modal bin (Figure 12-13). The **inflection points** of a bell-shaped distribution lie one standard deviation away from the mean.

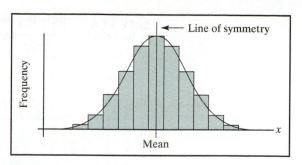

Figure 12-13 A normal distribution is bell shaped.

Normal distributions follow the **Empirical** or **68-95-99.7 Rule**, which states that

- About 68% of the data lie within one standard deviation of the mean.
- About 95% of the data lie within two standard deviations of the mean.
- About 99.7% of the data lie within three standard deviations of the mean.

END *of* CHAPTER EXERCISES

Mean and Standard Deviation

1. What is the mean for each data set in Table 12-10? Which of the two data sets has a greater standard deviation? *Try answering these questions by just inspecting the data.*

TABLE 12-10

Set 1	5	10	15	20	25	30	35	40	45
Set 2	17	19	21	23	25	27	29	31	33

2. What is the mean for each data set in Table 12-11? Which of the two data sets has a greater standard deviation? *Try answering these questions by just inspecting the data.*

TABLE 12-11

Set 1	200	300	400	500	600	700	800	900	1,000
Set 2	200	300	400	450	600	750	800	900	1,000

3. Which should have the smaller standard deviation from its yearly mean: daily temperatures in Portland, Oregon, or daily temperatures in Chicago, Illinois? Explain.

4. Arrange in ascending order based upon standard deviation from its yearly mean:
   a. Hours of daylight in Montreal, Canada
   b. Hours of daylight in Havana, Cuba
   c. Hours of daylight in Baltimore, United States

5. Compare the histograms displayed in Figure 12-14.

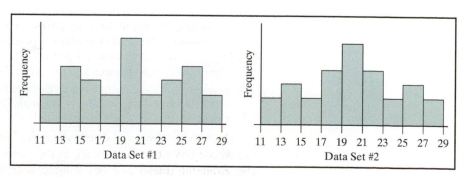

Figure 12-14

   a. Explain why both data sets have about the same mean and median.

   b. Which data set has greater standard deviation?

6. Compare the histograms displayed in Figure 12-15.

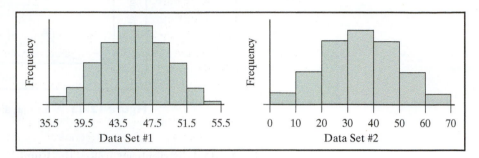

Figure 12-15

   a. For each data set, estimate the mean and median.

   b. Which data set has the greatest standard deviation?

7. Scientists drilled wells at a Florida broiler chicken farm to collect random samples of groundwater.[7] Potassium levels were measured in each water sample (Table 12-12).

TABLE 12-12

Sample #	1	2	3	4	5	6	7	8	9	10	11	12
Potassium (mg/L)	1.5	1.8	1.6	1.3	0.7	0.7	0.6	0.6	0.4	0.4	0.4	0.1

Source: U.S. Geological Survey

   a. Compute the mean potassium level for the 12 samples. Then make a table (such as Table 12-1) to determine the sum of the squared deviations from the mean. (Use four decimal places for $\bar{x}$, $x - \bar{x}$, and $(x - \bar{x})^2$ in your table.) Finally, compute the standard deviation. Include correct units.

   b. Which potassium measurement is furthest from the mean? Which is closest to the mean? How do you know?

8. In the same 12 groundwater samples described in the previous exercise, scientists also measured nitrate levels (Table 12-13).

TABLE 12-13

Sample #	1	2	3	4	5	6	7	8	9	10	11	12
Nitrate (mg/L)	18	19	20	21	9.4	10	11	26	5.2	5.4	5.0	4.9

Source: U.S. Geological Suvey

   a. Compute the mean nitrate level in the 12 samples. Make a table to determine the sum of the squared deviations from the mean. (Use two decimal places for $\bar{x}$, $x - \bar{x}$, and $(x - \bar{x})^2$ in your table.) Finally, compute the standard deviation. Include units.

   b. Comparing standard deviations of two data sets is difficult when the data measure different things, and when magnitudes of data values are quite different. To adjust for these differences, we can normalize the standard deviations by dividing each by its mean. The end result is called the **coefficient of variation**. Compute the coefficient of variation for both the potassium and nitrate concentrations (see the previous exercise for potassium concentrations). Express the coefficient of variation in percentage form. Use the results to decide which data set varies the most.

z-Scores

9. Suppose that a sample of stream velocities has a mean of $\bar{x} = 2.10$ miles per hour (mph) and a standard deviation of $s = 0.60$ mph. Determine the z-scores for the following individual velocities.

   a. $x = 1.8$ mph

   b. $x = 5.0$ mph

   c. $x =$ the mean

   d. x is 2.6 standard deviations below the mean.

10. Suppose that a sample of tree heights has a mean of $\bar{x} = 24.50$ m with a standard deviation of $s = 2.40$ m. Determine the z-scores for the following individual heights.

   a. $x = 24.5$ m

   b. $x = 29.3$ m

   c. $x = 20.9$ m

   d. x is 2.75 standard deviations above the mean.

11. Suppose that a sample of air temperature measurements has a mean of $\bar{x} = 18.64°C$ and a standard deviation of $s = 0.08°C$.

   a. Which Celsius reading has a z-score of $z = 0.75$?

   b. Which Celsius reading has a z-score of $z = -3$?

   c. Which Celsius reading has a z-score of $z = 0$?

   d. In general, which air temperatures have positive z-scores? Negative z-scores?

12. Suppose that a collection of water samples from a lake has a mean phosphorus concentration of $\bar{x} = 60.0\ \mu g/L$ with a standard deviation of $s = 4.5\ \mu g/L$.

   a. Which phosphorus concentration has a z-score of $z = 0$?

   b. Which phosphorus concentration has a z-score of $z = 0.5$?

   c. Which phosphorus concentration has a z-score of $z = -2.5$?

   d. If the phosphorus concentrations are positively skewed, what can you say about the z-score for the median?

13. Which is smaller, Iowa's nuclear energy production or its hydroenergy production? Explain your answer by making use of z-scores (see Table 12-5 and Table 12-6).

14. Which is larger, Arizona's nuclear energy production or its hydroenergy production? Explain your answer by making use of z-scores (see Table 12-5 and Table 12-6).

15. Illinois produces the most nuclear energy. Which state produces the second most? Is that state's nuclear energy production an outlier? (See Table 12-5.)

16. Washington State produces the most hydroenergy. Which state produces the least amount (of the states listed in Table 12-6)? Is that state's hydroenergy production an outlier?

Chebychev's Rule and the Empirical Rule

17. Which of the following statements are true about *any* data set?

   a. About 75% of the data lie within two standard deviations of the mean.

   b. At least 75% of the data lie within two standard deviations of the mean.

   c. About 25% of the data lie more than two standard deviations from the mean.

   d. No more than 25% of the data lie more than two standard deviations from the mean.

18. Which of the following statements are true about a *bell-shaped* data set?

   a. About 95% of the data lie within two standard deviations of the mean.

   b. At least 95% of the data lie within two standard deviations of the mean.

   c. About 5% of the data lie more than two standard deviations from the mean.

   d. No more than 5% of the data lie more than two standard deviations from the mean.

19. Estimate the mean of each normal distribution; then use the Empirical Rule to estimate the standard deviation.

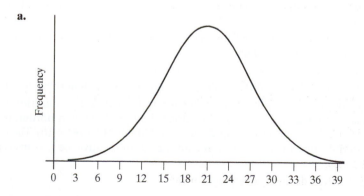

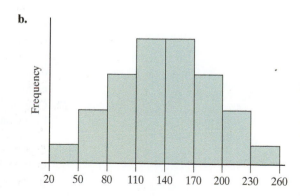

20. Estimate the mean of each normal distribution; then use the Empirical Rule to estimate the standard deviation.

a.

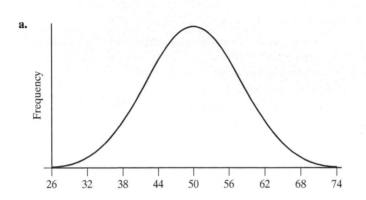

b.

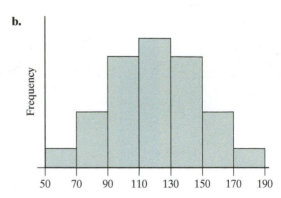

Applications

21. Motor oil recycling rates (in pounds per capita) for all 36 counties in Washington State are given in Table 12-14. The data are based on a three-year mean from 1997 to 1999.[8]

TABLE 12-14

County	Rate (lb/capita)	County	Rate (lb/capita)	County	Rate (lb/capita)
Adams	0.71	Island	1.48	Pierce	0.81
Asotin	5.63	Jefferson	3.41	San Juan	4.81
Benton	1.60	King	1.78	Skagit	0.66
Chelan	2.06	Kitsap	1.45	Skamania	2.47
Clallam	2.50	Kittitas	1.70	Snohomish	1.21
Clark	2.16	Klickitat	3.33	Spokane	1.34
Cowlitz	2.96	Lewis	2.29	Stevens	3.05
Douglas	1.35	Lincoln	0.22	Thurston	1.86
Ferry	1.63	Mason	1.15	Walla Walla	0.66
Franklin	2.14	Okanogan	0.82	Whatcom	1.67
Grant	0.09	Pacific	1.87	Whitman	0.27
Grays Harbor	0.62	Pend Oreille	1.27	Yakima	2.35

Source: Washington State Dept of Ecology

a. Using technology, compute the mean and standard deviation.

b. On graph paper, draw a frequency histogram of the recycling rates. Label and scale correctly.

c. What percent of the counties lie within two standard deviations of the mean? Does your answer concur with Chebychev's Rule?

d. Use z-scores to identify any counties that might be outliers.

e. Suggest two reasons why the recycling rates might be so high in some counties.

22. One impact of global warming is the reduction in the extent and duration of ice that forms on lakes in winter. Lake Superior, along the U.S.–Canadian border, is typically covered with ice from mid December through mid March (Figure 12-16). The number of days when Lake Superior was ice covered during 39 consecutive ice seasons is provided in Table 12-15.[9]

Figure 12-16 Lake Superior in April 2003. Ice, white in this photo, is confined to sheltered bays with waters less than 75 meters deep. *Photo courtesy of Liam Gumley, Space Science and Engineering Center, University of Wisconsin–Madison.*

TABLE 12-15

Ice Season	Days of Ice	Ice Season	Days of Ice	Ice Season	Days of Ice
1955–1956	87	1970–1971	115	1983–1984	48
1956–1957	137	1971–1972	118	1984–1985	118
1957–1958	106	1972–1973	82	1985–1986	116
1958–1959	97	1973–1974	115	1986–1987	81
1959–1960	105	1974–1975	97	1987–1988	116
1960–1961	118	1975–1976	104	1988–1989	123
1961–1962	118	1976–1977	146	1989–1990	112
1962–1963	136	1977–1978	126	1990–1991	99
1963–1964	91	1978–1979	141	1991–1992	102
1966–1967	107	1979–1980	111	1992–1993	118
1967–1968	96	1980–1981	123	1993–1994	63
1968–1969	114	1981–1982	118	1994–1995	62
1969–1970	125	1982–1983	83	1996–1997	132

Source: National Snow and Ice Data Center

a. Enter the ice days into a single list on your technology device. Then compute the mean and standard deviation.

b. Make a relative frequency histogram of the ice data on graph paper. Label and scale correctly.

c. Would you say that the Empirical Rule applies to this data set? How about Chebychev's Rule?

d. Compute the value of the skew using Bowley's formula (see Chapter 11). Does the skew value agree with the general shape of the histogram?

e. Would you agree with the statement, "Near Lake Superior, extremely cold winters are more common than extremely warm winters"? Explain.

23. Suppose that for a sample of 58 businesses, the amount of cardboard discarded each week was normally distributed with a mean of 220 pounds and a standard deviation of 30 pounds.

a. What percent of the businesses discarded between 190 and 250 pounds per week?

b. How many businesses discarded between 160 and 280 pounds per week?

c. How many businesses discarded more than 280 pounds per week?

d. Each week, 84% of the businesses discard cardboard amounts below which weight?

24. Arkshell clams (also known as blood clams) are harvested commercially off the coast of Virginia. A sample of two species of these clams (*Noetia ponderosa* and *Anadara ovalis*) had shell heights normally distributed with a mean of 20.3 mm and a standard deviation of 1.4 mm.[10]

a. For the Arkshell sample, find the seven numbers $\bar{x} - 3s$, $\bar{x} - 2s$, $\bar{x} - s$, $\bar{x}$, $\bar{x} + s$, $\bar{x} + 2s$, and $\bar{x} + 3s$.

b. Make a diagram of a bell curve. Label the seven numbers that you found in part (a) along the horizontal axis of the bell curve. Also mark the location of the inflection points.

c. Determine the percentages of the data that lie between consecutive numbers that you marked on your bell curve in part (b). Label these six percentages on your diagram.

d. Which shell heights lie two to three standard deviations from the mean? In a sample of 300 blood clams, how many clam shells would have these heights?

SCIENCE IN DEPTH

Impermeable Surfaces and Urban Runoff

Walk around your neighborhood and look at the variety of surfaces found in a few square blocks. There are manicured lawns of nonnative grasses, strips of grass next to the sidewalks, flower beds, shrubs and trees of various sizes, and backyard gardens. There are surfaces made of concrete and asphalt such as roads, sidewalks, driveways, parking lots, and alleys. The roofs of houses, garages, apartments, commercial buildings, and other structures also make up a large part of the surface area of the neighborhood. Now imagine a rainstorm; what happens to the rainwater that falls on each of those surfaces?

Permeable or **pervious** surfaces such as uncompacted lawns, shrubbery, and gardens allow rainfall to penetrate into the ground, and recharge the **groundwater** that lurks below in sand, gravel, and cracks in bedrock. Highly permeable materials such as light, sandy soils with trees and lawns can have rainfall infiltration rates of 5–10 centimeters (2–4 inches) of water per hour or higher.[11] Soils with such high permeabilities can accommodate very heavy rainfall events. Dense, clay-rich soils may have permeabilities of less than 1 centimeter (a third of an inch) per hour. Very little water can be absorbed by these soils.

Impermeable or **impervious** surfaces such as roads, sidewalks, rooftops, compacted lawns, and dense soils produce **runoff**.[12] Runoff includes both excess water that cannot be accommodated by soils and water that never touches a natural surface! Runoff can be emptied into the sewer system and wind up at the municipal sewage treatment plant. Alternatively, runoff may be captured by a separate stormwater system and dumped into local water bodies such as lakes, rivers, or seas. Increased urbanization results in more impermeable surface area and more runoff, which necessitates larger and more costly sewer and stormwater systems. Large storm events in a highly urbanized area can produce too much runoff for a sewer system to handle, resulting in the dumping of raw sewage into nearby water bodies. (See Figure 8-4 in Chapter 8 concerning combined sewer overflows.)

Increased urbanization also results in more contamination of runoff.[13] Vehicles contribute a variety of toxic heavy metals, such as zinc, cadmium, lead, and arsenic, to roads and parking lots. These metals originate from wear of brake pads, engines, and tires, and from tailpipe emissions. Vehicles also leak organic compounds such as gasoline, diesel fuel, lubricants, and antifreeze. Stormwater runoff can also be contaminated by pesticides and fertilizers from lawns and playfields. And pet waste can contribute a significant load of pathogens to urban runoff. Genetic testing of the bacterium *E. coli* in one urban stream demonstrated that the main source of this bacterium was household cats.[14] Urbanization and increased amounts of impervious surface also result in the loss of habitat and the reduction of biodiversity.

Figure 12-17 An open, vegetated swale engineered to capture and retain runoff. Compare to a typical street with curbs. *Photo: Langkamp/Hull.*

Fortunately, there are a number of successful strategies for reducing impervious surfaces, while simultaneously reducing pollutants and runoff. Many neighborhood streets are two lanes wide (plus parking) despite the low volumes of traffic they accommodate. These streets can be reduced to one-lane streets with pullouts or one-way access, with the excess lane converted to shallow ditches or **swales** on either side of the roadway (Figure 12-17). Curbs are removed so that runoff can wash directly into the swales, which are landscaped to act as temporary retention ponds and infiltration centers. The swales also help restore lost habitat.

Capping a building with an **eco-roof** (green roof) is another strategy for mitigating the impervious surfaces associated with houses, apartments, and other buildings. Eco-roofs have a thin layer of soil on top of a waterproof membrane and drainage layer. Small, low-maintenance plants that can tolerate the unusual rooftop environment are planted on the eco-roof. The vegetation, soil, and drainage layers provide temporary storage of rainfall and slow the volume of runoff delivered to storm and sewer systems. The vegetation also recycles rainfall back into the atmosphere by **transpiration**.

 CHAPTER PROJECT:
URBAN RUNOFF SCORECARD

Perhaps the neighborhood where you live is full of lush green lawns, wide driveways and streets, and spacious parks. Or maybe you live in an highly urbanized area with tall apartment buildings, narrow streets, and few green spaces. Or maybe you live in an area with a little of both. How would you rate the environmental quality of your neighborhood?

In the companion project for this chapter found at (**enviromath.com**), you will assess the "runoff quality" of a city block near your school or home by focusing on the amounts of permeable and impermeable surface area. You will compare surface areas for your block with the class means and develop an "Urban Runoff Index" to evaluate the environmental quality of your block. And you will discuss some simple strategies for reducing urban runoff and its pollution. How much runoff does your neighborhood create?

NOTES

[1]International Association of Energy Efficient Lighting (IAEEL), "CFL Sales Up and Up," *IAEEL Newsletter* 9, no. 24, http://www.iaeel.org/

[2]Heather M. Stapleton et al., "Polybrominated Diphenyl Ethers in House Dust and Clothes Dryer Lint," *Environmental Science and Technology* 39 (2005): 925–931.

[3]U.S. Dept of Energy, Energy Information Administration, http://www.eia.doe.gov/

[4]U.S. Dept of the Interior, *Estimated Use of Water in the United States in 1990*, USGS National Circular 1081, Table 28, http://water.usgs.gov/watuse/tables/hytab.st.html

[5]City of San Jose, Environmental Services Dept, *Recycled Water Quality Information for the San Jose/Santa Clara Water Pollution Control Plant: 2004*, http://www.ci.san-jose.ca.us/esd/

[6]U.S. Dept of the Interior, *Analysis of Current Meter Data at Columbia River Gaging Stations, Washington and Oregon*, by J. Savini and G. L. Bodhaine, U.S. Geological Survey Water Supply Paper No. 1869-F (1971): 59. More information about these data can be found in Data Set #010 at the QELP Web site: http://www.seattlecentral.edu/qelp/sets/010/010.html

[7]U.S. Dept of the Interior, *Effects of Waste Disposal Practices on Ground-Water Quality at Five Poultry (Broiler) Farms in North-Central Florida, 1992–93*, by H. H. Hatzell, U.S. Geological Survey Water Resources Investigation Report 95-4064 (1995).

[8]Washington State Dept of Ecology, *Solid Waste in Washington State: 9th Annual Status Report*, by Ellen Caywood, Publication no. 00-07-037 (2000), http://www.ecy.wa.gov/biblio/0007037.html

[9]National Snow and Ice Data Center/World Data Center for Glaciology, *Great Lakes Daily Ice Observations at NOAA Water Level Gauge Sites*, Gauge Station 9004, http://nsidc.org/data/g00945.html

[10]U.S. Dept of Commerce, "Population Structure of the Arkshell Clam," by Katherine McGraw, Sally Dennis, and Michael Castagna, NOAA technical report (1996). As cited in J. Susan Milton, *Statistical Methods in the Biological and Health Sciences*, 3rd ed. (New York: McGraw-Hill, 1999), 189.

[11]Barrett Kays, "Relationship of Forest Destruction and Soil Disturbance to Increased Flooding in the Suburban North Carolina Piedmont," *Metropolitan Tree Improvement Alliance (METRIA) Proceedings* 3, 1980, 118–125, http://www.ces.ncsu.edu/fletcher/programs/nursery/metria/metria3/

[12]U.S. Environmental Protection Agency, *Urban Storm Water Best Management Practices Study*, EPA-821-R-99-012, August 1999, http://www.epa.gov/ost/stormwater/

[13]National Resources Defense Council, *Stormwater Strategies: Community Responses to Runoff Pollution*, http://www.nrdc.org/water/pollution/storm/stoinx.asp

[14]King County, Washington, "Swimming Beach Monitoring Program," http://dnr.metrokc.gov/wlr/waterres/swimbeach/beachFAQ.htm

13

Normal Distributions

The organization *Redefining Progress* has created a measure of the **ecological balance** for each country in the world, to monitor whether or not nations are living within their ecological means. The ecological balance formula takes the biological capacity of each country and subtracts the ecological footprint—the demands placed on nature to produce the resources and absorb the wastes for the people in the country. In the year 2000, the United States had a biocapacity of 4.61 hectares per person and a footprint of 9.57 hectares per person. That left the United States with an ecological balance of −4.96 hectares per person, meaning that the average person in the United States requires almost 5 hectares (over 12 acres) of land *in excess of what's available in the United States* to sustain his or her consumption level. How does the United States compare with other countries in the world? The frequency histogram in Figure 13-1 illustrates the ecological balance for 137 countries.[1] The United States, with its large ecological deficit, is located in the second bar from the left.

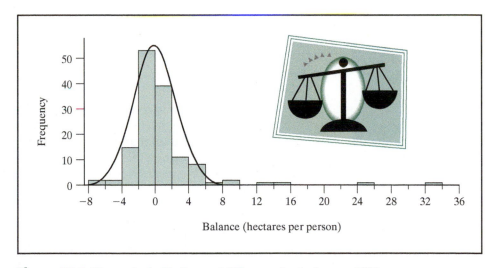

Figure 13-1 The ecological balance of 137 countries in the year 2000.

Four countries had large, positive ecological balances in 2000: Congo (13.52), Papua New Guinea (14.94), Mongolia (24.13), and Gabon (33.05). If we ignore these outliers, we can approximate the frequency distribution with a normal (bell-shaped) curve. Many environmental data sets have frequency distributions that are normal or nearly normal: stream velocities, wheat yields, human birth weights, and shoreline habitat for a collection of lakes. Beyond the Empirical Rule, what ties these distributions together?

The answer lies with the **standard normal distribution**, which allows us to work with all normal distributions using the same mathematics. Recall that the Empirical Rule tells us the proportions of normal distributions that lie within one, two, and three standard deviations of the mean. In this chapter we extend this rule to find proportions of normal distributions that lie between *any* two standard deviation values. We then conclude with a discussion about confidence intervals and margins of error—numbers that are always mentioned with opinion polls. Surprisingly, margins of error are based on the mathematics of normal distributions, too!

THE STANDARD NORMAL DISTRIBUTION

In Chapter 12 we investigated several data sets that had normal (bell-shaped) distributions, and we learned that the Empirical Rule tells us the percentages of the data that lie within one, two, or three standard deviations of the mean. Those percentages are 68%, 95%, and 99.7%, respectively. But how do we find the percentage of normal data that lies, for example, within 1.5 standard deviations of the mean? Or more than 2.5 standard deviations above the mean?

Before answering these more difficult questions, we revisit two of the data sets from Chapter 12 that have bell-shaped distributions. The first data set, concerning river velocities, has a mean of $\bar{x} = 3.38$ m/s and a standard deviation of $s = 0.21$ m/s. The second, about wheat yields at an experimental farm, has a mean of $\bar{x} = 3.949$ bushels with a standard deviation of $s = 0.458$ bushels. The bell curves that approximate these data sets are displayed in Figure 13-2. Notice that each horizontal axis has been scaled using both the original units of measure and z-scores.

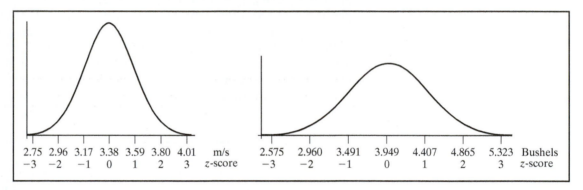

Figure 13-2 Bell curves approximating stream velocities (left) and wheat yields (right).

With the horizontal axes marked with a "z-scale," the two bell-shaped curves in Figure 13-2 are quite similar. They can be made identical by adjusting the vertical scales so that the total area under the curves is equal to 1 or 100% (although we don't explicitly show this). The single, resulting bell curve is called the **standard normal curve** and provides a "template" for all normally distributed data sets (Figure 13-3).

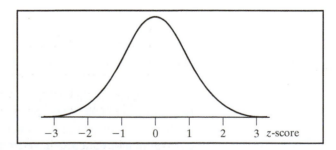

Figure 13-3 This bell curve is called the standard normal curve and can be used to model any data set that is normally distributed. The area under the curve is equal to 1 or 100%.

Using the standard normal curve, we can restate the Empirical Rule in terms of z-scores (sometimes called **standard scores**). Furthermore, the percentages given in the Empirical Rule can be interpreted as areas—a fact that we'll find enormously convenient. For example, because about 68% of normally distributed data lie within one standard deviation of the mean, the *area* under the standard normal curve between $z = -1$ and $z = 1$ equals about 68% or 0.68 (Figure 13-4).

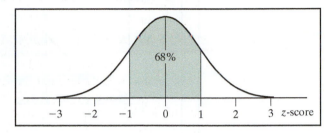

Figure 13-4 By the Empirical Rule, the area under the standard normal curve from $z = -1$ to $z = 1$ is about 68% or 0.68.

Also by the Empirical Rule, about 95% of the area under the standard normal curve lies between $z = -2$ and $z = 2$, with about 99.7% of the area between $z = -3$ and $z = 3$ (Figure 13-5).

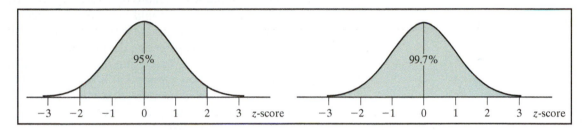

Figure 13-5 Under the standard normal curve, 95% of the area lies between $z = -2$ and $z = 2$ (left) and 99.7% of the area lies between $z = -3$ and $z = 3$ (right).

Using the symmetry of the normal curve, we can determine the area that lies between integer-valued z-scores such as $z = -1$ and $z = 2$ (Figure 13-6). The area between $z = -1$ and $z = 0$ is about 34% (half of 68%), and the area between $z = 0$ and $z = 2$ is about 47.5% (half of 95%). Thus the total area between $z = -1$ and $z = 2$ is about 81.5%.

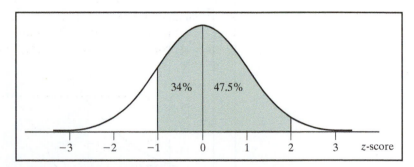

Figure 13-6 The area under the standard normal curve from $z = -1$ to $z = 2$ can be found by summing two smaller areas.

We now investigate the techniques used to find the area under the standard normal curve between *any* two z-scores, even between those with noninteger values. Let's look at a practical example.

EXAMPLE 13-1 SUV Fuel Economy

The city fuel economy ratings of 224 different sport utility vehicle (SUV) models are approximately normally distributed with a mean of $\bar{x} = 16.6$ mpg and a standard deviation of $s = 3.3$ mpg (Figure 13-7).[2]

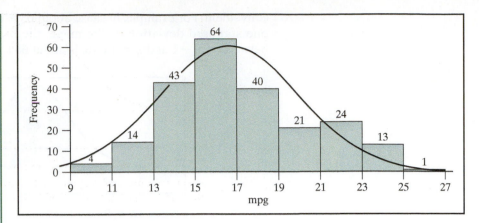

Figure 13-7 A frequency histogram of SUV fuel economy ratings is approximated with a bell curve.

What proportion of the SUVs have fuel economy ratings between 16.6 mpg and 22.3 mpg? If we had all the ratings in front of us, we could simply count the number of SUVs that fall within these bounds and divide by 224. Without the individual fuel economy ratings we could use the frequencies given in Figure 13-7 to make a rough estimate. (Our estimate is 95/224 = 42%; what's yours?)

A different approach to estimating the proportion assumes that the data can be modeled with a normal curve. We see that the shape of the histogram in Figure 13-7 is roughly bell shaped, so that assumption is reasonable. To get started, we convert the lower and upper fuel economy ratings into their z-scores. The z-score of 16.6 mpg (the mean) is $z = 0$. For 22.3 mpg, the z-score is

$$z = \frac{22.3 \text{ mpg} - 16.6 \text{ mpg}}{3.3 \text{ mpg}} = 1.73$$

Note that we have rounded the z-score to two decimal places—the reason for this will be clear shortly.

Our question now becomes: What proportion of the SUVs have fuel economy ratings with z-scores between $z = 0$ and $z = 1.73$? Those z-scores are indicated on the SUV histogram in Figure 13-8.

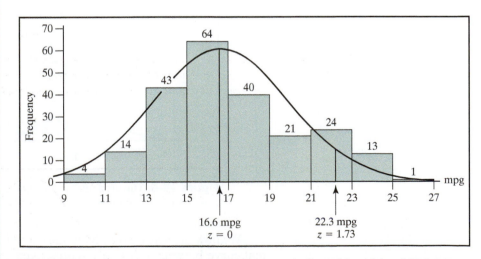

Figure 13-8 The SUV histogram with the z-scores indicated for 16.6 and 22.3 mpg.

The proportion of SUVs with ratings between $z = 0$ and $z = 1.73$ is equivalent to the shaded area under the standard normal curve displayed in Figure 13-9. We cannot determine this shaded area using the Empirical Rule, because that rule only states percentages for z-scores such as 0, ±1, ±2, and ±3. Furthermore, finding areas

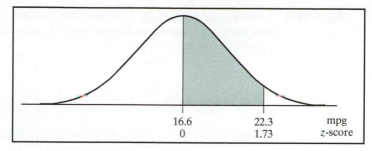

Figure 13-9 The proportion of SUVs that have fuel economies between 16.6 mpg and 22.3 mpg is determined by finding the area under the standard normal curve from $z = 0$ to $z = 1.73$.

under curves is, in general, quite difficult (often requiring techniques in calculus). Fortunately, statisticians have determined many areas under standard normal curves and organized the results in tables such as Table 13-1.

Table 13-1 lists areas between $z = 0$ and a positive z-score labeled $z = c$. The table is designed to be used with z-scores that have two decimal places of precision. To look up the area corresponding to the shaded region in Figure 13-9, we use $c = 1.73$. In the table, scan down the left-hand column until you see the row starting with **1.7**, then go across to the column with the header **.03**. (The cell in the table is lightly shaded.) The area is 0.4582, meaning that 45.82% of the area under the standard normal curve lies between $z = 0$ and $z = 1.73$. In terms of our problem, about 46% of SUVs have z-scores between $z = 0$ and $z = 1.73$, which is equivalent to saying that about 46% of SUVs get between 16.6 mpg and 22.3 mpg. This estimate is quite close to our earlier estimate based on the frequencies labeled on the histogram (about 42%).

Table 13-1 can be somewhat challenging to use, so we'll work through several additional problems to give you a feel for its structure. Let's estimate the proportion of SUVs that get between 9.7 mpg and 16.6 mpg. To use Table 13-1, we need to first convert these fuel economy ratings to z-scores. The z-score of 16.6 mpg (the mean) is $z = 0$, while the z-score of 9.7 mpg is

$$z = \frac{9.7 \text{ mpg} - 16.6 \text{ mpg}}{3.3 \text{ mpg}} = -2.09$$

Again we round the z-score to two decimal places, so that it is easier to use Table 13-1. The area under the standard normal curve bounded by $z = -2.09$ and $z = 0$ will give us the sought-after proportion of SUVs (Figure 13-10).

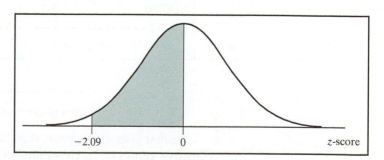

Figure 13-10 The proportion of SUVs getting between 9.7 mpg and 16.6 mpg can be estimated by finding the area under the standard normal curve from $z = -2.09$ to $z = 0$.

TABLE 13-1 Areas under a Standard Normal Distribution: This table gives the area under the standard normal curve between the mean $z = 0$ and a positive z-score $z = c$.

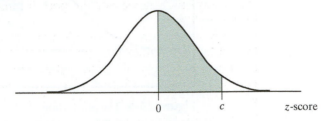

c	.00	.01	.02	.03	.04	.05	.06	.07	.08	.09
0.0	.0000	.0040	.0080	.0120	.0160	.0199	.0239	.0279	.0319	.0359
0.1	.0398	.0438	.0478	.0517	.0557	.0596	.0636	.0675	.0714	.0753
0.2	.0793	.0832	.0871	.0910	.0948	.0987	.1026	.1064	.1103	.1141
0.3	.1179	.1217	.1255	.1293	.1331	.1368	.1406	.1443	.1480	.1517
0.4	.1554	.1591	.1628	.1664	.1700	.1736	.1772	.1808	.1844	.1879
0.5	.1915	.1950	.1985	.2019	.2054	.2088	.2123	.2157	.2190	.2224
0.6	.2257	.2291	.2324	.2357	.2389	.2422	.2454	.2486	.2517	.2549
0.7	.2580	.2611	.2642	.2673	.2704	.2734	.2764	.2794	.2823	.2852
0.8	.2881	.2910	.2939	.2967	.2995	.3023	.3051	.3078	.3106	.3133
0.9	.3159	.3186	.3212	.3238	.3264	.3289	.3315	.3340	.3365	.3389
1.0	.3413	.3438	.3461	.3485	.3508	.3531	.3554	.3577	.3599	.3621
1.1	.3643	.3665	.3686	.3708	.3729	.3749	.3770	.3790	.3810	.3830
1.2	.3849	.3869	.3888	.3907	.3925	.3944	.3962	.3980	.3997	.4015
1.3	.4032	.4049	.4066	.4082	.4099	.4115	.4131	.4147	.4162	.4177
1.4	.4192	.4207	.4222	.4236	.4251	.4265	.4279	.4292	.4306	.4319
1.5	.4332	.4345	.4357	.4370	.4382	.4394	.4406	.4418	.4429	.4441
1.6	.4452	.4463	.4474	.4484	.4495	.4505	.4515	.4525	.4535	.4545
1.7	.4554	.4564	.4573	.4582	.4591	.4599	.4608	.4616	.4625	.4633
1.8	.4641	.4649	.4656	.4664	.4671	.4678	.4686	.4693	.4699	.4706
1.9	.4713	.4719	.4726	.4732	.4738	.4744	.4750	.4756	.4761	.4767
2.0	.4772	.4778	.4783	.4788	.4793	.4798	.4803	.4808	.4812	.4817
2.1	.4821	.4826	.4830	.4834	.4838	.4842	.4846	.4850	.4854	.4857
2.2	.4861	.4864	.4868	.4871	.4875	.4878	.4881	.4884	.4887	.4890
2.3	.4893	.4896	.4898	.4901	.4904	.4906	.4909	.4911	.4913	.4916
2.4	.4918	.4920	.4922	.4925	.4927	.4929	.4931	.4932	.4934	.4936
2.5	.4938	.4940	.4941	.4943	.4945	.4946	.4948	.4949	.4951	.4952
2.6	.4953	.4955	.4956	.4957	.4959	.4960	.4961	.4962	.4963	.4964
2.7	.4965	.4966	.4967	.4968	.4969	.4970	.4971	.4972	.4973	.4974
2.8	.4974	.4975	.4976	.4977	.4977	.4978	.4979	.4979	.4980	.4981
2.9	.4981	.4982	.4982	.4983	.4984	.4984	.4985	.4985	.4986	.4986
3.0	.4987	.4987	.4987	.4988	.4988	.4989	.4989	.4989	.4990	.4990

At first glance, it appears that Table 13-1 can only be used for positive z-scores (positive values of c). However, because of the symmetry of the standard normal curve, the area between $z = -c$ and $z = 0$ is equivalent to the area from $z = 0$ to $z = c$. So finding the area between $z = -2.09$ and $z = 0$ is equivalent to finding the area between $z = 0$ and $z = 2.09$.

Table 13-1 indicates that for $c = 2.09$ the area equals 0.4817. Thus, about 48.17% of the SUVs have fuel economy ratings between 9.7 mpg and 16.6 mpg. Note that we are stating the answer with a high degree of precision for illustration purposes only. Practically speaking, we would report the answer with less precision—something like 48.2% or 48%.

Now we ask one last question about the SUV fuel economy ratings. What is the rating that divides the upper 25% of all SUVs from the lower 75%? This rating will equal the value of the 3rd quartile or Q3. A diagram of the standard normal curve is shown in Figure 13-11. Note that this is an "inverse problem" in which we are given an area and need to determine the fuel economy rating (or corresponding z-score).

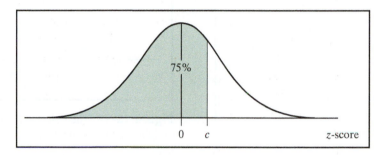

Figure 13-11 To determine Q3, find the z-score with 75% of the area under the standard normal curve on its left.

Because 50% of the area lies to the left of $z = 0$, we need a value for c that gives an area between $z = 0$ and $z = c$ equal to 25% or 0.2500. A portion of Table 13-1 is displayed next with the two entries shaded that "sandwich" the area of 0.2500. The z-score for the area of 0.2500 lies between $z = 0.67$ and $z = 0.68$. We chose $z = 0.67$ because 0.2500 lies a bit closer to 0.2486 than to 0.2517.

c	.00	.01	.02	.03	.04	.05	.06	.07	.08	.09
0.5	.1915	.1950	.1985	.2019	.2054	.2088	.2123	.2157	.2190	.2224
0.6	.2257	.2291	.2324	.2357	.2389	.2422	.2454	.2486	.2517	.2549
0.7	.2580	.2611	.2642	.2673	.2704	.2734	.2764	.2794	.2823	.2852

So what is the SUV fuel economy rating x when its z-score is $z = 0.67$? To answer this, substitute the given information into the z-score formula:

$$z = \frac{x - \overline{x}}{s}$$

$$0.67 = \frac{x - 16.6\,\text{mpg}}{3.3\,\text{mpg}}$$

Then multiply each side by 3.3 to clear fractions, and solve the resulting equation for x to get the rating:

$$0.67(3.3\,\text{mpg}) = x - 16.6\,\text{mpg}$$

$$18.811\,\text{mpg} = x$$

So an SUV getting about 18.8 mpg or higher will have a fuel economy rating in the top 25%. In Figure 13-7 there are 59 SUVs that get 19 mpg or higher out of 224 total, which is 59/224 = 26.3% of all SUVs. The value predicted by modeling with the standard normal curve agrees quite well with the value estimated by using the binned data in the histogram.

EXAMPLE 13-2 Male Chinese Birth Weights

The measurements of physical features of organisms are often normally distributed. For example, the weights of 9,465 male Chinese newborns measured in 1950 and 1951 were normally distributed with an approximate mean of 109.9 oz (ounces) and a standard deviation of 13.6 oz.[3] A diagram of the normal distribution of birth weights is given in Figure 13-12.

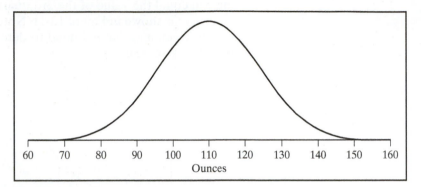

Figure 13-12 The bell curve approximation of male Chinese birth weights, with $\overline{x} = 109.9$ oz and $s = 13.6$ oz.

We now work through four examples involving the normal distribution of the birth weights.

a. Determine the percentage of newborns weighing between 120 and 130 ounces.

Solution We first convert these weights into z-scores, and then shade the corresponding area under the standard normal curve (Figure 13-13).

$$z = \frac{120\,\text{oz} - 109.9\,\text{oz}}{13.6\,\text{oz}} \approx 0.74$$

$$z = \frac{130\,\text{oz} - 109.9\,\text{oz}}{13.6\,\text{oz}} \approx 1.48$$

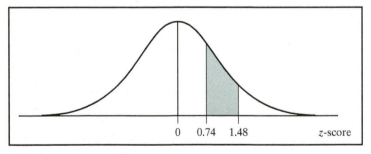

Figure 13-13 The proportion of newborns weighing between 120 oz and 130 oz is equal to the area under the standard normal curve from $z = 0.74$ to $z = 1.48$.

One strategy to find the shaded area is to take the area between $z = 0$ and $z = 1.48$ and subtract the area between $z = 0$ and $z = 0.74$. Reading Table 13-1, we determine that these areas are 0.4306 and 0.2704, respectively. Thus the proportion of male Chinese babies with weights between 120 oz and 130 oz is

$$0.4306 - 0.2704 = 0.1602 = 16.02\%$$

We estimate that about 16% of the newborns sampled (or approximately 1,514 of the 9,465 newborns) weighed between 120 oz and 130 oz.

b. Determine the proportion of the sample newborns that have weights below 90 oz.

Solution We begin by converting 90 oz into its z-score and shading the appropriate area under the standard normal curve. See the following z-score computation and Figure 13-14.

$$z = \frac{90 \text{ oz} - 109.9 \text{ oz}}{13.6 \text{ oz}} \approx -1.46$$

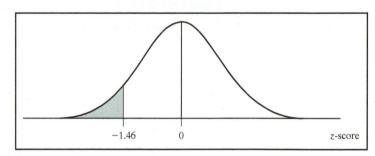

Figure 13-14 The proportion of newborns weighing below 90 oz is estimated by finding the area to the left of $z = -1.46$.

To find the shaded area using Table 13-1, we take the entire area to the left of $z = 0$ (area $= 0.5$ or 50%) and subtract the area between $z = -1.46$ and $z = 0$ (which is equal to the area between $z = 0$ and $z = 1.46$). The shaded area is $0.5 - 0.4279 = 0.0721$. Our answer suggests that about 7.21% of the sample, 682 newborns in total, had weights below 90 oz.

c. Find the proportion of the Chinese male newborn population that have weights equal to 112 oz.

Solution Actually, this is not a well-defined problem because the precision of the number 112 is not clear. Does 112 mean *exactly* 112 with infinite precision in the mathematical sense, or does it mean *approximately* 112 in a practical, scientific sense, where the weight has been rounded to the nearest 1 ounce? Let's look at how each meaning will affect our answer.

If we consider that 112 oz means exactly 112 ounce (i.e., 12.000000000. . . .), then the chance that a newborn weighs *exactly this weight* should be extremely small, if not zero. In fact, if we use the approach described up to this point, we would first convert 112 to its z-score

$$z = \frac{112 \text{ oz} - 109.9 \text{ oz}}{13.6 \text{ oz}} \approx 0.15$$

and then find the area under the standard normal curve between that z-score and itself ($z = 0.15$ to $z = 0.15$). Clearly, this infinitely thin region has zero area!

If we think of the number 112 oz as being rounded to the nearest 1 oz, then the true weight lies between 111.5 oz and 112.5 oz. The area below the standard normal curve that lies between these two weights is small but not zero. In the homework exercises, you can show that this area is about 0.0275, or slightly less than 3%.

d. Suppose we visit China today and randomly sample one male newborn and measure his weight at 140 oz. How unusual is it for a newborn to weigh this amount?

Solution To answer this question, we assume that male newborns in China today have weights with the same mean and standard deviation as they did in the early 1950s. If we consider a weight of *exactly* 140 oz, or a weight between 139.5 oz and 140.5 oz, then the chance is either zero or close to zero, as we discovered in part (c).

As it turns out, it's rare to select a newborn of *any* given weight (e.g., 86 oz, 112 oz, or 140 oz) because there are so many possible weights to choose from, even when rounded to the nearest ounce. But this begs the question: How do we know when a particular weight, say 140 oz, is unusually high or low?

We answer this question by finding the probability of having a baby weigh 140 oz *or more*. (We could again consider rounding and use a lower weight of 139.5 oz, but we'll ignore this minor difference in this and future computations.) This probability corresponds to the cumulative frequency of weights greater than or equal to 140 oz. As before, we determine this proportion by finding the area under the standard normal curve that lies above the z-score for 140 oz, which is about $z = 2.21$. See Figure 13-15.

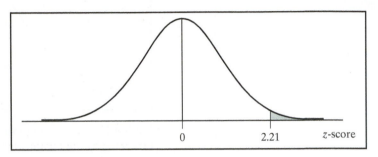

Figure 13-15 How unusual is it for a newborn to weigh 140 ounces? The answer is given by the area to the right of $z = 2.21$.

From Table 13-1, the area between $z = 0$ and $z = 2.21$ is 0.4864; thus the shaded area is $0.5 - 0.4864 = 0.0136$ or 1.36%. So we see that a weight of 140 oz, which lies 2.21 standard deviations above the mean, is actually quite rare—only 1.36% of the newborns weigh more than that amount.

By randomly choosing one male newborn from the entire Chinese population and getting such an extremely high weight, we might conclude that something is amiss. Perhaps the sample was not random (we searched for a heavy-looking baby) or perhaps the data value was recorded in error. But there could be an environmental explanation, too. Perhaps the mean of 109.9 oz and the standard deviation of 13.6 oz, the statistics from the early 1950s, *do not describe the distribution of weights today*! We might hypothesize that Chinese male newborns weigh more today than they did in the 1950s. Can you suggest a few reasons why this might be true?

TRANSFORMATIONS TO NORMAL

In data sets with positive or negative skew, the values can often be transformed so that the resulting distribution is close to normal. This transformation is advantageous because we can then apply the powerful properties of the Empirical Rule to the new distribution, or model the distribution with a standard normal curve. There are many techniques to transform data, all depending on the type of skew (negative or positive) and the magnitude of the skew. Though many of these transformations are beyond the scope of this text, in this section we investigate transforming positive-skewed data using logarithms. The following example explains how this works.

EXAMPLE 13-3 San Diego Ozone

Ozone (O_3) is a compound of oxygen found high in the stratosphere ("good" ozone blocking harmful ultraviolet radiation) and low in the troposphere ("bad" ozone that is the main part of photochemical smog). See Figure 13-16. Ozone causes inflammation of airways that can reduce lung capacity and result in permanent damage to lung tissue. The U.S. Environmental Protection Agency has established a federal limit of 12 parts per thousand for the average one-hour ozone concentration in cities.

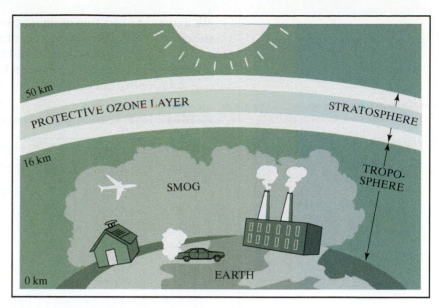

Figure 13-16 Ozone in the stratosphere blocks harmful ultraviolet rays, while ozone nearer to Earth is a major component of smog.

The California Air Resources Board analyzed air samples taken on Overland Avenue in downtown San Diego for 70 consecutive days during the summer of 1998.[4] Maximum one-hour ozone concentrations for 69 of those days (on one day there was no value reported) are given in Table 13-2. The concentrations

TABLE 13-2

23	40	43	49	57	65	81
27	40	44	50	57	67	85
31	40	44	50	57	67	86
34	40	45	50	57	67	90
36	41	45	51	60	67	94
37	42	46	52	62	68	97
38	42	47	53	63	71	105
38	42	47	54	65	75	115
38	42	47	54	65	75	125
39	43	48	54	65	76	

Source: California Air Resources Board

are given in parts per thousand (ppt). A histogram of the data (Figure 13-17) indicates that the ozone concentrations have a strong positive skew.

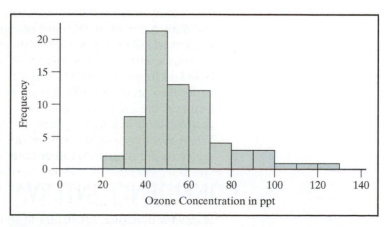

Figure 13-17 Histogram of ozone concentrations showing a strong positive skew.

To make the histogram closer to normal, the ozone concentration data can be transformed by taking the logarithm of each value (Table 13-3). The histogram of the logged concentrations (Figure 13-18) indicates that the transformed data are close to normal (though not perfectly bell shaped). The units of measure in the transformed data are log(ppt).

TABLE 13-3 **Transformed ozone concentrations in log(ppt)**

1.36	1.60	1.63	1.69	1.76	1.81	1.91
1.43	1.60	1.64	1.70	1.76	1.83	1.93
1.49	1.60	1.64	1.70	1.76	1.83	1.93
1.53	1.60	1.65	1.70	1.76	1.83	1.95
1.56	1.61	1.65	1.71	1.78	1.83	1.97
1.57	1.62	1.66	1.72	1.79	1.83	1.99
1.58	1.62	1.67	1.72	1.80	1.85	2.02
1.58	1.62	1.67	1.73	1.81	1.88	2.06
1.58	1.62	1.67	1.73	1.81	1.88	2.10
1.59	1.63	1.68	1.73	1.81	1.88	

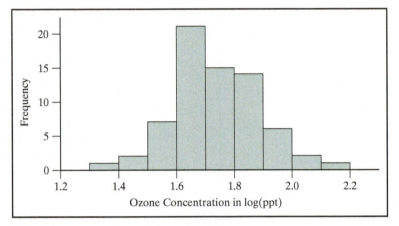

Figure 13-18 The histogram of the logged or transformed data is approximately normal.

Working with logged data that are approximately normal, we can apply the Empirical Rule or use the standard normal curve and Table 13-1. We will explore the details of this topic in one of the homework exercises and in the project linked to this chapter.

It may seem kind of strange working with transformed data rather than the original data. After all, do the units of log(parts per thousand) really make any sense? In science, transformations such as these are very common. In Chapter 1 you read that measurements of acidity (pH scale) and earthquake magnitude (moment scale) are based on logged values. In Chapter 5 and Chapter 6, we transformed bivariate data in order to get the best-fitting exponential and power functions. Finally, something as simple as the size of a circular lake can be given in terms of diameter (units of length) or area (units of length^2). These alternate scales are used primarily for *convenience*. When measuring ozone concentrations, it may be more convenient to use units of log(parts per thousand) rather than parts per thousand.

CONFIDENCE INTERVALS

Recall the discussion in Chapter 11 about sampling a school's water system to determine the mean lead concentration. Results of environmental inquiries like this will state the population mean using a **confidence interval**, such as 150 ± 10 parts per billion (ppb).

A confidence interval of 150 ± 10 ppb predicts that the true population mean (the mean lead concentration for all of the school's water) will lie between 140 ppb and 160 ppb. Confidence intervals are used to predict population parameters because it is highly unlikely that a sample statistic could ever *exactly pinpoint* the true value of the population parameter. The 10 ppb "leeway," either up or down from 150 ppb, is called the **margin of error**. The margin of error measures the precision of the prediction.

In addition to predicting population means, confidence intervals are used to predict population proportions. Examples of confidence intervals for proportions are frequently encountered in the news, especially with results from opinion polls. For example, a poll might claim that 54% of the U.S. public favors tougher environmental regulations, with a maximum sampling error of ±3 percentage points. This poll predicts that the true population proportion (the proportion of all U.S. residents favoring tougher environmental regulations) lies between 51% and 57%. The margin of error for this poll would equal 3%.

Confidence intervals are based upon results from samples. But all samples, even when collected properly, have inherent variability, and thus it is possible that a confidence interval will fail to contain the true population mean or population proportion. One way to increase the predictive success for a confidence interval is with a larger margin of error. For example, we might state that the mean lead concentration in our school is 150 ± 75 ppb. Because this confidence interval is wider than the interval 150 ± 10 ppb, it has a higher chance of containing the true mean lead concentration. In a similar manner, a confidence interval of 150 ± 2 ppb would have a smaller chance of containing the population mean.

Perhaps you see the dilemma. It is best to have a confidence interval that is precise (i.e., with a *small* margin of error), but it should also have a high chance of containing the population parameter, which can be obtained using a *large* margin of error. As a tradeoff between precision and certainty, many scientists use a **95% confidence interval**. To understand the meaning of this phrase, suppose a large number of samples of the same size are collected from a population, and a 95% confidence interval is constructed for each sample. We expect that 95% of these intervals will contain the true population parameter.

Confidence Intervals for Means

We now learn how to compute the 95% confidence interval when estimating the mean of a population. The theory behind this method (which is explained in more depth in elementary statistics texts) requires that two conditions be met. First, the sample must be a simple random sample. Recall that a simple random sample is collected through a process in which every sample of a given size has an equal chance of being selected. Second, the population must be somewhat normally distributed, or the sample size must be large. The term *somewhat normally distributed* generally means "peak shaped and symmetric." What is typically meant by *large* is a sample size of $n \geq 30$.

A 95% confidence interval for the mean takes the form $(\bar{x} - E, \bar{x} + E)$. This interval has $\bar{x}$ (the sample mean) at its center and extends an equal distance E, which is the margin of error, to the right and left of the mean (Figure 13-19). The 95% confidence interval can also be written in the form $\bar{x} \pm E$.

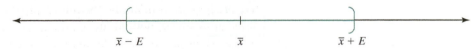

$$\bar{x} - E \qquad \bar{x} \qquad \bar{x} + E$$

Figure 13-19 A diagram of the confidence interval that contains the population mean.

The margin of error, E, depends upon the sample size and the variation or spread in the data. The margin of error when estimating a population mean is given by the formula

Margin of error for means: $E = t \cdot \dfrac{s}{\sqrt{n}}$

In the preceding formula, s represents the sample standard deviation, n is the sample size, and t is a number found in a table such as Table 13-4. Note that the value of t is dependent upon sample size n. Because t-values quickly level off and eventually approach the number 1.960, it is common to use $t = 1.960$ for sample sizes greater than or equal to 30.

TABLE 13-4

n	t	n	t	n	t
2	12.706	12	2.201	22	2.080
3	4.303	13	2.179	23	2.074
4	3.182	14	2.160	24	2.069
5	2.776	15	2.145	25	2.064
6	2.571	16	2.132	26	2.060
7	2.447	17	2.120	27	2.056
8	2.365	18	2.110	28	2.052
9	2.306	19	2.101	29	2.048
10	2.262	20	2.093	≥ 30	1.960
11	2.228	21	2.086		

EXAMPLE 13-4 Mercury in CFLs

Recall from Chapter 12 our example concerning mercury in compact fluorescent lights (CFLs). Suppose that the sample of $n = 9$ CFLs was a simple random sample, and that the population of all CFL mercury weights is normally distributed. We found that the mean of the sample is $\overline{x} = 5.00$ mg and the standard deviation is $s = 0.25$ mg. Predict the mean mercury weight in all CFLs using a 95% confidence interval.

Solution In Table 13-4, we find that when $n = 9$, $t = 2.306$. We now substitute these values into the margin of error formula to determine E:

$$E = t \cdot \frac{s}{\sqrt{n}}$$

$$= 2.306 \cdot \frac{0.25 \text{ mg}}{\sqrt{9}}$$

$$= 0.1921666667 \text{ mg}$$

$$\approx 0.19 \text{ mg}$$

For the margin of error, we round off using the same number of decimal places as found in the sample mean—in this case two decimal places.

The 95% confidence interval for the population mean can now be computed as follows:

$$(\overline{x} - E, \overline{x} + E) = (5.00 \text{ mg} - 0.19 \text{ mg}, 5.00 \text{ mg} + 0.19 \text{ mg})$$

$$= (4.81 \text{ mg}, 5.19 \text{ mg})$$

We are 95% confident that the interval from 4.81 milligrams to 5.19 milligrams contains the true population mean.

How would a larger sample influence the size of the margin of error in the previous example? Suppose that the simple random sample in the previous example contained $n = 90$ CFLs. Using the same sample standard deviation $s = 0.25$ g, and a t-value of 1.960, we find that the margin of error is much smaller:

$$E = 1.960 \cdot \frac{0.25 \text{ mg}}{\sqrt{90}} \approx 0.05 \text{ mg}$$

The smaller margin of error should come as no surprise—by using a larger sample, we should gather more information about the population and thus have a more precise estimate of the true value of the population mean. Using this revised margin of error, we would be 95% confident that the population mean is between 4.95 mg and 5.05 mg.

When a 95% confidence interval is used to estimate a population mean, scientists often say they are working at the **95% confidence level** or at the **5% error level**. In some scientific studies, confidence levels different than 95% are used, with 90% being a typical lower level of confidence (error level = 10%) and 99% a typical higher level of confidence (error level = 1%). For different levels of confidence, there are different tables giving values of t, and thus different values for the margin of error E. In this text we only use 95% confidence levels with t-values given by Table 13-4.

When working with a random sample, and the sample size is large ($n \geq 30$), the margin of error formula can be used when the population distribution takes on any shape (normal or not). It is therefore "safer" statistically to have sample sizes greater than or equal to 30. Unfortunately, environmental scientists often obtain small samples because of sampling costs, environmental hazards, or some other limiting constraint. For example, measuring the mercury contents in a CFL requires access to expensive laboratory equipment and trained technicians, and so a sample size of $n = 9$ might be the largest sample that is feasible. Consider a real study in which 12 Hispanic women were sampled to measure how the banned pesticide DDT, found in the fish they ate, was passed through breast milk to their nursing children.[5] In this type of study, a larger sample might be difficult to obtain because of privacy reasons, or simply because there were too few Hispanic women in the study region who had the given attributes (fish diet, nursing infant, and exposure to DDT).

Confidence Intervals for Proportions

A sample proportion can be used to create a 95% confidence interval that predicts a population proportion. Using the symbol $\hat{p}$ ("p-hat") to denote a sample proportion, the 95% confidence interval for the population proportion takes the form $(\hat{p} - E, \hat{p} + E)$ or $\hat{p} \pm E$ (Figure 13-20).

$\hat{p} - E$	$\hat{p}$	$\hat{p} + E$

Figure 13-20 A diagram of the confidence interval that contains the population proportion.

The margin of error, E, for the confidence interval for proportions is given by the formula below. As with the margin of error formula for means, n represents sample size and the value of t is found in Table 13-4.

$$\textbf{Margin of error for proportions:} \quad E = t \cdot \sqrt{\frac{\hat{p}(1 - \hat{p})}{n}}$$

There are two sample conditions that must be met for the confidence interval to be statistically valid. First, the sample must be a simple random sample. Second, the sample size should be large enough to satisfy both $n\hat{p} \geq 5$ and $n(1 - \hat{p}) \geq 5$. Again, reasons for these conditions are too advanced for this text, but interested readers will find more thorough explanations in statistics texts.

Figure 13-21 Near Sheldon, Illinois, a pesticide is applied to a field. *Source: USDA Agricultural Research Service.*

EXAMPLE 13-5 Pesticide Residues in Lettuce

The United States Department of Agriculture randomly tests thousands of fruit and vegetable samples for pesticide residues as part of its Pesticide Data Program (Figure 13-21). In the year 2000, 740 lettuce samples were collected, washed, and tested for pesticide residues; 18 of the samples (about 2.4%) had detectable levels of the pesticide diazinon. Detected amounts ranged in value from 0.003 ppm to 0.021 ppm; the EPA tolerance level is 0.7 ppm.[6] If the sample was a simple random sample, estimate the percentage of lettuce in the United States that contains diazinon, using a 95% confidence interval.

Solution We use the sample size of $n = 740$ and sample proportion of $\hat{p} = 2.4\%$ $= 0.024$ to verify that the conditions for the margin of error have been met:

$$n\hat{p} = 740(0.024) = 17.76 \geq 5 \quad \checkmark$$

$$n(1 - \hat{p}) = 740(1 - 0.024) = 722.24 \geq 5 \quad \checkmark$$

The t-value from Table 13-4 is $t = 1.960$, and so the margin of error for the 95% confidence interval is

$$E = t \cdot \sqrt{\frac{\hat{p}(1 - \hat{p})}{n}}$$

$$= 1.960 \sqrt{\frac{0.024(1 - 0.024)}{740}}$$

$$= 0.0110273394$$

$$\approx 1.1\%$$

Note that we have rounded off the margin of error to the same number of decimal places used in the sample proportion $\hat{p}$.

The 95% confidence interval is $\hat{p} \pm E = 2.4\% \pm 1.1\%$. This interval indicates that we are 95% confident that the percentage of lettuce in the United States with detectable amounts of diazinon (even after washing) is between 1.3% and 3.5%.

CHAPTER SUMMARY

All normal (bell-shaped) curves can be rescaled so that the horizontal axis is given in terms of z-scores, and the total area under the curve equals 1. The resulting bell curve is the **standard normal curve**, which has a mean of 0 and standard deviation of 1. Areas under the standard normal curve can be found using Table 13-1.

Analysis of normally distributed data sets begins with z-scores or areas, then uses the standard normal curve. For example, suppose air quality measurements are normally distributed with a mean of $\mu = 160$ ppm and a standard deviation of $s = 20$ ppm. Finding the proportion of the samples between 160 ppm and 185 ppm is equivalent to finding the area under the standard normal curve between $z = 0$ and $z = 1.25$ (Figure 13-22). Table 13-1 indicates that the proportion is 39.44%.

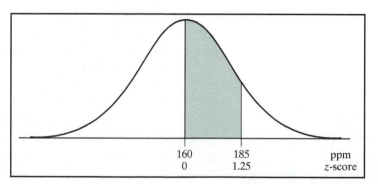

	160	185	ppm
	0	1.25	z-score

Figure 13-22 The proportion of the samples between 160 ppm and 185 ppm equals the area under the standard normal curve from $z = 0$ to $z = 1.25$.

When a proportion or area is given, Table 13-1 can be read "backward" to find a z-score or actual value. Data sets with positive skew (long tail to the right) can be **transformed to normal** using logarithms before analysis takes place.

Confidence intervals use sample statistics to predict population parameters. The theory behind confidence intervals requires that the sample be a simple random sample, and that the population is somewhat normally distributed or the sample size is large ($n \geq 30$).

A confidence interval that estimates the population mean takes the form $\bar{x} \pm E$. E is called the **margin of error** and is given by the formula $E = t \cdot \frac{s}{\sqrt{n}}$. A confidence interval to estimate a population proportion is of the form $\hat{p} \pm E$, where the margin of error is $E = t \cdot \sqrt{\frac{\hat{p}(1 - \hat{p})}{n}}$. When working at the 95% confidence level, the value for t in the preceding formulas is given by Table 13-4.

END *of* CHAPTER EXERCISES

Finding Areas under the Standard Normal Curve

1. Use the Empirical Rule to find the area under the standard normal curve, given the following information. Include a sketch with your work.
   a. The area lies between $z = -3$ and $z = 0$.
   b. The area lies to the right of $z = 1$.
   c. The area lies between $z = -2$ and $z = 3$.
2. Use the Empirical Rule to find the area under the standard normal curve, given the following information. Include a sketch with your work.
   a. The area lies between $z = 0$ and $z = 2$.
   b. The area lies between $z = -2$ and $z = -1$.
   c. The area lies to the left of $z = 2$.
3. Use Table 13-1 to find the area under the standard normal curve, given the following information. Include a sketch with your work.
   a. The area falls between $z = 0$ and $z = 1.25$.
   b. The area falls between $z = 0$ and $z = 2.2$.
   c. The area falls between $z = -2.28$ and $z = 0$.
4. Use Table 13-1 to find the area under the standard normal curve, given the following information. Include a sketch with your work.
   a. The area falls between $z = 0$ and $z = 1.81$.
   b. The area falls between $z = 0$ and $z = 1.3$.
   c. The area falls between $z = -0.24$ and $z = 0$.
5. Use Table 13-1 to find the area under the standard normal curve, given the following information. Include a sketch with your work.
   a. The area lies to the right of $z = 2.61$.
   b. The area lies between $z = -2.2$ and $z = 1.62$.
   c. The area lies between $z = 0.6$ and $z = 2.8$.
6. Use Table 13-1 to find the area under the standard normal curve, given the following information. Include a sketch with your work.
   a. The area lies between $z = -1$ and $z = 2.66$.
   b. The area lies to the left of $z = -1.75$.
   c. The area lies between $z = -3.05$ and $z = -1.5$.
7. Use Table 13-1 to estimate the z-score $z = c$ in each diagram.
   a.

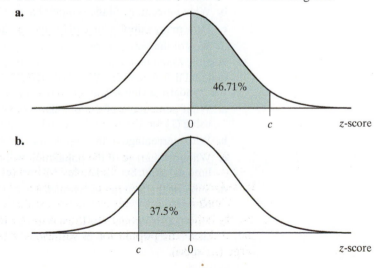

   b.

c.

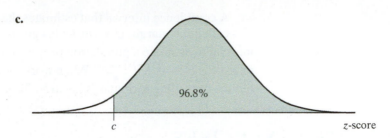

8. Use Table 13-1 to estimate the z-score $z = c$ in each diagram.

a.

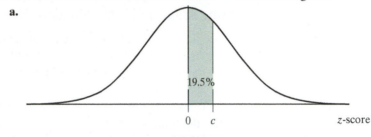

b.

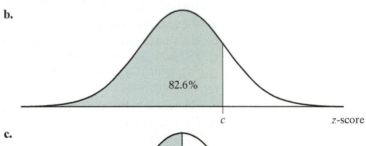

c.

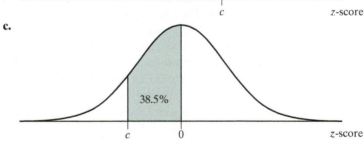

Applications of Normal Distributions

9. What percentage of SUVs have fuel economy ratings between 15.8 mpg and 21.4 mpg? See Example 13-1, and make use of the standard normal curve.

10. At the end of part (c) in Example 13-2, we claimed that about 2.75% of newborn Chinese males weighed between 111.5 oz and 112.5 oz. Show the steps involved to compute this percentage.

11. Suppose that 64 water samples have pH values that are normally distributed with a mean of 5.5 and a standard deviation of 0.4.

   a. What percentage of the samples have pH values between 5.5 and 5.8?

   b. What percentage of the samples have pH values between 4.9 and 5.5?

   c. How many samples have pH values greater than 5.7?

   d. 80% of the samples are below which pH value?

12. A study found that in a population of herring, *Pomolobus aestivalis*, the lengths of the individual fish were normally distributed (Figure 13-23). The mean length was 54.0 mm and the standard deviation was 4.0 mm.[7]

   a. What percentage of the population have lengths greater than 58 mm? Answer this using (1) the Empirical Rule and (2) Table 13-1. How do your answers compare?

   b. What percentage of the population will measure *exactly* 54 mm?

   c. What percentage of the population will measure 54 mm, rounded to the nearest millimeter? *Hint:* See Example 13-2 part (c).

   d. Suppose that you randomly catch a herring. What is the chance that its length is less than 62 mm?

   e. How unusual would it be to catch a random herring with a length of 43 mm? If you caught such a fish, would you be suspicious of the claim that the mean length was 54.0 mm? Explain.

Figure 13-23 Herring racks in the village of Tununak, Alaska. *Courtesy of U.S. Fish and Wildlife Service/photo by Jerry L. Hout.*

13. The ability of unsaturated soils to retain water is critical to soil ecosystems. A team of soil scientists investigated the water retention properties of soil cores sampled from a field. The water content of the soil cores (measured in cubic meters of water per cubic meter of soil) was determined to be approximately normal with $\bar{x} = 0.27\,\text{m}^3/\text{m}^3$ and $s = 0.04\,\text{m}^3/\text{m}^3$.[8]

   a. What percentage of the soil cores will have water contents less than 0.34 m^3/m^3?

   b. What percentage of the soil cores will have water contents between 0.21 m^3/m^3 and 0.35 m^3/m^3?

   c. 40% of soil cores will have water contents below which value?

14. During the 137-year period from 1863 to 1999, annual rainfall for Reading, Pennsylvania, was normally distributed with a mean of $\bar{x} = 41.5$ inches and a standard deviation of $s = 6.5$ inches.[9]

   a. What percentage of these years had annual rainfall under 25 inches? Comment on your answer.

   b. The 30 years with the most rain had annual rainfall amounts above which number?

   c. The 20 years with the least rain had which rainfall amounts?

15. Coal-fired power plants produce over half of the electricity generated in the United States, lessening America's dependence on foreign fuel sources. However, emissions from these plants are harmful to human health and are a major contributor to environmental problems such as acid rain, smog, and global warming. The 1998 rates of carbon dioxide (CO_2) production for the 100 largest coal-fueled electric power plants in the United States are given in Table 13-5.[10] The units of measure are pounds of CO_2 produced per megawatt-hour of energy produced (lb/MWh). The mean and standard deviation of the CO_2 rates are 2,223.1 lb/MWh and 211.3 lb/MWh, respectively.

TABLE 13-5

1,665	1,963	2,036	2,111	2,162	2,233	2,282	2,353	2,409	2,468
1,744	1,997	2,050	2,113	2,169	2,243	2,295	2,356	2,434	2,470
1,744	2,001	2,059	2,116	2,170	2,247	2,298	2,367	2,439	2,479
1,869	2,006	2,064	2,117	2,183	2,249	2,310	2,372	2,440	2,487
1,914	2,008	2,075	2,120	2,189	2,257	2,313	2,373	2,446	2,519
1,918	2,015	2,081	2,128	2,192	2,263	2,315	2,375	2,448	2,528
1,929	2,016	2,092	2,140	2,195	2,265	2,320	2,388	2,451	2,553
1,950	2,021	2,098	2,145	2,196	2,268	2,335	2,394	2,454	2,561
1,956	2,029	2,102	2,147	2,205	2,275	2,342	2,396	2,463	2,636
1,958	2,031	2,110	2,158	2,206	2,276	2,349	2,402	2,467	2,982

Source: U.S. Environmental Protection Agency

   a. Tally the data into 8 to 10 bins, and make a frequency histogram. Be sure to label the axes.

   b. Comment on the shape of the histogram. Are there any outliers? Explain how you know.

c. Assuming that the data are normally distributed, what percentage of the data will lie at least 1.5 standard deviations above the mean *in theory*?

d. Check whether your answer to part (c) actually holds for the data set. In other words, what percentage of the data are at least 1.5 standard deviations above the mean?

e. The coal-fired power plant in Marion, Illinois (Figure 13-24) ranked number 281 in coal energy production in 1998 and so is not included in Table 13-5. Its CO_2 production rate was 2,590 lb/MWh. What was the Marion plant's z-score, as compared to the plants listed in the table? In theory, what percentage of the data should lie below that z-score? What percentage of the data are actually smaller than the Marion plant's CO_2 emission rate?

Figure 13-24 A coal-powered plant in Marion, Illinois. This view includes coal and sludge conveyor belts and smokestacks. *Photo courtesy of the U.S. Dept of Energy.*

Transformation to Normal

16. The 32 states in the United States that produce nuclear energy have production values listed in Table 13-6. All values have units of megawatt-hours (MWh). See Chapter 12 for further details.

TABLE 13-6

13,050	19,800	3,000	1,300	8,000	23,600	57,800	3,600
20,600	21,800	7,900	5,100	7,500	25,900	42,900	23,800
11,300	24,800	14,200	21,600	4,100	10,700	14,000	5,700
32,700	71,900	4,900	12,100	23,800	6,100	15,900	11,200

Source: U.S. Dept of Energy

a. Use technology to create a histogram of the data set. (If necessary, see Chapter 3 or Chapter 11 for assistance.) Sketch the histogram on your homework paper.

b. Your histogram should indicate that the nuclear energy data have a large positive skew. To transform the histogram into one that is more normal, take the logarithms of the Table 13-6 values and store them in a different column. Create a histogram of the transformed (logged) data using technology. Sketch on your homework paper. How normal are the transformed data?

c. Calculate the one-variable statistics for the transformed (logged) data. What are the values of the mean ($\bar{x}$) and standard deviation (s) of the transformed data? Include units of measure.

d. Find the interval ($\bar{x} - 3s, \bar{x} + 3s$), using the mean and standard deviation of the logged values.

e. Express the interval that you found in part (d) using unlogged energy production values.

f. Do any of the nuclear energy production values from Table 13-6 lie more than three standard deviations from the mean, based on the interval that you found in part (e)? In other words, are any of the production values outliers?

95% Confidence Intervals

17. Assume that a random sample was gathered from a population that was normally distributed. Given the following sample statistics, find the margin of error and a 95% confidence interval for the population mean. Include units in each answer.

    a. $\bar{x} = 20.5$ lb, $s = 2.2$ lb, $n = 10$

    b. $\bar{x} = 45$ km, $s = 5$ km, $n = 24$

    c. $\bar{x} = 3.19$ cfs, $s = 0.15$ cfs, $n = 50$

18. Suppose that a study is conducted to estimate the mean amount of hazardous waste produced by hospitals. A 95% confidence interval that estimates the population mean is found to equal (210 kg/day, 260 kg/day). Determine the sample mean, $\bar{x}$, and margin of error, E.

19. In 1994, physicians measured blood lead levels in 373 bridge workers employed by painting contractors in eight states. The lead levels had a mean of 27.2 μg/dL (micrograms per deciliter) and a standard deviation of 16.1 μg/dL.[11]

    a. Determine a 95% confidence interval for the mean. Assume that the 373 workers make up a random sample.

    b. In 2000, a health objective of the U.S. Centers for Disease Control and Prevention was the elimination of lead exposures for occupations in which blood lead levels exceeded 25 μg/dL. Interpret your 95% confidence interval in terms of this objective.

20. A sample of 106 Yellowfin tuna (*Thunnus albacares*) were caught off Hawaii in 1998 to determine the mean concentration of mercury in this fish. The sample had a mean concentration of 0.210 ppm and a standard deviation of 0.112 ppm.[12] Determine a 95% confidence interval for the mean, and interpret your answer. Assume that the 106 tuna make up a simple random sample.

21. In a study on sea stars, a sample of $n = 67$ sea star arm lengths had a mean of 6.98 cm and a standard deviation of 2.00 cm.[13]

    a. Determine the margin of error for a 95% confidence interval of the mean.

    b. If sample size were quadrupled, and the sample mean and standard deviation stayed the same, by what factor would the margin of error be reduced?

    c. Suppose that you want to estimate the mean arm length of a sea star so that the margin of error is less than 0.10 cm. What is the minimum sample size needed?

22. The National Water and Climate Center monitors snowpack (snow on the ground) levels throughout the year in many western states of the United States (Figure 13-25). The snowpack acts as a natural water reservoir during winter months. Snowpack melts during spring and summer, providing valuable water for ecosystems as well as for farming and hydropower. Snowpack totals for Vail Mountain in Colorado on April 1 are provided for five different years (Table 13-7).[14]

TABLE 13-7

Year	Total Snowpack (inches of water)
1980	28.9
1985	22.8
1990	16.6
1995	19.9
2000	18.8

Source: U.S. Dept of Agriculture

Figure 13-25 A researcher collecting snow data. *Scott Bauer/USDA Agricultural Research Service.*

a. Assume that the five snowpack amounts represent a simple random sample of all April 1 snowpack totals for Vail Mountain. Determine the sample mean and sample standard deviation for the snowpack totals.

b. Further assume that yearly snowpack totals at Vail Mountain are normally distributed. Determine a 95% confidence interval of the mean for all April 1 snowpack totals.

c. Comment on the assumptions made in parts (a) and (b).

23. Assume that a simple random sample was gathered from a population. Use the following sample statistics to verify that $n\hat{p} \geq 5$ and $n(1 - \hat{p}) \geq 5$; then, find the margin of error and a 95% confidence interval for the population proportion.

    a. $\hat{p} = 74\%, n = 40$

    b. $\hat{p} = 38\%, n = 15$

    c. $\hat{p} = 35.4\%, n = 1,000$

24. In 1999, a study was conducted by the Centers for Disease Control and Prevention to estimate the percentage of U.S. residents 18 years or older that were currently smoking. The 95% confidence interval for that percentage is (22.9%, 24.1%).[15] Determine the sample proportion, $\hat{p}$, and the margin of error, E.

25. StarLink® corn is a genetically engineered variety that was approved as a source of animal feed but never approved for human consumption. However, in September 2000, StarLink corn was found in taco shells in U.S. supermarkets. A subsequent study by the U.S. Department of Agriculture found that about 9% of 110,000 corn samples were contaminated with StarLink corn.[16]

    a. With such a large sample size, should the margin of error be large or small? After you think about this question, compute the margin of error, E.

    b. Given that the percentage of samples contaminated with StarLink was "about 9%," is it appropriate to state the 95% confidence interval as 9% ± E?

26. Review Example 12-5 in Chapter 12 regarding the 66 velocity measurements for the Columbia River. Assume that the sample was a simple random sample.

    a. What percentage of the measurements in the sample are below 3.4 m/s?

    b. Verify that $n\hat{p} \geq 5$ and $n(1 - \hat{p}) \geq 5$ for the sample proportion found in part (a).

    c. Using a 95% confidence interval, estimate the percentage of all velocity measurements at the same Columbia River location that are below 3.4 m/s. Interpret your answer.

27. In a Gallup poll conducted in March 2004, 55% of those sampled said that the U.S. government was doing "too little" in terms of protecting the environment.[17] The sample size was $n = 526$ people. Assume that the sample was a simple random sample and that all people surveyed were U.S. residents.

    a. Verify that $n\hat{p} \geq 5$ and $n(1 - \hat{p}) \geq 5$.

    b. Estimate the percentage of all U.S. residents (in March 2004) who thought that the U.S. government was doing "too little" to protect the environment. Use a 95% confidence interval and interpret your answer.

    c. To decrease the margin of error to 3%, what is the minimum number of people that need to be polled? You may assume that the sample proportion, $\hat{p}$, will still be equal to 55%.

28. In years 2001 and 2002, 42% of 2,000 people polled in Germany said it was important that electricity was generated from water, wind, or other green energy sources.[18] Assume that the sample was a simple random sample.

    a. Was this a large enough sample? How do you know?

    b. Using a 95% confidence interval, estimate the percentage of all Germans who think that it is important that electricity be generated by green power.

    c. To decrease the margin of error to 1%, what is the minimum number of people that need to be polled? You may assume that the sample proportion, $\hat{p}$, will remain equal to 42%.

SCIENCE IN DEPTH

Hazardous Household Waste

Hazardous waste is material that is toxic, flammable, corrosive, reactive, or explosive. Examples of hazardous waste include toxic dioxin, corrosive battery acids, and reactive chemicals like sodium hydroxide (lye). Many types of businesses produce hazardous waste. Some are small businesses such as dry cleaners, auto repair shops, and photo processing centers. Others are larger firms which may generate enormous quantities of hazardous waste, such as chemical manufacturers, electroplating companies, and petroleum refineries.

The Resource Conservation and Recovery Act (RCRA, pronounced "wreck-ruh") is the main Federal law that regulates hazardous and other waste to ensure that they are managed properly.[19] RCRA waste is assigned a federal hazardous waste code and regulated because of permitting standards or because it was shipped subject to hazardous transportation requirements.

Household waste (also called municipal waste) was excluded from the RCRA list by Congress, but there are components of household waste that are hazardous. Fluorescent light bulbs (including energy-saving compact fluorescents), old-fashioned thermometers, mirrors, and several kinds of batteries contain toxic mercury. Carcinogenic benzene can be found in household products such as cleaners, nail polish remover, and varnishes.

The most common method of disposal of municipal waste is to bury the waste in a **landfill**. A landfill is a large pit lined with plastic sheeting or other barriers to contain the liquid effluent. The pit is filled with waste and then covered. Landfills can be reclaimed as green space, parks, golf courses, or even ski runs, such as Wilmot Mountain in Wisconsin. Methane gas from the decomposition of organic matter in the landfill is generated, occasionally in quantities large enough to be captured as a fuel source. However, much landfill waste decomposes at an incredibly slow rate because of the lack of oxygen in the tightly packed debris. Excavation of former landfills can yield 10-year old heads of lettuce and 20-year-old guacamole.[20]

An alternative method of disposal of municipal waste is **incineration**. The waste is burned at high temperatures and the heat generated can be used to produce electricity and hot water for local heating. Large quantities of carbon dioxide and nitrogen oxides (greenhouse gases) and smaller quantities of toxic substances like dioxin are emitted. The remaining solid combustion byproduct is referred to as **ash**. The volume of ash produced is very small compared to the original refuse, which helps extend the life of landfills. Disposal of the ash is a concern, however, as the ash is often rich in toxic heavy metals such as lead and cadmium.

While disposal of incinerator ash may be an environmental issue, ash is not on the RCRA list of toxic wastes. Disposal of ash in landfills has been a common practice, but as landfills run out of room, incinerator operators look for other disposal venues. In 1986, a private waste-disposal company, working for the city of Philadelphia, transported 13,000 tons of incinerator ash to the Bahamas on the Norwegian transport ship *Khian Sea*.[21] However, the Bahamas rejected the shipment and the *Khian Sea* began a two-year voyage to nowhere, with unsuccessful stops in Florida, Haiti (where part of the ash was dumped on the beach), Senegal, Yugoslavia, Indonesia, and the Philippines. The *Khian Sea* eventually docked in Singapore, minus its deadly cargo, whose fate remains unknown.

CHAPTER PROJECT:
TAXES ON TOXICS

Information on hazardous waste can be found at the U.S. Environmental Protection Agency's Biennial Reporting System (BRS).[22] BRS represents the only nationally consistent reporting of information on hazardous waste generation and management activities in the United States. BRS contains enormous amounts of data submitted by hazardous waste generators and handlers—it's easy to investigate who is generating the pollutants in your neighborhood!

In the companion project to Chapter 13 found at the text's Web site (**enviromath.com**), you will analyze data on RCRA waste from a state of your choice. You will examine whether or not county RCRA waste values are normally distributed, and decide whether transforming the data using logarithms improves normality. And you will develop a system of rewards for green counties and punishments for polluters based on z-scores.

NOTES

[1]Ecological footprint data from Redefining Progress, *Ecological Footprint of Nations 2004*, by Jason Venetoulis, Dahlia Chazan, and Christopher Gaudet, March 2004, www.RedefiningProgress.org; Biocapacity data for year 2000 obtained by request from Redefining Progress.

[2]U.S. Environmental Protection Agency, U.S. Dept of Energy, *Model Year 2001 Fuel Economy Guide*, http://www.fueleconomy.gov/

[3]Robert R. Sokal and F. James Rohlf, *Biometry* (San Francisco: W. H. Freeman, 1969), 60. Based on data published by J. Millis and Y. P. Seng, "The Effect of Age and Parity of the Mother on Birth Weight of the Offspring," *Ann. Human Genetics* 19 (1954): 58–73.

[4]California Air Resources Board, "Hourly Ozone Measurements in San Diego: Overland Avenue, 1998," http://www.arb.ca.gov/html/aqe&m.htm

[5]Washington State Dept of Health, Office of Environmental Health Assessment Services, *DDT and DDE Transmission Through Breast Milk: Yakima River Basin*, by Koenraad Mariën, 1998.

[6]U.S. Dept of Agriculture, Pesticide Data Program, *Annual Summary: Calendar Year 2000*, Appendix E, 12, http://www.ams.usda.gov/science/pdp/index.htm

[7]Thomas Glover and Kevin Mitchell, *An Introduction to Biostatistics* (New York: McGraw-Hill, 2003), 24.

[8]P. J. Shouse et al., "Spatial Variability of Soil Water Retention Functions in a Silt Loam Soil," *Soil Science* 159 (1995): 1–8.

[9]U.S. Historical Climate Network: http://www.ncdc.noaa.gov/ol/climate/research/ushcn/ushcn.html. For more information, consult QELP Data Set #049: http://www.seattlecentral.edu/qelp/sets/049/049.html

[10]U.S. Environmental Protection Agency, *2000 Emissions & Generation Resource Integrated Database (EGRID 2000)*, http://www.epa.gov/cleanenergy/egrid/index.htm; Note: Size of plants based upon annual energy produced, with all 100 plants producing more than 7 MWh in 1998.

[11]U.S. Dept of Health and Human Services, Centers for Disease Control and Prevention, "Lead Toxicity Among Bridge Workers, 1994," *Morbidity and Mortality Weekly Report*, December 15, 1995, http://www.cdc.gov/mmwr/

[12]Anne M. L. Kraepiel et al., "Sources and Variations of Mercury in Tuna," *Environmental Science & Technology* 37 (2003): 5551–5558.

[13]Glover and Mitchell, *An Introduction to Biostatistics*, 304.

[14]U.S. Dept of Agriculture, National Water and Climate Center, "Colorado Historical Snowpack Data," http://www.co.nrcs.usda.gov/snow/data/sc_list.html

[15]U.S. Centers for Disease Control and Prevention, *Morbidity and Mortality Weekly Report*, October 12, 2001, http://www.cdc.gov/mmwr/preview/mmwrhtml/mm5040a1.htm

[16]Anthony Shadid, "Ten Percent of U.S. Corn Contaminated by StarLink," *Boston Globe,* May 3, 2001.

[17]Gallup Poll, March 8–11, 2004. As cited at PollingReport.com, http://www.pollingreport.com/

[18]Reuters, "Fewer Germans Aware of Power Prices in 2002—Poll," *PlanetArk World Environmental News*, April 12, 2002, http://www.planetark.com/avantgo/dailynewsstory.cfm?newsid=18874.

[19]U.S. Environmental Protection Agency, Resource Conservation and Recovery Act Overview, http://www.epa.gov/enviro/html/rcris/

[20]William Rathje and Cullen Murphy, *Rubbish! The Archaeology of Garbage* (Tucson: University of Arizona Press, 2001).

[21]Kenny Bruno, "Philly Waste Go Home," *Multinational Monitor*, January/February 1998, http://multinationalmonitor.org/mm1998/mm9801.03.html

[22]U.S. Environmental Protection Agency, Hazardous Waste Data, http://www.epa.gov/epaoswer/hazwaste/data/#brs

Appendix: Unit Conversions

DISTANCE

1 centimeter (cm) = 10 millimeters (mm) $\approx$ 0.3937 inches (in)

1 meter (m) = 100 cm $\approx$ 39.37 inches $\approx$ 3.281 feet $\approx$ 1.094 yards

1 kilometer (km) = 1,000 m $\approx$ 0.621 miles

1 mile (mi) = 5,280 feet = 1,760 yards $\approx$ 1.609 kilometers

AREA

Units of area often come in the form length^2.

$$1 \text{ meter}^2 \approx 10.764 \text{ feet}^2$$

$$1 \text{ acre} = 43,560 \text{ ft}^2 \approx 0.405 \text{ hectare}$$

$$1 \text{ hectare (ha)} = 100 \text{ m} \times 100 \text{ m} = 0.1 \text{ km} \times 0.1 \text{ km} = 0.01 \text{ km}^2 \approx 2.47 \text{ acres}$$

$$1 \text{ km}^2 = 1,000 \text{ m} \times 1,000 \text{ m} = 1 \text{ million m}^2 = 100 \text{ ha} \approx 0.386 \text{ mile}^2$$

$$1 \text{ mile}^2 = 640 \text{ acres} \approx 2.59 \text{ km}^2$$

VOLUME

Units of volume often come in the form length^3.

$$1 \text{ cm}^3 = 1 \text{ cc (cubic cm)} = 1 \text{ milliliter} \left(\text{mL or } \frac{1}{1,000} \text{ liter} \right) \approx 0.0338 \text{ fluid ounces}$$

$$1 \text{ liter} = 1,000 \text{ mL} = 1,000 \text{ cm}^3 \approx 1.057 \text{ quarts (liquid)}$$

$$1 \text{ gallon (gal)} = 4 \text{ quarts} = 8 \text{ pints} = 16 \text{ cups} \approx 3.785 \text{ liters} \approx 0.1337 \text{ ft}^3$$

$$1 \text{ m}^3 = 1,000 \text{ liters} \approx 264.172 \text{ gallons} \approx 35.315 \text{ ft}^3$$

$$1 \text{ barrel of oil} = 42 \text{ gallons of oil} \approx 158.987 \text{ liters of oil} \approx 0.158987 \text{ m}^3 \text{ of oil}$$

$$1 \text{ bushel of grain} \approx 0.03524 \text{ m}^3 \approx 1.244 \text{ ft}^3$$

$$1 \text{ acre-foot of water} = 1 \text{ acre} \times 1 \text{ foot} = 43,560 \text{ ft}^3 \approx 325,851 \text{ gallons}$$

DISCHARGE

Discharge is volume divided by time and is a common measure of streamflow or irrigation rate.

$$1 \frac{\text{ft}^3}{\text{sec}} \text{ (cubic feet per sec or cfs)} \approx 7.481 \frac{\text{gal}}{\text{sec}}$$

$$1 \frac{\text{m}^3}{\text{sec}} \approx 35.315 \frac{\text{ft}^3}{\text{sec}}$$

$$\frac{1 \text{ acre-foot}}{1 \text{ day}} \approx 226.29 \frac{\text{gallons}}{\text{minute}} \text{ (gallons per minute or gpm)}$$

MASS AND WEIGHT (ON EARTH)

Mass (the amount of matter) is *not* the same as weight (the force on an object due to gravity). Units of mass include the gram or kilogram (metric) while units of weight include the Newton (metric) and the pound (English). We can, however, equate the kilogram (mass) with the pound (weight), even though these measures belong to different families of units, as long as we stay on Earth!

$$1 \text{ dry ounce (oz)} \approx 28.35 \text{ grams (gm or g)}$$

$$1 \text{ pound (lb)} = 16 \text{ dry ounces (oz)} \approx 0.454 \text{ kilograms (kg)}$$

$$1 \text{ kilogram (kg)} = 1{,}000 \text{ grams} = 1 \text{ liter of water} \approx 2.205 \text{ pounds (on Earth)}$$

$$1{,}000 \text{ kilograms} = 1 \text{ tonne (metric ton)} \approx 2{,}204.6 \text{ pounds}$$

$$2{,}000 \text{ pounds} = 1 \text{ ton (U.S. ton or "short" ton)} \approx 0.9072 \text{ tonnes}$$

DENSITY

Density (also known as specific gravity) is mass divided by volume or mass per volume. Lead and gold are dense, whereas chicken feathers are "light" (low density). The density of pure water at room temperature is 1 gram per cubic centimeter (1 gram per milliliter).

$$\frac{1 \text{ g}}{1 \text{ cm}^3} = \frac{1{,}000 \text{ kg}}{1 \text{ m}^3}$$

The term *density* can also refer to concentration, such as parts per million.

FORCE

Force is mass times acceleration. When the acceleration is due to gravity, gravitational force is often called "weight" (see previous section).

$$1 \text{ dyne} = 1 \text{ g} \times 1 \frac{\text{cm}}{\text{sec}^2}$$

$$1 \text{ newton (Nt or N)} = 1 \text{ kg} \times 1 \frac{\text{m}}{\text{sec}^2} = 10^5 \text{ dynes}$$

PRESSURE

Pressure is force divided by the area on which the force is acting. One atmosphere is the typical pressure on the Earth's surface caused by the weight of gases in the atmosphere. If you dive to the bottom of a 30-foot-deep swimming pool, you can feel an extra 1 atmosphere of pressure due to the weight of the water.

$$10 \frac{\text{dyne}}{\text{cm}^2} = 1 \frac{\text{newton}}{\text{m}^2} = 1 \text{ pascal (Pa)}$$

$$\frac{1 \text{ pound}}{1 \text{ in}^2} \text{ (pounds per square inch or psi)} \approx 0.068 \text{ atmospheres}$$

$$1 \text{ bar} = 10^5 \text{ pascals (Pa)} \approx 0.987 \text{ atmospheres}$$

WORK, MOMENT, ENERGY, AND HEAT

These four physical concepts have the same units. Work or moment is a force multiplied by the distance through which the force acts. If you have used a tire wrench to take off lug nuts, you are familiar with both moment and work! Energy is the capacity or ability to do work and comes in many forms, such as kinetic energy or potential energy. Heat is the amount of energy required to change the temperature of an object.

$$1 \text{ erg} = 1 \text{ dyne} \times 1 \text{ cm}$$

$$1 \text{ joule (J)} = 1 \text{ newton} \times 1 \text{ meter} = 1 \text{ N-m} = 10^7 \text{ ergs}$$

$$1 \text{ calorie (cal, with small c)} \approx 4.1868 \text{ joules}$$

$$1{,}000 \text{ calories} = 1 \text{ kilocalorie (kcal)} = 1 \text{ Calorie (capital C; "the food" calorie)}$$

$$1 \text{ British thermal unit (BTU)} \approx 1{,}055.056 \text{ joules} = 1.055 \text{ kJ}$$

Power is energy per time (see the next section). Therefore,

$$\text{power} = \frac{\text{energy}}{\text{time}}$$

$$\text{power} \times \text{time} = \left(\frac{\text{energy}}{\text{time}}\right) \times \text{time} = \text{energy}$$

Somewhat perversely, the amount of energy produced or consumed is often expressed in units of "power-time." Watt is a common unit of power (see the next section); therefore,

$$1 \text{ joule} = 1 \text{ watt} \times 1 \text{ sec} = 1 \text{ watt-sec}$$

$$3{,}600 \text{ joules} = 1 \text{ watt} \times 1 \text{ hour} = 1 \text{ watt-hour} = 1 \text{ Wh}$$

$$1{,}000 \text{ watt-hours} = 1 \text{ kilowatt-hour} = 1 \text{ kWh}$$

POWER

Power is work or energy divided by time and is the rate at which energy is produced or consumed. Power is work per time or energy per time.

$$1 \text{ watt} = \frac{1 \text{ joule}}{1 \text{ sec}} \approx 3.413 \frac{\text{BTU}}{\text{hour}}$$

$$1 \text{ horsepower} \approx 0.7457 \text{ kilowatts (kW)}$$

TEMPERATURE

There are three common scales of temperature; Celsius (sometimes mislabeled as "centigrade"), Fahrenheit, and Kelvin. Plain water freezes at 0°C (32°F or 273.15 K) and boils at 100°C (212°C and 373.15 K). No degree symbol is used for Kelvin!

$$\text{Celsius} = \frac{5}{9}(\text{Fahrenheit} - 32°)$$

$$\text{Fahrenheit} = \frac{9}{5}\text{Celsius} + 32°$$

$$\text{Kelvin} \approx \text{Celsius} + 273.15$$

ENERGY CONTENT

How much energy is contained in common fuels if they're burned efficiently? Note that these relationships are not "unit equivalents"; pounds and gallons do not belong to the same category as BTUs. Some conversions may vary.

$$1 \text{ cubic foot of natural gas} = 1{,}026 \text{ BTU}$$

$$1 \text{ pound of coal} = 10{,}340 \text{ BTU}$$

$$1 \text{ gallon of propane} = 91{,}000 \text{ BTU}$$

$$1 \text{ gal gasoline} = 0.898 \text{ gal crude oil} = 0.8921 \text{ gal heating oil}$$

$$1 \text{ gal gasoline} = 0.828 \text{ gal fuel oil} = 124{,}000 \text{ BTU}$$

RADIATION AND DOSAGE

The Curie and the Becquerel are units of radiation, whereas the rad, Gray, rem, and Sievert are units of exposure or dose.

$$1 \text{ becquerel (Bq)} = \frac{1 \text{ radioactive decay event}}{1 \text{ second}} = \frac{1}{\text{second}}$$

$$1 \text{ curie(Ci)} = \frac{37 \times 10^9 \text{ decay events}}{1 \text{ second}} = 37 \text{ GBq}$$

$$100 \text{ radiation absorbed doses} = 100 \text{ rad} = 1 \text{ gray (Gy)} = 1 \frac{\text{joule}}{\text{kilogram}}$$

$$100 \text{ roentgen equivalent mammal} = 100 \text{ rem} = 1 \text{ sievert (Sv)}$$

MOLE

Some familiar terms that tell us "how many" include a dozen (12) doughnuts or a ream (500 sheets) of paper. Chemists have their own term for "how many" atoms or molecules. This term is the **mole** and is a very large number, approximately 6.022×10^{23}, almost a trillion trillion. If you have one mole of carbon atoms, or a mole of water molecules, you have 6.022×10^{23} carbon atoms or water molecules. Because individual atoms and molecules are so small, a mole of carbon or water isn't a lot of mass; a mole of carbon is 12 grams (less than half an ounce), and a mole of water is about 18 grams.

Moles show up occasionally when discussing concentration (see Chapter 1). For example, the amount of nitrate in seawater can be expressed as "microMoles of nitrate per liter of water" or $\mu M/L$. A liter of water contains about 56 moles of water. Therefore, one microMole in one liter equals

$$\frac{1 \text{ microMole}}{\text{liter}} = \frac{10^{-6} \text{ Moles}}{56 \text{ Moles}} = \frac{10^{-6} \text{ \sout{Moles}}}{56 \text{ \sout{Moles}}} = 1.79 \times 10^{-8}$$

$$1.79 \times 10^{-8} = \frac{1.79 \times 10^{-8} \times 1 \text{ billion}}{1 \text{ billion}} = \frac{17.9}{1 \text{ billion}} = 17.9 \text{ ppb}$$

One microMole per liter equals about 18 parts per billion.

PREFIXES

Prefix	Symbol	Value	Exponential Equivalent	Example Using Prefix
peta	P	1 quadrillion	10^{15}	petagram (Pg)
tera	T	1 trillion	10^{12}	terabyte (Tbyte)
giga	G	1 billion	10^{9}	gigawatt (GW)
mega	M	1 million	10^{6}	mega-annum (Ma)
kilo	k	1 thousand	10^{3}	kilopascal (kPa)
milli	m	1 thousandth	10^{-3}	milliliter (mL)
micro	μ	1 millionth	10^{-6}	micrometer (μm)
nano	n	1 billionth	10^{-9}	nanogram (ng)
pico	p	1 trillionth	10^{-12}	picosecond (psec)

Answers to Odd-Numbered Exercises

Chapter 1 Exercises

1. Clam shell. *Answers may vary.* 5.5 cm long and 4.1 cm wide. Choice of axes may vary.
3. Reservoir evaporation. 0.07 inches per day less than the measured value. Accuracy seems moderate.
5. MTBE in groundwater. Precision seems very high, within one part out of a billion parts.
7. Gallons of gasoline. 10,000 miles per year at 20 miles per gallon equals 500 gallons of gasoline.
9. Food consumption. 1 pound of food per meal three times a day equals 1,095 pounds per year.
11. Kilogram and joule (among others)
13. Volume
15. Forest transect

$$13{,}267 \text{ feet} \times \frac{1 \text{ mile}}{5{,}280 \text{ feet}} \approx 2.51 \text{ miles}$$

17. Wetlands

$$1.45 \text{ km}^2 \times \frac{100 \text{ ha}}{1 \text{ km}^2} = 145 \text{ ha}$$

19. Pickup power

$$137 \text{ horsepower} \times \frac{0.7457 \text{ kilowatts}}{1 \text{ horsepower}} \approx 102.16 \text{ kW}$$

compared with 99.18 kW.

21. Earth's circumference

$$40{,}000 \text{ km} \times \frac{1 \text{ mile}}{1.609 \text{ km}} \approx 24{,}860 \text{ miles}$$

23. Discharge

$$\frac{3.5 \text{ gallons}}{\text{minute}} \times \frac{60 \text{ minutes}}{1 \text{ hour}} = \frac{210 \text{ gallons}}{1 \text{ hour}}$$

$$\frac{210 \text{ gallons}}{1 \text{ hour}} \times \frac{1 \text{ m}^3}{264.172 \text{ gallons}} = 0.795 \frac{\text{m}^3}{\text{hour}}$$

25. River sediment

$$\frac{2.4 \text{ million cubic yards}}{89 \text{ years}} = \frac{26{,}966 \text{ yd}^3}{1 \text{ year}}$$

$$\frac{26{,}966 \text{ yd}^3}{1 \text{ year}} \times \frac{1 \text{ year}}{365 \text{ days}} = 73.9 \frac{\text{yd}^3}{\text{day}}$$

27. Meters per second
29. Megawatt (MW)
31. One-millionth (10^{-6}) of a meter
33. Mammals versus insects. 4.65×10^3 mammals and 1.025×10^6 insects. 2–3 orders of magnitude.
35. PM10. $0.000025 \frac{\text{g}}{\text{m}^3}$ is $2.5 \times 10^{-5} \frac{\text{g}}{\text{m}^3}$, whereas $0.00015 \frac{\text{g}}{\text{m}^3}$ is $1.5 \times 10^{-4} \frac{\text{g}}{\text{m}^3}$. One order of magnitude.
37. $-6, 8, 5,$ and -3
39. 6.238 and -3.453
41. Wildfire acreages. 5.004 and 6.224. 1 order of magnitude.
43. Indonesian earthquake magnitude

$$\log\left((3.5 \times 10^{29} \text{ dyne} \times \text{cm})^{\frac{2}{3}} \Big/ 10^{10.7} \right) \approx 9.0$$

45. Mine drainage
$\text{pH} = -\log(1.6 \times 10^{-5} \text{ moles/liter}) \approx 4.8$
The lake water is rather acidic.

Chapter 2 Exercises

1. Red List

$$\frac{15{,}042 \text{ species}}{10{,}731 \text{ species}} \approx 1.40$$

3. Land areas

$$\text{California's area} = \frac{4 \text{ million km}^2}{10} = 400{,}000 \text{ km}^2$$

5. California power plants
   a.

Plant Name	Normalized (tons/TBTU)
Santa Clara	0.058
SCA	0.521
Scattergood	0.054
South Bay	0.059
SPA	0.059
Walnut	0.058

   b. With one exception (SCA), all the power plants have very similar ratios.
   c. SCA
7. San Francisco automobiles

Year	Normalized (cars/person)
1930	0.293
1940	0.353
1950	0.375
1960	0.445
1970	0.541
1980	0.634
1990	0.656
2000	0.698

a. Automobiles/person

b. Per capita car ownership has increased more slowly over the last few decades.

9. Forest areas

a.

Country	% Protected	Country	% Protected
Cameroon	6.2	Equatorial Guinea	0.0
C.A.R.	20.9	Gabon	3.7
Congo	4.6	Zaire	6.5

b. Zaire

c. Central African Republic

11. Water bottles

0.84×1 billion bottles/year
$= 840$ million bottles/year discarded

13. U.S. birth rate

$$\frac{4{,}137{,}000 \text{ births}}{281{,}422{,}000 \text{ people}} \approx 0.0147 = 14.7 \text{ ppt} = 14.7\text{‰}$$

15. Lead in trout

$$\frac{1 \text{ milligram}}{1 \text{ kilogram}} = \frac{10^{-3} \text{ gram}}{10^3 \text{ gram}} = \frac{1}{1 \text{ million}} = 1 \text{ ppm}$$

$$\frac{1}{1 \text{ million}} \times \frac{1{,}000}{1{,}000} = \frac{1{,}000}{1 \text{ billion}} = 1{,}000 \text{ ppb}$$

17. Prairie dogs

$$\frac{30 \text{ dogs} - 10 \text{ dogs}}{10 \text{ dogs}} \times 100\% = 200\%$$

$$\frac{300 \text{ dogs} - 100 \text{ dogs}}{100 \text{ dogs}} \times 100\% = 200\%$$

19. Toxics

$$\frac{7.1 \text{ billion lb} - 7.7 \text{ billion lb}}{7.7 \text{ billion lb}} \times 100\% = -7.79\%$$

21. England and Wales air pollution

$$\frac{225 \text{ incidents} - \text{initial}}{\text{initial}} = -0.54$$

initial amount ≈ 489 incidents

23. Idaho wheat farms

$$\frac{789 \text{ acres} - 1{,}259 \text{ acres}}{1{,}259 \text{ acres}} \times 100\% = -37.3\%$$

$$\frac{1{,}690 \text{ acres} - 1{,}259 \text{ acres}}{1{,}259 \text{ acres}} \times 100\% = 34.2\%$$

25. Laboratory analysis

$$\frac{159 \text{ ppb} - 150 \text{ ppb}}{150 \text{ ppb}} \times 100\% = 6.0\%$$

27. Daily water use

$$x = \frac{(50 \text{ gallons})(115{,}000 \text{ people})}{1 \text{ person}} = 5.75 \text{ million gallons}$$

29. Areas

$$x = \frac{100 \text{ ha} \times 1 \text{ mi}^2}{0.386 \text{ mi}^2} \approx 259.07 \text{ ha}$$

31. Brownfield reclamation

$$x = \frac{3.2 \text{ acres} \times 100\%}{28.7 \text{ acres}} \approx 11.1\%$$

33. PBDE in fish

$$x = \frac{3{,}078 \times 1 \text{ billion}}{1 \text{ trillion}} = 3{,}078 \times 10^{-3} = 3.078 \text{ ppb}$$

35. Desert survey

a. $x = \dfrac{120 \text{ ironwood/ha} \times 270 \text{ palo verde/ha}}{315 \text{ ironwood/ha}}$

≈ 103 palo verde/ha

b. $y = \dfrac{120 \text{ ironwood/ha} \times 165 \text{ saguaro/ha}}{315 \text{ ironwood/ha}}$

≈ 63 saguaro/ha

37. Canyon treefrog

$$\frac{(61 \text{ marked tadpoles in population})(68 \text{ tadpoles in recapture})}{36 \text{ marked tadpoles in recapture}}$$
$$= N \approx 115 \text{ tadpoles}$$

Boot Spring was probably a closed habitat; the tadpoles were recaptured after a short period of time and the tadpoles probably didn't travel that far.

39. New York pickerel

$$\frac{232 \text{ marked in population}}{16 \text{ marked in sample}} = \frac{N}{329 \text{ pickerel total in sample}}$$
$$N \approx 4{,}771 \text{ pickerel}$$

Most lakes are closed environments for most fish; that will help the quality of the population estimate.

41. Tumbling dice

a. 36 possible outcomes

1-1	2-1	3-1	4-1	5-1	6-1
1-2	2-2	3-2	4-2	5-2	6-2
1-3	2-3	3-3	4-3	5-3	6-3
1-4	2-4	3-4	4-4	5-4	6-4
1-5	2-5	3-5	4-5	5-5	6-5
1-6	2-6	3-6	4-6	5-6	6-6

b. 6 successful outcomes

1-1	2-1	3-1	4-1	5-1	**6-1**
1-2	2-2	3-2	4-2	**5-2**	6-2
1-3	2-3	3-3	**4-3**	5-3	6-3
1-4	2-4	**3-4**	4-4	5-4	6-4
1-5	**2-5**	3-5	4-5	5-5	6-5
1-6	2-6	3-6	4-6	5-6	6-6

$$P(\text{sum} = 7) = \frac{6}{36} = 0.1\overline{6} \approx 17\%$$

c. 4 successful outcomes. The probability is

$$P(\text{sum} = 9) = \frac{4}{36} = 0.\overline{1} \approx 11\%$$

d. 21 combinations. The probability is

$$P(\text{sum} > 6) = \frac{21}{36} = 0.58\overline{3} \approx 58\%$$

e. 4 combinations. The probability is

$$P(\text{sum} \neq 5) = 1 - P(\text{sum} = 5)$$
$$= 1 - \frac{4}{36} = 0.\overline{8} \approx 89\%$$

43. Mississippi River floods

a. $P(\text{flood}) = \dfrac{26 \text{ years with floods}}{99 \text{ years possible}} = 0.\overline{26} \approx 26\%$

b. $P(\text{no flood}) = 1 - P(\text{flood})$
$= 1 - 0.\overline{26} = 0.\overline{73} \approx 74\%$

45. Cascadia earthquakes

a. $R = \dfrac{3,200 \text{ years}}{6 \text{ intervals}} \approx 533 \dfrac{\text{years}}{\text{interval}}$

b. We should not expect another great earthquake 200 years from now, as the actual recurrence intervals probably varied quite a bit.

47. Colorado wildfires
36 large wildfires started in 135 days.

$$R = \dfrac{135 \text{ days}}{35 \text{ intervals}} \approx 3.9 \text{ days/interval}$$

Chapter 3 Exercises

1. Greenhouse gases
Estimates may vary.

Gas	CO_2	CH_4	N_2O	Total
%	66%	18%	16%	100%

3. Toxic release inventory

a.

State	Toxics	Percentage
Nevada	1,000	28.3%
Utah	956	27.0%
Arizona	744	21.0%
Alaska	535	15.1%
Texas	302	8.5%

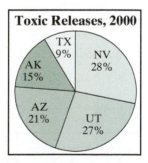

Toxic Releases, 2000

b. Nevada and Utah
c. The values in the table support the assertion.

5. Recycled materials

a.

Category	1986	1986 %	1998	1998 %
Papers	391,994	87.1%	821,994	37.9%
Metals	9,528	2.1%	318,710	14.7%
Organics	0	0.0%	815,809	37.6%
Plastics	349	0.1%	9,871	0.5%
Glass	48,013	10.7%	113,338	5.2%
Others	352	0.1%	87,657	4.0%

1986 WA Recycling

1998 WA Recycling

b. Organics, plastics, metals, and others are not apparent on the 1986 chart but are shown on the 1998 chart.

c. Organics, plastics, and others have increased as percentages of the total from 1986 to 1998. Paper and glass have declined.

7. Ecological footprint
a. Hectares per person (ha per capita)
b. Grains like rice and wheat, trees for lumber, fruit-bearing trees, and grass and forage
c. About 8.8 ha/person to 0.8 ha/person or 11 to 1

9. U.S. energy consumption

a.

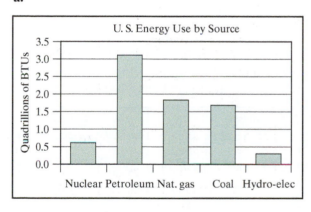

U. S. Energy Use by Source

b. Only hydroelectricity is renewable.
c. Solar, geothermal, and wind are missing.

11. Bangladeshi females
a. 5 years
b. The 80+ bin
c. Approximately 63 million females

13. Mexico population pyramid
a. 5 years
b. Bin 80+
c. The 25–30 age interval
d. Approximately 4 to 4.5 million people

15. Tennessee annual precipitation
  a. Results may vary depending upon the choice of bins.

Inches	Frequency
25–30	1
30–35	0
35–40	5
40–45	11
45–50	10
50–55	8
55–60	3
60–65	3
65–70	2
70–75	1

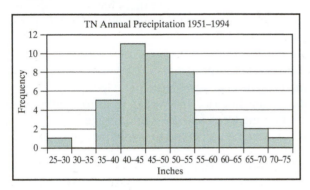

TN Annual Precipitation 1951–1994

  b. 30–35 inches per year
  c. 40–45 inches per year
  d. Six values fall below and 27 values exceed the modal bin.

17. Annual precipitation in Pennsylvania

Reading			Montrose		
Inches/Year	Frequency	Rel. Freq.	Inches/Year	Frequency	Rel. Freq.
25–30	2	0.015	25–30	0	0.000
30–35	19	0.144	30–35	6	0.171
35–40	27	0.205	35–40	6	0.171
40–45	39	0.295	40–45	9	0.257
45–50	33	0.250	45–50	10	0.286
50–55	10	0.076	50–55	2	0.057
55–60	1	0.008	55–60	2	0.057
60–65	0	0.000	60–65	0	0.000
65–70	1	0.008	65–70	0	0.000

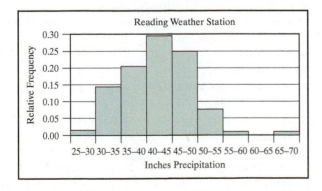

Reading Weather Station

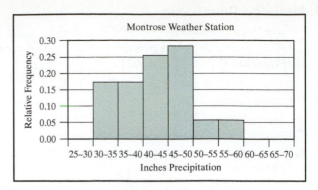

Montrose Weather Station

  a. The modal bin for Reading is 40–45 inches per year. The modal bin for Montrose is 45–50 inches.
  b. About the same (0.29 and 0.28)
  c. Reading has the longer period of record and does have more extreme drought and deluge than Montrose.

19. Wisconsin gray wolves
  a. Time (in years) and number of wolves
  b. 15 and 330
  c. 1994. Hunting and trapping of wolves may have been banned around 1994. The deer population might have increased.

21. Recycling aluminum
  a.

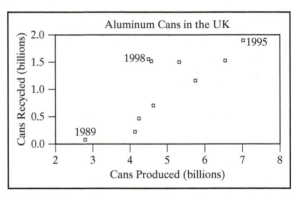

Aluminum Cans in the UK

  b. The higher the amount of cans produced, the higher the number of cans recycled.

23. UK aluminum cans

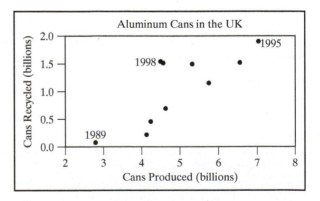

Aluminum Cans in the UK

Chapter 4 Exercises

1. Writing linear functions
  a. GT = global temperature in °F; t = time in years since 2001. $GT = 58.2 + 0.03t$.
  b. F = U.S. total fertility rate; t = years since 2001. $F = 2.034 - 0.022t$.

3. Two tables
  a. Example A: $m = 400$ people/year

  Example B: $m = -550 \dfrac{\text{people/km}^2}{\text{km}}$

  b. Example A: $P = 2,000 + 400t$
  Example B: $P = 5,000 - 550d$

  c.

Example A		Example B	
t	P	d	P
0	2,000	0	5,000
1	2,400	1	4,450
2	2,800	2	3,900
3	3,200	3	3,350
4	3,600	4	2,800

5. Two tables
  a. Example A: $m = 750 \dfrac{\text{kWh}}{\$1,000}$

  Example B: $m = -55 \dfrac{\text{million gallons}}{\text{year}}$

  b. Example A: $E = 10,500 + 750I$
  Example B: $C = 2,925 - 55t$

  c.

Example A		Example B	
I	E	t	C
0	10,500	0	2,925
2	12,000	5	2,650
4	13,500	10	2,375
6	15,000	15	2,100
8	16,500	20	1,825

7. Balancing units
The right side of the equation reduces to millions of hectares when years are cancelled; those are the same units as on the left side of the equation.

$$F(\text{Mha}) = 3,510(\text{Mha}) - 11.2\left(\frac{\text{Mha}}{\text{yr}}\right) \times t(\text{yr})$$

$$\text{Mha} = \text{Mha}$$

9. U.S. population
  a. Let P = U.S. population in millions of people and t = months since April 1, 2000.
  b. 0.2267 million people/month. The slope indicates that the United States was gaining, on average, 226,700 people per month during this time period.
  c. $P = 281.4 + 0.2267t$
  d. In $t \approx 82$ months, or February 1, 2007

e. See graph. Extrapolation to 240 months is a bit extreme.

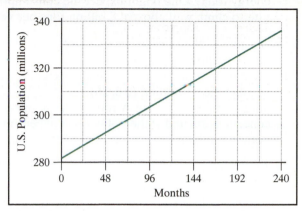

11. Soil thickness in northern Midwest states
  a. $S = 0.003t$
  b. The slope tells us the rate that soil forms: 0.003 inches/year
  c.

Years	Inches	Years	Inches
0	0	6,000	18
1,000	3	7,000	21
2,000	6	8,000	24
3,000	9	9,000	27
4,000	12	10,000	30
5,000	15		

  d. $t \approx 42$ years

13. Energy consumption as a function of wealth
  a. Let E represent energy consumption (quadrillion BTUs) and GDP represent Gross Domestic Product (billion U.S.$). $E = 0.011GDP - 2.25$.
  b.

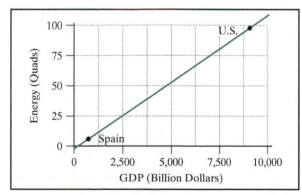

  c. Units are $\dfrac{\text{quadrillion BTUs}}{\text{billion U.S.\$}}$. The slope indicates that energy consumption climbs 0.011 quadrillion BTUs (11 trillion BTUs) for every 1 billion dollar increase in GDP.
  d. The United State uses more energy (10,741 BTU/$) as compared to Spain (7,884 BTU/$).

15. Coal on planet Earth
  a. 4.8 billion tons per year   **b.** $A = 1,081 - 4.8t$
  c. 841 billion tons   **d.** $t \approx 184$ years, or the year 2186
  e. Annual world coal consumption has risen tremendously because of the increase in the world's

population and because people are using more energy today. One might expect the consumption rate to climb well above 4.8 billion tons per year.

17. Water quality at a Florida chicken farm
    a. See the following graph.
    b. Most of the data look quite linear, except for the one significant outlier: (0.6, 26).
    c. *Solutions may vary.* Let N represent the nitrate concentration and P the potassium concentration, both in milligrams per liter. $N = 12.2P + 2$.

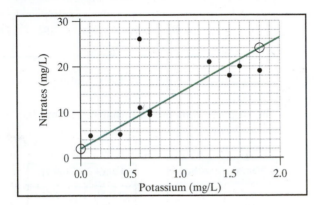

    d. $N = 9.517P + 4.8981, r = 0.71$. The correlation coefficient tells us that potassium concentration is a fairly good predictor of nitrate concentration.
    e. Approximately 15.4 mg/L

19. Hawaii-Emperor chain
    a. See the following graph.
    b. The data are very linear, except for the point corresponding to the island of Kimmei. It appears that Kimmei's estimated age is too low.
    c. *Solutions may vary.* Let D represent distance in kilometers, and a the age in millions of years. $D = 75a + 250$.

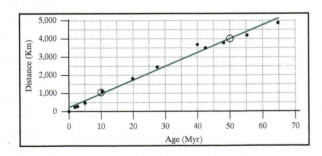

    d. $D = 75.75a + 184.84$
    e. $r = 0.9944$. The size of the correlation coefficient indicates that the data are very linear, since the value is close to 1.0. The sign indicates that the regression line has a positive slope.
    f. 21.86 million years old

21. Columbia River velocities
    a. velocity $= -0.1164$ depth $+ 1.6801$
    b. Depth is a good predictor of velocity, because $r = -0.94$ (which is very close to -1).

c. The slope tells us how rapidly the velocity decreases as depth increases; for every 1-foot increase in depth, the velocity decreases by 0.1164 feet per second.
d. Horizontal intercept: 14.4 ft. This represents the hypothetical depth at which velocity diminishes to 0.
e. The closer to the bottom, the greater the frictional forces or "drag" on the water caused by rocks, vegetation, debris, and so on.

Chapter 5 Exercises

1. Writing exponential functions
   a. Let t represent years since 2000 and P the population in millions. $P = 99.9(1.012)^t$.
   b. Let t represent years after 1988 and P the world production of CFCs in metric tons. $P = 1,074,465(0.734)^t$.

3. Two tables
   a. Example A: $M = 1.2$ Example B: $M = 0.95$
   b. Example A: $r = 0.2 = 20\%$
      Example B: $r = -0.05 = -5\%$
   c. Example A: $A = 20(1.2)^t$
      Example B: $P = 120,000(0.95)^t$

Example A		Example B	
t	A	t	P
0	20	0	120,000
1	24	1	114,000
2	28.8	2	108,300
3	34.56	3	102,885
4	41.472	4	$\approx$ 97,741

5. Solving exponential equations
   a. $x = \dfrac{\log(40)}{\log(4)} \approx 2.661$
   b. $W = \dfrac{\log(10)}{\log(1.0625)} \approx 37.981$
   c. $t = \dfrac{\log(150/52)}{\log(3)} \approx 0.964$
   d. $x = \dfrac{\log(0.06)}{\log(4)} \approx -2.029$

7. Biodiesel
   a. Let B represent biodiesel production in millions of gallons and t the number of years after 1999. $B = 5(1.431)^t$
   b.

t	B	t	B
0	5	6	42.9
1	7.2	7	61.4
2	10.2	8	87.9
3	14.7	9	125.8
4	21.0	10	180.0
5	30.0	11	257.6

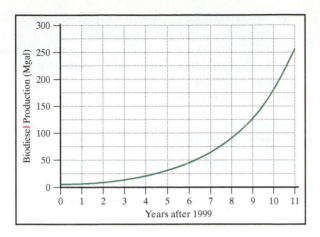

Biodiesel Production (Mgal) vs. Years after 1999

c. Yes. Another 5 years

9. Shrinking populations
   a. Belarus: $P(t) = 10.4(0.9983)^t$;
   Hungary: $P(t) = 10.1(0.9971)^t$
   b. Hungary
   c. About 0.47 million or 470,000 people

11. Vehicle miles in United States
    a. $M = 1.0289, r = 0.0289 = 2.89\%$
    b. Let VM represent vehicle miles in billions and t the number of years since 1988. $VM = 1{,}511(1.0289)^t$.
    c. Approximately 6,648 billion miles
    d.

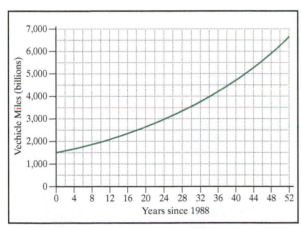

Vehicle Miles (billions) vs. Years since 1988

    e. Graph: approximately 24 years. Equation: 24.3 years.
    f. Vehicle miles will most likely grow exponentially because the U.S. population is growing exponentially, and limitations in space seem to be offset by people's increasing desire to travel in vehicles.

13. Timber harvests in California
    a. $M = 0.9520$. Let t represent the number of years since 1996 and H the harvest in billions of board feet. $H = 2.27(0.9520)^t$.
    b. 4.8%
    c. About 147,000 homes

15. Genetic diversity of domestic livestock
    a. $DLB = 5{,}000(0.95)^t$ where DLB = number of domestic livestock breeds and t = years, with $t = 0$ referring to the year when $DLB = 5{,}000$.
    b. About 13.5 years. Every 13.5 years, the number of domestic livestock breeds will be reduced by 50%.

17. The 3 fastest-growing U.S. counties
    a. Douglas: 0.854%, Loudoun: 0.794%, Forsyth: 0.764%
    b. Douglas: $P = 175{,}766(1.00854)^t$,
    Loudoun: $P = 169{,}599(1.00794)^t$,
    Forsyth: $P = 98{,}407(1.00764)^t$
    c. Douglas: about 82 months, Loudoun: about 88 months, Forsyth: about 91 months

19. Atmospheric CO_2 at Mauna Loa, Hawaii
    a. $r = 0.004$ or 0.4%
    b. In about 62 years
    c. About 174 years
    d. In general, the Northern Hemisphere experiences cooler weather from October to April and warmer weather from April to October. From October to April, there is more plant decay than growth—this leads to a net increase in atmospheric CO_2. From April to October, the opposite happens—there is more growth than decay, and net atmospheric CO_2 drops.

21. Galapagos Islands cactus finches
    a. See the graph. The data look somewhat exponential, although there is not a smooth decrease in population values.

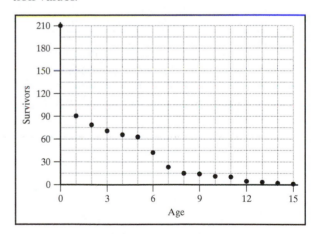

Survivors vs. Age

    b.

Age	Log(survivors)	Age	Log(survivors)
0	2.32	8	1.18
1	1.96	9	1.15
2	1.89	10	1.04
3	1.85	11	1.00
4	1.81	12	0.60
5	1.79	13	0.48
6	1.62	14	0.30
7	1.36	15	0.00

c. See the graph. $\log(y) = -0.12x + 2.2$.

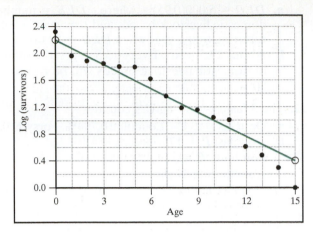

Log (survivors) vs. Age

d. $y = 158(0.759)^x$

e. $y = 205(0.727)^x$. This equation has a significantly higher value for y_0 than does the straightedge method. The multiplier is similar to but slightly smaller than the one found in the straightedge method.

f. $r = -0.273$ or -27.3%

23. DDT in the food chain

a.

x	y	y	$\log(y)$
1	0.04/1,000,000	4×10^{-8}	-7.4
2	0.23/1,000,000	2.3×10^{-7}	-6.6
3	2.07/1,000,000	2.07×10^{-6}	-5.7
4	13.8/1,000,000	1.38×10^{-5}	-4.9

b. The plot of the $(x, \log(y))$ points looks very linear. See the graph. An exponential model would be a very good fit to the original (x, y) data, because a linear model is a very good fit to the transformed data.

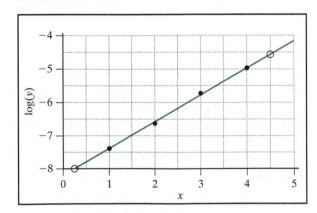

log(y) vs. x

c. $\log(y) = 0.8235x - 8.2059$ converts to $y = 6.2 \times 10^{-9}(6.66)^x$, where y is the DDT concentration and x the consumer level. Written with DDT concentrations in parts per million: $y = 0.0062(6.66)^x$.

d. For each 1-step increase in consumer level, DDT concentration increases by 6.66 or approximately a factor of 7.

e. $y = 0.0052(7.19)^x$, where y represents the DDT concentration in parts per million.

f. $r = 0.999$. The correlation coefficient indicates that food-chain level is an excellent indicator of DDT concentration.

g. About 100 ppm

Chapter 6 Exercises

1. Power functions

a.

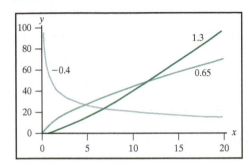

b. The first two equations pass through the origin, but not the function with the negative exponent.

c. $c = 1$

3. Solving power functions

a. $x = 0.523$

b. $x = 12.426$

c. $x = 0.038$

5. Phosphorus discharge

a. 6,231 pounds per year and 18,629 pounds per year

b. 45.5 square miles

7. Paleoclimatology

a. 323 centimeters or 127 inches. *Answers may vary*: There are a few places in the United States where the mean annual precipitation exceeds 127 inches, including southeast Alaska, windward Hawaii, and the Olympic Peninsula of Washington.

b. The average area is 20.5 square centimeters, with an expected mean annual precipitation of 140 centimeters.

9. Glacier dimensions

a.

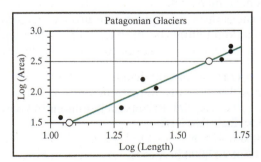

Patagonian Glaciers — Log (Area) vs. Log (Length)

b. *Answers may vary.* The slope m is 1.818 and the y-intercept is -0.454. The linear equation is

$$\log(\text{area}) = -0.454 + 1.818\,\log(\text{length})$$

c. *Answers may vary.* The power function is

$$\text{area} = 0.35\,\text{length}^{1.818}$$

d. The power of 1.818 is close to that of 2, matching expectations.

11. Australian marsupials

a.

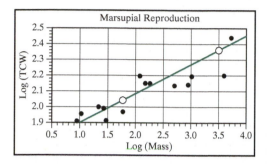

b. *Answers may vary.* The slope m is 0.188 and the y-intercept is 1.71. The linear equation through the transformed data is

$$\log(\text{tcw}) = 1.71 + 0.188\,\log(\text{mass})$$

The corresponding power function is

$$\text{tcw} = 51\,\text{mass}^{0.188}$$

c. As the size of female dasyurids increases, so does the time from conception to weaning. For small dasyurids, a small change in mass yields a large change in time to weaning, but many large dasyurids have similar weaning times.

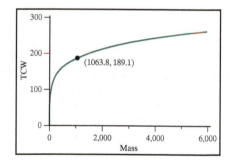

d. 237 days

13. Makaopuhi lava lake

a. Time is the independent variable. The best-fitting regression model is

$$T = 1.15762(t)^{0.51699}$$

The correlation coefficient r is 0.9976. A power function is an excellent model.

b. $T = 1.2$ feet. We expect very thin crust after only one day of cooling.

c. About 47.8 feet for the layer after 1,333 days. The observed thickness is 52 feet. This prediction is still quite reasonable, as the actual data on which the model is based extend only to about 500 days.

d. The power of 0.517 is consistent with cooling by conduction.

15. Smallmouth bass

a. The best-fitting power law regression is

$$\text{weight} = 45.61(\text{age})^{1.555}$$

The correlation coefficient is $r = 0.9036$

b. The model is a good fit despite bass of the same age having a variety of weights.

c. 134 grams

17. Tonawanda Creek discharge

a. 100 cfs and 939 cfs

b. *Results may vary.*

Bin	Frequency	BMP	RCF
100–200	17	150	31
200–300	6	250	14
300–400	2	350	8
400–500	2	450	6
500–600	1	550	4
600–700	2	650	3
700–800	0	750	1
800–900	0	850	1
900–1,000	1	950	1

c. The shape is consistent with a power function with a negative power.

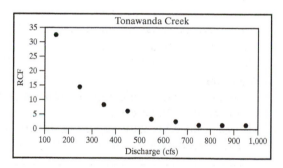

d. The regression equation is

$$RCF = 767{,}996\,BMP^{-1.9733}$$

The power $c \approx -2$ falls in the range -1 to -5.

e. r is -0.9738, with a negative sign consistent with a decreasing power function.

f. Discharges show a good power law frequency distribution.

19. Landslides

a. *Results may vary.*

BMP	RCF	BMP	RCF
1,000	3,163	11,000	259
3,000	1,909	13,000	196
5,000	1,300	15,000	141
7,000	1,028	17,000	80
9,000	652	19,000	35

b. The equation is $RCF = 160{,}191{,}465 BMP^{-1.4433}$. The correlation coefficient is -0.8978

c. Landslides show a drop-off in frequency; however, a power law seems less than ideal as a model.

d. The equation is $RCF = 4{,}485(0.9998)^{BMP}$. The correlation coefficient is -0.9929.

e. The exponential decay model

21. River bank failure

a.

BMP	RCF	log(BMP)	log(RCF)
2,500	187	3.40	2.27
7,500	75	3.88	1.88
12,500	43	4.10	1.63
17,500	21	4.24	1.32
22,500	16	4.35	1.20
27,500	9	4.44	0.95
32,500	5	4.51	0.70

b.

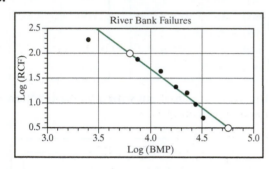

The slope is -1.58 and the y-intercept is 8.00. The linear equation is
$\log(RCF) = 8.00 - 1.58 \times \log(BMP)$. The power function is $RCF = 10^8 BMP^{-1.58}$.

c. The RCF is 4.9. One bank failure in this size range or greater occurs approximately every 7 years.

Chapter 7 Exercises

1. Finding sequence values

 a. 6 **b.** 54 **c.** -3 **d.** 36

3. Evaluating sequences A and B

 a. 42 **b.** -6 **c.** 0 **d.** 162

5. Sequence F

n	0	1	2	3	4	5
f(n)	10	18	26	34	42	50

The initial term value is 10. To get any term value, add 8 to the previous term value. Math:
$f(n) = f(n-1) + 8; f(0) = 10$

7. Sequence H

The initial term value is 2,000. To get any term value, multiply the previous term value by 0.75 and then add 50.

n	0	1	2	3	4
h(n)	2,000	1,550	1,212.5	959.375	769.53125

9. Affine difference equation

a.

n	0	1	2	3	4	5
a(n)	60	56	53.6	52.16	51.296	50.7776

b.

n	0	1	2	3	4	5
a(n)	40	44	46.4	47.84	48.704	49.2224

c.

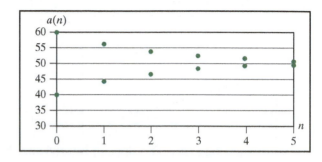

d. As n gets bigger, both sequences approach the number 50.

11. Finding solution equations

 a. Linear; solution equation is $u(n) = 50 + 10n$

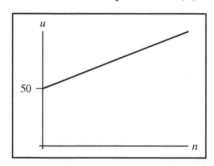

 b. Exponential; solution equation is $u(n) = 2{,}000(1.25)^n$

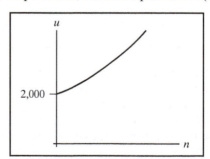

c. Exponential; solution equation is $u(n) = 100(0.8)^n$

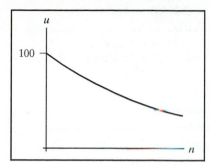

d. Linear; solution equation is $u(n) = 25 - 3n$

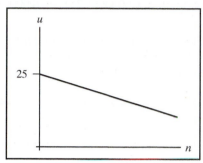

13. Consider: $a(n) = a(n-1) + 5$.

a. The solution equation is $a(n) = 10 + 5n$.

b. The solution equation is $a(n) = 15 + 5n$.

c.

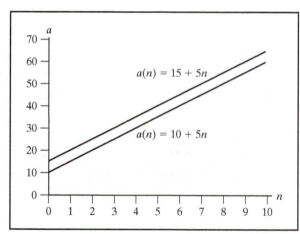

d. Use the difference equation $a(n) = a(n-1) + 5$ and initial condition $a(0) = 10$ to generate all sequence values from $a(1) = 15$ to $a(4) = 30$. Alternatively, use the solution equation and plug in $n = 4$ directly: $a(4) = 10 + 5(4) = 30$.

15. Modeling with affine difference equations

a. Let $p(n)$ represent the population after n months; $p(n) = 1.05p(n-1) - 20$.

b. 2,544 individuals

17. Meat consumption in China

This statistic tells us how meat consumption in any given year is related to meat consumption in the previous year. We can write a difference equation $c(n) = c(n-1) + 2$, where $c(n)$ is meat consumption (in kg per person) in year n.

19. North Cascade glaciers

a. $T(n) = T(n-1) - 0.30$

b.

Year	n	$T(n)$
1984	0	60
1985	1	59.7
1986	2	59.4
1987	3	59.1
1988	4	58.8
1989	5	58.5

c. $T(n) = 60 - 0.30n$. In 2001 the thickness is 54.9 meters. The loss is 8.5% of its original thickness. So the claim "about 10%" is reasonable.

21. Renewable energy

a. Industrialized: $M \approx 1.02419$ and $r \approx 0.02419$
Less industrialized: $M \approx 1.01755$, $r \approx 0.01755$

b. Industrialized: $E(n) = 1.02419E(n-1)$, $E(0) = 335,929$
Less industrialized: $E(n) = 1.01755E(n-1)$, $E(0) = 924,052$

c. Industrialized: $E(7) = 397,110$. Less industrialized: $E(7) = 1,043,727$

d. In about 155 years. Factors that might make this assumption unlikely: political changes, technological advances, and so on.

23. Mexico City population

a. $p(n) = (p(n-1) + 0.75)(1.017)$

n	$p(n)$	n	$p(n)$
0	20	4	24.52
1	21.10	5	25.70
2	22.22	6	26.90
3	23.36		

b. Similarities: Both take into account the linear and exponential growth of the population, through a two-stage process. Differences: They switch the order of these stages. Neither is more correct; both use reasonable assumptions.

25. Deer population

a. $P(n) = 1.04P(n-1) - 200$, affine

b.

n	$P(n)$	n	$P(n)$	n	$P(n)$	n	$P(n)$
0	2,000	4	1,490	8	894	12	197
1	1,880	5	1,350	9	730	13	5
2	1,755	6	1,204	10	559	14	-195
3	1,625	7	1,052	11	382		

c.

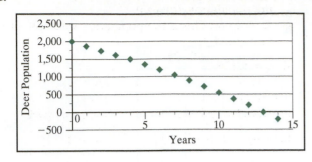

d. 80 deer

e. 46 to 47 deer

Chapter 8 Exercises

1. Solve the system: $60 = a + b$ and $58.634 = 1.025a + b$.
   Rewrite the first equation as $b = 60 - a$, and substitute into the second equation to get $-1.366 = 0.025a$. Solve to get $a = -54.64$. Substitute to find that $b = 114.64$.

3. Affine solution equations
   a. Substitute points $(0, 400)$ and $(1, 1900)$ to get the equations $400 = a + b$ and $1,900 = 2.5a + b$. Solve equations to get $a = 1,000$ and $b = -600$; $p(n) = 1,000(2.5)^n - 600$.
   b. Substitute points $(0, 510)$ and $(1, 337.5)$ to obtain equations $510 = a + b$ and $337.5 = 0.75a + b$. Solve equations to get $a = 690$ and $b = -180$; $p(n) = 690(0.75)^n - 180$.
   c. Substitute points $(0, 75)$ and $(1, 2.25)$ to obtain equations $a + b = 75$ and $1.03a + b = 2.25$. Solve these equations to get $a = -2,425$ and $b = 2,500$; $p(n) = -2,425(1.03)^n + 2,500$.

5. Finding solution equations
   a. exponential; $u(n) = 4(0.64)^n$
   b. linear; $v(n) = 100 - 6n$
   c. affine; substitute points $(0, 7)$ and $(1, 5.4)$ to get the equations $a + b = 7$ and $0.2a + b = 5.4$. Solving yields the solution equation $w(n) = 2(0.2)^n + 5$.

7. Population model
   a. For $p(0) = 600$, $p(n) = -400(0.80)^n + 1,000$.
   For $p(0) = 800$, $p(n) = -200(0.80)^n + 1,000$.
   For $p(0) = 1,300$, $p(n) = 300(0.80)^n + 1,000$.
   b.

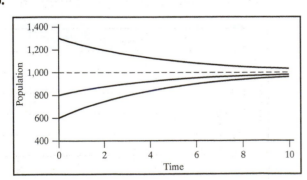

   c. The equilibrium value is stable, because all populations move toward that value.

9. Finding equilibrium values
   a. $E = 40$; $40 = 2.2(40) - 48$ ✓
   b. $E = 0$; $0 = 3(0)$ ✓
   c. $E = 80$; $80 = 0.75(80) + 20$ ✓
   d. $E = \pm 5$; $5 = 5 + 5^2 - 25$ ✓ and
   $-5 = -5 + (-5)^2 - 25$ ✓

11. Classifying equilibrium values
    a. $E = 8,000$; unstable
    b. $E = 120$; stable

13. Equilibrium values and affine solution equations
    a. $E = 90$; $u(n) = 210(1.5)^n + 90$
    b. $E = 2,000$; $v(n) = -1,898(0.98)^n + 2,000$
    c. $E = -8,000$; $w(n) = 10,000(1.025)^n - 8,000$

15. Forest plantation
    a. Let T equal the number of live trees and n the number of years; $T(n) = 0.75T(n-1) + 500$
    b. $T(n) = 4,000(0.75)^n + 2,000$
    c.

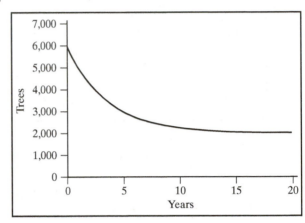

    d. The tree population will level off at 2,000 trees; 500 trees harvested each year.

17. Microloan example
    a. $E = 20,000$; if the loan balance in one month is $20,000, then the loan balance will remain at $20,000 for the following month.
    b. $u(n) = -18,500(1.0025)^n + 20,000$
    c. Using the difference equation, create a table to find that after 10 months the balance is $1,032.26. Using the solution equation, the balance after 10 months is $-18,500(1.0025)^{10} + 20,000 = 1,032.26$.

19. Amalgam fillings
    a. Solve $M^{15} = 0.5$ to get $M = (0.5)^{1/15} \approx 0.9548$.
    b. $120 \times 365 = 43,800$ micrograms; 0.0438 grams
    c. Let A represent the amount of mercury in the body in micrograms and n the number of years; $A(n) = 0.9548A(n-1) + 43,800$.
    d. $E = 969,026.5487 \approx 969,000$ micrograms. No matter what the current level of mercury in the body, the long-term mercury level will approach 969,000 micrograms.
    e. $A(n) = -469,000(0.9548)^n + 969,000$

21. Rework salmon hatcheries example
    The initial condition and difference equation are $p(0) = 2,000$ and $p(n) = 0.75p(n-1) + k$. The equilibrium value is $E = 4k$. The solution equation is $p(n) = (2,000 - 4k)(0.75)^n + 4k$. Use the point

(15, 6000) to find $k = 1,513.54$. The hatchery must add slightly more than 1,500 salmon each year.

23. Grameen Shakti microenergy
    a. $u(0) = \$382.50$
    b. $u(n) = 1.01u(n-1) - k$
    c. $E = 100k$
    d. $u(n) = (382.50 - 100k)(1.01)^n + 100k$
    e. Use the point $(36, 0)$ to determine the monthly payment of $k = \$12.70$.
    f. The amount due after 35 months is $12.77. The last payment equals $12.89.

Chapter 9 Exercises

1. Graphing a logistic difference equation
   a. $K = 1,000, r_{\max} = 9\%$
   b.

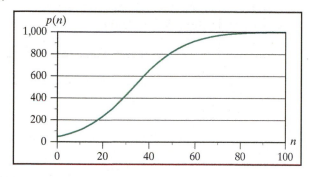

3. Writing logistic difference equations
   a. $p(n) = 1.02p(n-1) - \dfrac{0.02}{60}p(n-1)^2$
   b. $p(n) = 1.075p(n-1) - \dfrac{0.075}{4,000}p(n-1)^2$
   c. $p(n) = 1.05p(n-1) - \dfrac{0.05}{20,000}p(n-1)^2$
   d. $p(n) = 1.003p(n-1) - \dfrac{0.003}{2,000}p(n-1)^2$

5. Graph in Figure 9-3
   a. slope $= -0.00000001 = -10^{-8}$
   b. $r = 0.06 - 0.00000001p$
   c. The y-intercept is 0.06, which is $r_{\max}$. This agrees with the graph.
   d. Yes, it is the same.

7. Linearly decreasing population growth rates
   a.

p = Population	r = Growth Rate
0	0.1250
100	0.1125
200	0.1
300	0.0875
400	0.075
500	0.0625
600	0.05
700	0.0375
800	0.025
900	0.0125
1,000	0

b. $r_{\max} = 0.125$ or 12.5%. This is the growth rate when the population equals 0.
c. $K = 1,000$. This is the population when the growth rate equals 0.
d. $r = 0.125 - \dfrac{0.125}{1,000}p$

9. World population
   a. About 10 billion people. *Answers may vary.*
   b. $p(n) = 1.016p(n-1) - \dfrac{0.016}{10}p(n-1)^2$
   c.

Year	n	p	Year	n	p
1850	0	1.26	2030	180	7.19
1880	30	1.89	2060	210	8.05
1910	60	2.73	2090	240	8.70
1940	90	3.77	2120	270	9.16
1970	120	4.94	2150	300	9.46
2000	150	6.12			

d. When $n = 305$ or in the year 2155.

11. G. F. Gause bacteria experiment
    a.

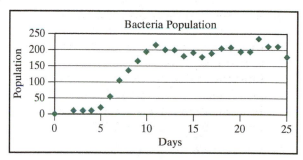

b. About 200 bacteria
c. $p(n) = 1.8p(n-1) - \dfrac{0.8}{200}p(n-1)^2$
d. The last 10–13 days in which the graph bounces up and down between 180 and 230 bacteria. It's possible that the test-tube environment had too much or too little food or water, or the waste products were not removed uniformly each day.
e. The population would outgrow the food supply, and the waste materials would pile up, so that the population would eventually start to die off.

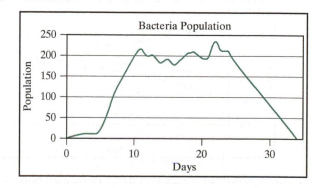

13. Population of England and Wales

a. $r_{max} = 0.02$, $K = 60$ million people. *Answers may vary.*

b. $p(n) = 1.02p(n-1) - \dfrac{0.02}{60}p(n-1)^2$, where p represents the population in millions, and n the number of years.

c.

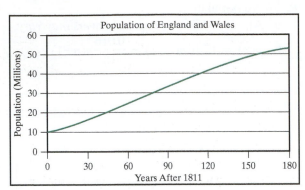

d. As people moved from farms to the city, families had fewer children. An increase in the general level of education and in the number of women working are also factors.

15. Finding equilibrium values

a. $E = 0$ and $E = 60$ **b.** $E = 20$ and $E = 40$

17. Bacteria population growing logistically

a.

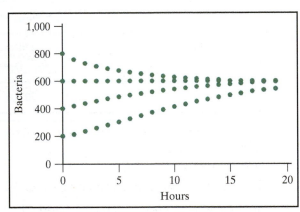

b. 600 bacteria   **c.** $E = 0$ and $E = 600$. If the bacteria population is at either equilibrium level, then it will stay at that level. $E = 600$ is the carrying capacity.

19. Fixed harvest of elk

a. $E = \dfrac{1,500 \pm \sqrt{1,250,000}}{2} \approx 1,309$ and 191

b.

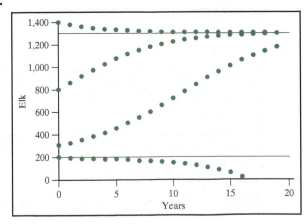

c. $E \approx 1,309$ is stable. $E = 191$ is unstable. The lower value tells us that if the population falls below 191, then the population will die off. This might be caused by decreased reproduction options, or decreased genetic diversity.

21. Pacific halibut population fixed harvest

a. $u(n) = 1.71u(n-1) - \dfrac{0.71}{80}u(n-1)^2 - 3$

b. The mass in the long term is about 75.5 million kg.

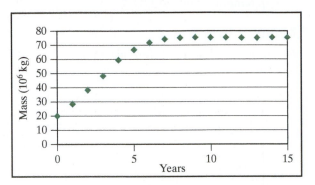

c. 5.625%

23. Pacific halibut population proportional harvest

a. $u(n) = 1.56u(n-1) - \dfrac{0.71}{80}u(n-1)^2$

b. The population mass levels off at about 63.1 million kg.

c. About 9%

d. The proportional harvest is 8.875%.

25. Periodic behavior of an insect population

a. $p(n) = 3.3p(n-1) - \dfrac{2.3}{1,000}p(n-1)^2$

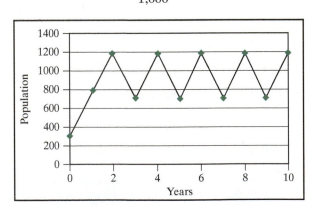

b. Between the approximate values of 688 and 1,182 insects.

27. Sensitivity to initial conditions

a. $p(n) = 4.0p(n-1) - \dfrac{3.0}{100}p(n-1)^2$, where p is the population and n is the number of years.

b.

n	Population	Population	Population
0	79	80	81
1	128.8	128.0	127.2
2	17.6	20.5	23.5
3	61.2	69.3	77.5
4	132.4	133.1	129.8
5	3.6	0.9	13.6
6	13.9	3.4	49.0
7	49.8	13.2	123.9
8	124.8	47.7	35.0
9	31.8	122.5	103.2
10	96.9	39.7	93.2

c. Yes, small differences in the initial population values create large differences in the projected population values, even by $n = 6$.

Chapter 10 Exercises

1. Regarding Example 10-1

n	$a(n)$	$b(n)$	$c(n)$	total
0	100	100	100	300
1	700	30	10	740
2	150	210	3	363
3	849	45	21	915
4	243	254.7	4.5	502.2
5	1,032.3	72.9	25.47	1,130.67

3. System of 3 difference equations

n	$a(n)$	$b(n)$	$c(n)$
0	20	10	30
1	30	40	50
2	70	90	80
3	160	170	150

5. System of 4 difference equations

n	$a(n)$	$b(n)$	$c(n)$	$d(n)$
0	2	5	3	4
1	7	13	2	13
2	5	40	12	37
3	25	112	8	118

7. Population with 3 age groups
  a. Every year, each member in group A will produce 2 offspring that survive for one full year. Group B members will each produce 3 offspring every year that survive for a full year. Group C members do not produce any offspring that survive a full year.
  b. Each year, 60% of group A and 25% of group B will live one additional year. No members in group C will live one additional year.

c.

	Group A	Group B	Group C
Annual Fertility Rate	2	3	0
Annual Survival Rate	0.6	0.25	0

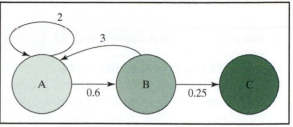

  d. $a(n) = 2a(n-1) + 3b(n-1)$
  $b(n) = 0.6a(n-1)$
  $c(n) = 0.25b(n-1)$

9. Life cycle and system of difference equations
  $a(n) = 2.0b(n-1) + 3.5c(n-1)$
  $b(n) = 0.36a(n-1)$
  $c(n) = 0.25b(n-1) + 0.88c(n-1)$

11. Sand smelt
  a. The annual fertility rates for females in groups A, B and C are 1.14, 3.65, and 6.97, respectively.
  b.

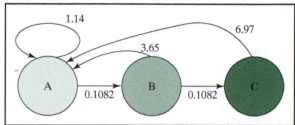

  $a(n) = 1.14a(n-1) + 3.65b(n-1) + 6.97c(n-1)$
  $b(n) = 0.1082a(n-1)$
  $c(n) = 0.1082b(n-1)$

  c. Total column below based on summing unrounded values.

n	$a(n)$	$b(n)$	$c(n)$	total
0	1,000	1,000	1,000	3,000
1	11,760	108	108	11,976
2	14,555	1,272	12	15,840
3	21,319	1,575	138	23,032
4	31,012	2,307	170	33,489
5	44,961	3,355	250	48,566
6	65,243	4,865	363	70,470
7	94,664	7,059	526	102,249

  d. $r = 45.1\%$
  e. $r = 45.1\%$. It appears that the sand smelt population's growth rate is stabilizing at 45.1%. This should be the long-term growth rate.
  f. 92.6%–6.9%–0.5%
  g. The revised fertility rates for groups A, B, and C are 0.68, 2.19, and 4.18, respectively. The new stable age distribution is about 89.0%–9.9%–1.1%.

13. PCH revisited
  All the growth rates equal 12.8%, which means that the

long-term growth rates for each stage will stabilize at this value. We also know that the total population will increase, in the long term, by 12.8% annually.

15. Waterbuck population
   a. $a(n) = 0.048b(n-1) + 0.081c(n-1)$
      $b(n) = 0.94a(n-1)$
      $c(n) = 1.00b(n-1) + 0.75c(n-1)$
   b. The percentage of females is 69.1%. Approximately 342 sighted were female.
   c. $a(0) = 114, b(0) = 114, c(0) = 114$
   d. By about 14.0% annually. The model predicts that after about $n = 25$ (year 2000) the total population would number under 10 waterbuck. Most likely when a herd gets this small, the chances for recovery are greatly diminished. We could assume that the waterbuck would go extinct in the park.

17. Pollution in Lake Erie and Lake Ontario
   a. Lake Erie's pollution level was 0.4489. The quantity flushed was 0.148137.
   b. Lake Ontario's pollution level was 1.1839. The quantity flushed was 0.201263.
   c. Lake Erie's pollution level will equal $0.4489 - 0.148137 = 0.300763$.
   d. Lake Ontario's pollution level will equal $1.1839 - 0.201263 + 0.148137 = 1.130774$.
   e. These answers match outputs from the system of difference equations.

19. Human exposure to lead
   a. $v(n) = 0.9714v(n-1) + 0.0111u(n-1)$

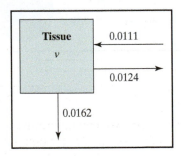

   b. $w(n) = 0.999965w(n-1) + 0.0039u(n-1)$

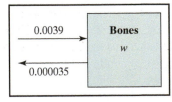

   c. $u(0) = 0 \ \mu g, v(0) = 0 \ \mu g$ and $w(0) = 0 \ \mu g$

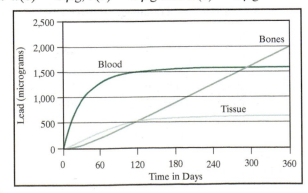

   d. The amounts of lead in the blood and the tissue grow quickly but then gradually level off. It appears that each of these amounts will reach its own equilibrium level. The amount of lead in the bones grows slowly at first, and then starts climbing in a linear manner. Eventually the amount of lead in the bones exceeds that of the blood and tissue; this is due to the fact that the bones only lose lead to the blood, and the flow rate is extremely slow (0.0035% each day).
   e. Yes, the blood lead reached the dangerously high concentration of 31.54 μg/dL.

21. Measles (Nigeria)
   a. $C(n) = 0.00003 \ S(n-1) \ C(n-1)$
   b. $S(n) = S(n-1) - 0.00003 \ S(n-1) \ C(n-1) + 360$
   c. $C(0) = 20; S(0) = 30{,}000$
   d.

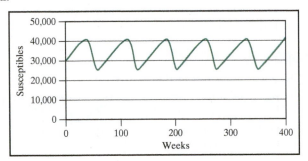

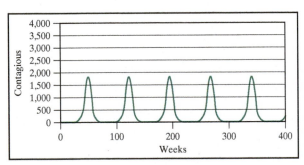

   e. At the peaks there are between 1,830 and 1,840 contagious children. The frequency of measles outbreaks is 73 to 74 weeks.

Chapter 11 Exercises

Note: If using technology other than the TI-83/84 to compute descriptive statistics, your results may differ slightly for Q1, Q3, boxplots, and the Bowley skew.

1. Discarded tires
   a.

	Fe	Mn	Al	Cr	Cu	Zn
Mean (mg/L)	6.984	0.732	0.336	0.094	0.182	1.012
Median (mg/L)	7.10	0.71	0.36	0.07	0.16	1.06

   b.

	Mn	Cu	Cr	Al	Zn	Fe
Range (mg/L)	0.08	0.14	0.17	0.33	0.43	0.96

The ranges indicate the amount of variation in each data set. The manganese data have the least amount of variation, whereas the iron data have the most.

c. 60%

3. Land biomass

Multiply the area percentage by the biomass to obtain the weighted biomass for each land type. Sum the weighted biomass values to get the mean biomass for Earth: approximately 3.8 kg C/m^2.

5. U.S. poverty rates

a. The number in each race is the number in poverty divided by the poverty rate. The total of 283 million people is about correct for the United States in years 2000–2001.

Race	Number in Race	% of U.S. Population
White (Non-Hispanic)	194,986,842	68.89%
Black	35,659,292	12.60%
American Indian and Alaska Native	3,226,667	1.14%
Asian and Pacific Islander	12,534,653	4.43%
Hispanic	36,613,953	12.94%
	Σ = 283,021,407	Σ = 100%

b. See the preceding table.

c. The overall U.S. poverty rate is about 11.57%.

d. The black poverty rate plays a larger factor because of the greater number of individuals of that race.

e. About 11.57%. This is the overall U.S. poverty rate, and matches the rate found in part (c).

7. Finding 5-number summaries

a. min = 26, Q1 = 36, median = 47, Q3 = 59.5, max = 69

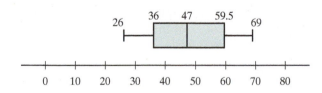

b. min = 15, Q1 = 17.5, median = 30, Q3 = 61, max = 69

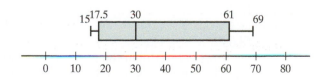

c. min = 3.4, Q1 = 3.8, median = 4.0, Q3 = 4.6, max = 5.2

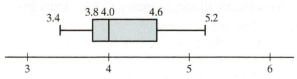

d. min = 100, Q1 = 120, median = 150, Q3 = 160, max = 160

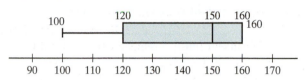

9. Water use

a. 129 gallons per person per day

b. Maryland

c. Both Louisiana and Ohio have equally small ranges (spreads).

d. Positively skewed. The Bowley skews for Maryland, Louisiana, and Ohio are 0.17, 0.33, and 0.18, respectively.

11. Soil density

a.

	Min	Q1	Median	Q3	Maximum
Before	0.10	0.145	0.20	0.48	0.81
After	0.15	0.56	0.77	0.96	1.13

b.

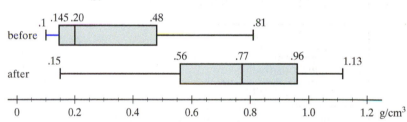

c. After logging takes place. We can determine this by inspecting the 5-number summary and boxplots—the five numbers for the post-logging data all lie to the right of those for the pre-logging data.

d. The range in the post-logging data (0.98 g/cm^3) is greater than in the pre-logging data (0.71 g/cm^3); thus the variation in the post-logging data is greater.

13. Frequency histogram

a. $n = 20$

b.

Bin (inches)	Frequency	Rel. Freq.
20–25	2	10%
25–30	4	20%
30–35	5	25%
35–40	6	30%
40–45	2	10%
45–50	1	5%

c. 35–40 inches

d. Q1 lies between 25 and 30 inches. The median lies between 30 and 35 inches. Q3 lies between 35 and 40 inches.

15. 4 histograms

    a. #1   **b.** #2   **c.** #3 and #4   **d.** #2   **e.** #1

17. Silver fir

    a. Mean = 15.25 in, min = 7 in, Q1 = 12 in, median = 15 in, Q3 = 18.5 in, max = 27 in, range = 20 in

    b. *Bins and frequencies may vary.*

Diameter (in)	Frequency
7–10	6
10–13	10
13–16	15
16–19	8
19–22	8
22–25	4
25–28	1

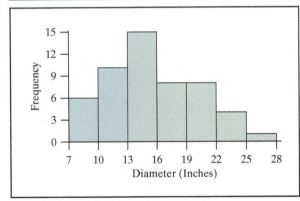

    c. The data are positively skewed, although just slightly. The Bowley skew value (0.077) indicates that the middle 50% of the data have positive skew.

    d. 13–16 inches. This is the most frequently occurring interval of diameters.

19. Sampling Atlantic surf clams

    a. Population: All surf clams living in Atlantic waters along the eastern United States. Sample: The 80 surf clams collected.

    b. A statistic

    c. No, they can only estimate the population mean with the sample mean.

    d. Researchers might sample shoreline with public access, where clam digging is more prevalent; this could result in fewer large clams in the sample. The sample mean would be smaller than the population mean.

21. Sampling light bulbs

    a.

Lumens	Watts	Efficiency	Lumens	Watts	Efficiency
210	25	8.4	1,200	75	16.0
220	25	8.8	1,600	100	16.0
820	60	13.7	1,500	100	15.0
830	60	13.8	1,750	120	14.6
1,170	75	15.6	1,800	120	15.0

    b. The sample mean is 13.69 lumens/watt, so we would estimate that the mean energy efficiency of all bulbs in the store is the same.

    c. 60%

    d. There are two bulbs with each wattage, perhaps from a systematic sample.

Chapter 12 Exercises

1. Means and standard deviation

Both sets have means of 25. Set 1 has the greater standard deviation.

3. Temperatures

Daily temperatures in Portland have a smaller standard deviation because Portland has smaller temperature fluctuations compared to Chicago.

5. Comparing 2 histograms

    a. Because both histograms are symmetric.

    b. Data Set 1

7. Potassium concentrations

    a. $\bar{x} = 0.84$ mg/L

x	$\bar{x}$	$x - \bar{x}$	$(x - \bar{x})^2$
1.5 mg/L	0.8417 mg/L	0.6583 mg/L	0.4334 mg²/L²
1.8 mg/L	0.8417 mg/L	0.9583 mg/L	0.9183 mg²/L²
1.6 mg/L	0.8417 mg/L	0.7583 mg/L	0.5750 mg²/L²
1.3 mg/L	0.8417 mg/L	0.4583 mg/L	0.2100 mg²/L²
0.7 mg/L	0.8417 mg/L	−0.1417 mg/L	0.0201 mg²/L²
0.7 mg/L	0.8417 mg/L	−0.1417 mg/L	0.0201 mg²/L²
0.6 mg/L	0.8417 mg/L	−0.2417 mg/L	0.0584 mg²/L²
0.6 mg/L	0.8417 mg/L	−0.2417 mg/L	0.0584 mg²/L²
0.4 mg/L	0.8417 mg/L	−0.4417 mg/L	0.1951 mg²/L²
0.4 mg/L	0.8417 mg/L	−0.4417 mg/L	0.1951 mg²/L²
0.4 mg/L	0.8417 mg/L	−0.4417 mg/L	0.1951 mg²/L²
0.1 mg/L	0.8417 mg/L	−0.7417 mg/L	0.5501 mg²/L²
		$\Sigma(x - \bar{x})$ = .0004 mg/L $\approx$ 0 mg/L	$\Sigma(x - \bar{x})^2$ = 3.4291 mg²/L²

$$s = \sqrt{\frac{3.4291\ \text{mg}^2/\text{L}^2}{11}} \approx 0.56\ \text{mg/L}$$

    b. 1.8 mg/L lies furthest from the mean, while 0.7 mg/L lies closest. We can tell this by comparing the absolute values of the deviations from the mean: $|x - \bar{x}|$.

9. Stream velocities

    a. $z = -0.5$

    b. $z = 4.8$

    c. $z = 0$

    d. $z = -2.6$

11. Temperatures

    a. $x = 18.7°C$

    b. $x = 18.4°C$

    c. $x = 18.64°C$

    d. Temperatures greater than the mean have positive z-scores, while those less than the mean have negative z-scores.

13. Iowa
Iowa's nuclear energy production. Nuclear: $z = -0.94$
Hydro: $z = -0.44$.

15. Nuclear energy
Pennsylvania. No, its z-score is $z = 2.54$, below the $z = 3$ cutoff level for outliers.

17. Which are true?
Both (b) and (d) are true.

19. Using the Empirical Rule
  a. $\bar{x} \approx 21, s \approx 6$
  b. $\bar{x} \approx 140, s \approx 40-50$

21. Motor oil recycling rates
  a. $\bar{x} = 1.816$ pounds per person, $s = 1.196$ pounds per person.
  b. *Bins may vary.*

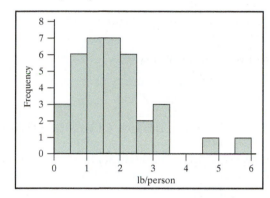

  c. 94.4%. Yes, this agrees with Chebychev's Rule.
  d. Asotin county, $z = 3.2$
  e. It's possible that those counties have mandatory recycling, very few industries, more recycling centers, or stronger recycling education.

23. A study of 58 small businesses
  a. 68%
  b. About 55 businesses
  c. About 1 business
  d. 250 pounds

Chapter 13 Exercises

1. Empirical Rule
  a. 49.85%

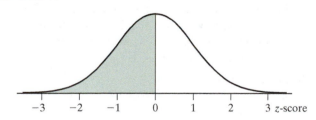

  b. 16%

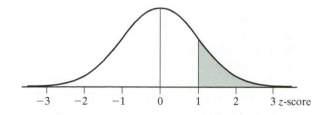

c. 97.35%

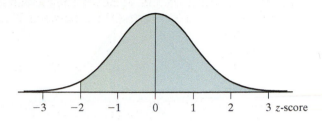

3. Standard normal curve
  a. 0.3944

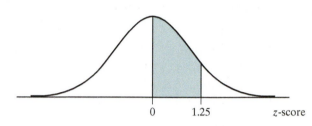

  b. 0.4861

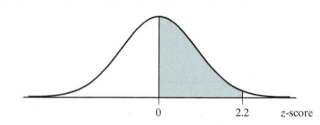

  c. 0.4887

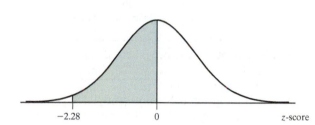

5. Standard normal curve
  a. 0.0045

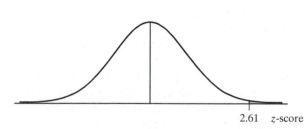

  b. 0.9335

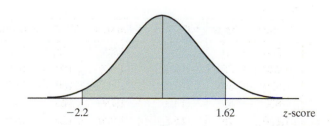

349

c. 0.2717

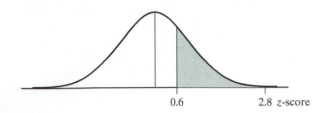

7. Estimate the z-score
  a. $c = 1.84$ **b.** $c = -1.15$ **c.** $c = -1.85$

9. SUVs
Approximately 52%

11. pH of water samples
  a. 27.34%   **b.** 43.32%   **c.** About 20 samples
  d. A pH of about 5.8 ($z = 0.84$)

13. Moisture in soils
  a. About 96%
  b. About 91%
  c. Soil moisture below 0.26 m³/m³ ($z = -0.25$)

15. Coal-fired power plants
  a. *Histograms will vary depending on binning strategy.*

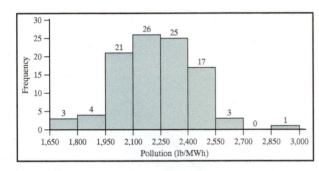

  b. The histogram has a decent bell shape, centered around 2,250 lb/MWh. The z-score of the plant that produces 2,982 lb/MWh is about 3.6, confirming that the plant is an outlier.
  c. About 7%
  d. 4% (which is reasonably close to 7%)
  e. $z \approx 1.74$. In theory, about 96%. In actuality, 98%.

17. Finding 95% confidence intervals for means
  a. $E = 1.6$ lb. The 95% CI is (18.9 lb, 22.1 lb) or 20.5 ± 1.6 lb.
  b. $E = 2$ km. The 95% CI is (43 km, 47 km) or 45 ± 2 km.
  c. $E = 0.04$ cfs. The 95% CI is (3.15 cfs, 3.23 cfs) or 3.19 ± 0.04 cfs.

19. Lead and bridge workers
  a. (25.6 µg/dL, 28.8 µg/dL)
  b. We are at least 95% certain that the mean blood lead level for all bridge workers is above 25 µg/dL, because the lower boundary of the 95% confidence interval is 25.6 µg/dL. Bridge workers should modify their exposure to lead to lower the mean blood level.

21. Sea stars
  a. $E = 0.48$ cm
  b. A factor of 2
  c. At least $n = 1,537$ sea stars

23. Finding 95% confidence intervals for proportions
  a. $40(0.74) = 29.6 \geq 5$ ✓ and $40(1 - 0.74) = 10.4 \geq 5$ ✓ $E = 0.14 = 14\%$. The 95% CI is (60%, 88%) or $74\% \pm 14\%$.
  b. $15(0.38) = 5.7 \geq 5$ ✓ and $15(1 - 0.38) = 9.3 \geq 5$ ✓ $E = 0.27 = 27\%$. The 95% CI is (11%, 65%) or $38\% \pm 27\%$.
  c. $1,000(0.354) = 354 \geq 5$ ✓ and $1,000(1 - 0.354) = 646 \geq 5$ ✓ $E = 0.0296 \approx 3.0\%$. The 95% CI is (32.4%, 38.4%) or $35.4\% \pm 3.0\%$.

25. StarLink® corn
  a. Small. $E \approx 0.002 = 0.2\%$.
  b. No, the margin of error is much smaller than the precision implied by the statement "about 9%."

27. Gallup poll
  a. $526(0.55) = 289.3 \geq 5$ ✓ and $526(1 - 0.55) = 236.7 \geq 5$ ✓
  b. $55\% \pm 4\%$. We can be 95% confident that the true population percentage is between 51% and 59%.
  c. A minimum of 1,057 people

Mathematics Index

2-cycle difference equation 216
4-cycle difference equation 217
5-number summary 257
68-95-99.7 Rule 292
95% confidence interval 315
95% confidence level 317

A

accuracy 3, 4
affine difference equations 173, 185
  equilibrium values 191
  solution equation 185, 196
approximating data with functions
  exponential 118
  linear 86
  power 139
approximation 5

B

bar chart 52
bell curve 290, 304
bimodal frequency distribution 263
bin 54, 144, 262
  width 54
  modal 54, 262
bivariate data 51, 60
Bowley skew 264
box-and-whisker plot 259
boxplot 259

C

calculator: see *using technology*
capture-recapture 24, 33
carrying capacity K 207
catch and release 33
categorical data 52
center value 255
chaos 218
  butterfly effect 219
Chebychev's Rule 289
circle chart 51
compound units 10
compounding, exponential 105
concentration 28
confidence intervals 314
  95% 315
  margin of error 315, 317
  of means 315
  of proportions 317
continuous equation 170, 230

convenience sample 269
correlation coefficient r
  exponential 121
  linear 90
  power 143
  using technology 91
correlation fallacies 91
cumulative frequency 146
cyclical difference equation model 215

D

damping term, logistic 208
data
  bivariate 51
  categorical 52
  exponential 118
  linear 78
  log transformation 118, 147
  nominal 52
  ordinal 53
  power 139
  shape of 261
  univariate 50
decay multiplier M 107, 168, 188
decay rate r 107, 168, 185
dependent variable 82
descriptive statistics 254
deviation from the mean 282
difference equation 164, 167
  affine 173
  chaotic 218
  cyclical 215
  damping term 208
  equilibrium value 185, 191, 212
  exponential 168
  harvest model 213
  linear 167
  logistic 206, 207, 208
  modeling 167, 171
  periodic 215
  solution equation 169, 185, 196
  systems 230
  using technology 174, 236
differential equations 172
discrete equation 169, 230
domain 83
doubling time 115

E

Empirical Rule 292